Edgar Leopold Layard

The Birds of South Africa

A Descriptive Catalogue of all the known Species occurring South of the 28th

Parallel of South Latitude

Edgar Leopold Layard

The Birds of South Africa
A Descriptive Catalogue of all the known Species occurring South of the 28th Parallel of South Latitude

ISBN/EAN: 9783337307707

Printed in Europe, USA, Canada, Australia, Japan

Cover: Foto ©berggeist007 / pixelio.de

More available books at **www.hansebooks.com**

THE

BIRDS OF SOUTH AFRICA.

A DESCRIPTIVE CATALOGUE

OF

ALL THE KNOWN SPECIES OCCURRING SOUTH OF THE
28TH PARALLEL OF SOUTH LATITUDE.

BY

EDGAR LEOPOLD LAYARD,

FELLOW OF THE ZOOLOGICAL SOCIETY, MEMBER OF THE ENTOMOLOGICAL
SOCIETY, AND OF VARIOUS FOREIGN SOCIETIES.

CAPE TOWN :
J. C. JUTA, WALE-STREET.

LONDON :
LONGMAN, GREEN, & Co., 39, PATERNOSTER ROW.

1867.

TO THE

LOVING COMPANION AND HELPMATE IN MY LABOURS,

MY WIFE,

THIS BOOK IS AFFECTIONATELY DEDICATED.

PREFACE.

THE following Catalogue of THE BIRDS OF SOUTH AFRICA was commenced eleven years ago, for my own information, and without any idea of its ever being before the public. On my arrival at the Cape, in December, 1854, I inquired in vain for any book which would give some insight into the ornithology of the country in which my lot had been cast. I was shown LE VAILLANT'S "Oiseaux D'Afrique," and Dr. A. SMITH's "Illustrations of the Zoology of South Africa;" but these visibly bore the stamp of incompleteness. I therefore commenced to form a catalogue for myself, and waded through whole libraries of works on South African travel, gleaning here and there a name ; but found myself still far from a complete list. Our Public. Library supplied me with the works of SHAW, LATHAM, and the older writers; and as descriptions and names of reputed South African birds came to hand, I wrote them on separate sheets of paper, adding from time to time my own notes on species obtained. I mention this in order to account for the broken and disjointed style which is apparent throughout the work. At length my MSS. grew to considerable proportions, and I found myself in possession of a mass of information, which many kind friends, with whom I had been corresponding on ornithological subjects, urged me to make public, and not lock away in the recesses of my tesk.

I have endeavoured to meet the requirements of South African readers by confining my descriptions of species to *general appearance*, and have not entered into scientific details: these, ornithologists will find in the authors quoted.

The classification followed is that of Mr. G. R. GRAY, in his " Genera of Birds;" and the characters of Orders, Tribes, Families, and Genera,* have been taken from that splendid work.

As regards identification of species, all those upon which I have felt any doubt have been submitted to the critical examination of my friends, Messrs. SCLATER and HARTLAUB, to whom my best thanks are due for their patient kindness in replying to the numerous questions with which I have troubled them. I am also much indebted to Mr. G. R. GRAY, of the British Museum, and to Professor NEWTON, of Cambridge, for much kind help and counsel.

At the Cape, my thanks are due to many whose names appear in these pages: to Mrs. BARBER, to Messrs. W. and T. ATMORE, W. CAIRNCROSS, D. ARNOT, J. CHAPMAN, C. J. ANDERSSON, H. JACKSON and his brother, my late lamented friend, ARTHUR JACKSON, of Nel's Poort, now, alas! no more, and many others.

With all these friends a most interesting correspondence has been maintained, from which much of my information is acquired. I think I have duly credited each with his share. I have but to add, that, owing to the enlightened influence of His Excellency, Sir GEORGE GREY, our then Governor, and His Excellency, Mr. RAWSON, now Governor

---

* With the exception of those founded by Swainson, whose own curt, vigorous descriptions ar used.

of the Bahamas, our then Colonial Secretary, the whole of this correspondence has been allowed to pass free through the Post-Office ; also, the Union Steam Company, carrying our mails, have in the most liberal manner conveyed all my little Zoological parcels to England free of charge.

I do not profess to put forth this Catalogue as complete. There are doubtless many species described in it, which a further acquaintance with works and type specimens contained in European museums, would sink into synonyms : it is a move forwards, and may serve as a foundation for the labours of others whose opportunities may be greater than my own.

No doubt many species on the Frontiers, particularly to the Eastward, are yet undescribed. I have never been able to visit these districts ; nor have I succeeded in getting any one to collect for me there.

As regards the identification of the eggs described, in every instance in which I have not myself taken them, the circumstances under which they have come into my possession have been detailed.

Many Birds are herein described that have no claim to be considered as South African, beyond the dictum of the old Authors. This has been done to serve as a beacon to others, that they may avoid the labour and trouble I have had in determining their true localities.

The Engraving, showing the various parts of a Bird, is copied from a plate in a German ornithological work, the name of which has escaped my memory.

ADDENDA ET CORIGENDA.

Page 15, line 29, for *Hæmatornis*, read *Hæmatornis*.
" 17 " 6 " *Focifer*, read *Vocifer*.
" 22 " 21 " *Tinnuncularias*, read *Tinnuncularius*.
" 34 " 24 " *Montaguii*, read *Montagui*.
" 44 " 11 " *J. Capensis*, read *S. Capensis*.
" 51 " 42 " *Velocifère*, read *Velocifèr*.
" 52 " 10 " CHRISTATA, read CRISTATUS.
" 52 " 19 " *coronata*, read *coronatus*.
" 64 " 23 " *Nutans*, read *Natans*.
" 76 " 21 " *plastrou*, read *plastron*.
" 95 " 14 " *supercilious*, read *superciliosus*.
" 99 " 31 " *Tardus*, read *Turdus*.
" 103 " 41 " *Minus*, read *Mimus*.
" 118 " 13 " *Levandiére*, read *Lavandière*.
" 119 " 31 " *Le Bergeronnet à guimpe*, read *l'aquimp*.
" 119 " 46 " *Le Levandiere*, read *La lavandière*.
" 121 " 19 " CATHROPÆ, read CALTHROPÆ.
" 127 " 26 " *Tardus*, read *Turdus*.
" 147 " 34 " *Flunicola*, read *Fluvicola*.
" 158 " 18 " *Æruginosus*, read *Æruginosus*.
" 170 " 32 " ŒNEA, read ÆNEA.
" 170 " 31 " *Longicanda*, read *Longicauda*.
" 173 " 27 " *Corocias*, read *Coracias*.
" 184 " 27 " *Euplectus*, read *Euplectes*.
" 200 " 1 " SABFLAVA, read SUBFLAVA.
" 212 " 1 " ALAUDA, read MEGALOPHONUS.
" 230 " 13 " *Piroquet*, read *Peroquet*.
" 233 " 2 " *Sud arc*, read *Sud Afr.*
" 337 " 12 " CŒRULESCENS, read CÆRULESCENS.

Nos. 463, 464, and 467 are, according to M. Verreaux (Proc. Zool. Soc., 1859), indentical.

SYSTEMATIC ARRANGEMENT

OF THE

ORDERS, SUB-ORDERS, TRIBES, SUB-TRIBES,
FAMILIES, SUB-FAMILIES, AND GENERA.

2

* Families III., TROCHILIDÆ; IV., MELIPHAGIDÆ; and V., CERTHIADÆ, are omitted. not being South African.

* Family II., Paradiseidæ, omitted in text, as not found in South Africa.

* Families I., CRACIDÆ; II., MEGAPODIDÆ; and VI., TINAMIDÆ, are omitted, not being South African.

Order I. ACCIPITRES. Linn.[*]

THIS Order embraces the Birds of Prey: they have the bill, of various length and form, more or less compressed; with the culmen suddenly hooked at the tip, and acute; the base more or less covered with a cere, in which are pierced the variously-formed nostrils; the wings lengthened and pointed; the feet strong, with the tarsi moderate, generally rounded and covered with scales of different shapes; the toes, three before, and one behind, all armed with strong claws, and their soles invariably rough.

The First Sub-Order,

ACCIPITRES DIURNI, or Diurnal Birds of Prey,

are distinguished from those that pursue their food in the twilight, or at night, by the lateral position of their eyes; the base of the upper mandible covered for nearly half its length with a prominent cere, in which the nostrils are placed; the tarsi moderate, scaled, and rarely covered with plumes.

The First Family,

VULTURIDÆ, or Vultures,

have the bill more or less strong, compressed, with the tip hooked, and acute; the wings lengthened and pointed; the tarsi short, rounded, and covered with reticulated scales; the toes moderate: the middle one the largest, and longer than the tarsi, the outer connected with the middle one by a membrane, the hind toe rather elevated, and all armed with strong obtuse claws.

Most of the birds of this family have the head and neck more or less clothed with down in the place of feathers, and the crop is prominent, naked, or covered with woolly hairs.

The First Sub-Family,

GYPAETINÆ, or Bearded Vultures,

have the head and neck completely clothed with feathers,

---

[*] The *Raptatores* of Illiger, the *Rapaces* of M. Temminck, or the *Raptores* of Mr. Swainson.

B

and the cere of base of the upper mandible entirely hidden by projecting bristles.

Genus GYPAETUS, Storr.

Bill strong, lengthened : upper mandible elevated near the end, which is hooked ; under mandible provided beneath with a bunch of setaceous bristles directed forwards. Nostrils oval, covered and defended by bristles. Feet short : the three anterior toes united to their base by a membrane ; the middle toe very long ; claws but slightly curved. Wings long ; the first quill rather shorter than the second ; the third longest.

No. 1. Gypaetus Meridionalis. Bp. Strikl.
Orni. Syn., p. 17.   The Bearded Vulture.   *Arend* of Colonists.

Top of head dirty white, bill black ; circle round the eyes, space between them and bill covered with black stiff hair, which extends on each side to the base of the lower mandible. A tuft of similar hair, of considerable length, projects outwards from under the bill. Back, wings, and tail dark blackish ash, each feather being light in the centre and darker on the edges, with white shafts and blotches. Shafts of wing and tail feathers white : tail wedge-shaped, under parts white, tinged and coated with a reddish substance, which can be scraped off. Length, 3' 10 ; wing, 2' 8"; tail, 1' 9".

This noble bird is locally distributed in the hilly parts of the colony. I have seen it often in Bain's Kloof, near Wellington ; also in the high mountains round Mr. Jackson's residence at Nel's Poort, near Beaufort. From this gentleman I have received sundry specimens, and he informs me that several pairs constantly breed in his neighbourhood ; also that they will kill lambs and sickly sheep. A pair of young birds was sent to the South African Museum from Graaff-Reinet by Mr. Ziervogel, the member for that division, who tells me it is a constant (though rare) resident there. Mr. Atmore found it about Blanco. He writes, May 25, 1864 :—

"We are going to lay wait for an 'Arend' to-morrow evening. He always sits on one particular yellow-wood tree in the forest; but he examines the premises very closely before he perches. What a beautiful flight they have, sailing about without even flapping a wing ; and when they stoop they come out of the sky like a lightning flash. There are several here, and we know of a nest; but they breed late, so no chance of eggs ; but we will give it a look up on our return trip."

Mr. G. R. Gray believes this species to be identical with the European *G. Barbatus*, Linn. ; indeed, he only admits the existence of one species.

The Second Sub-Family,

SARCORAMPHINÆ, or Condors,

have the bill lengthened, and rather slender, with the basal portion more or less covered with a soft cere; the apical part strong, much curved, and acutely hooked at the tip; the nostrils placed in the cere, with the opening large, exposed, oblong, and longitudinal; the wings lengthened and pointed; the tarsi long, and covered with small reticulated scales; the middle toe lengthened; the lateral ones short, equal, and united with the middle by a membrane; the hind toe generally short and weak.

Genus NEOPHRON, Savigny.

Bill very long and slender, with the cere covering two-thirds of its length, the apical portion rather arched, and acutely hooked at the tip, and the sides compressed; the nostrils placed near the middle of the bill, longitudinal and exposed. Wings lengthened and acute, with the third quill the longest. Tail moderate and wedge-shaped. Tarsi the length of the middle toe, plumed below the knee, and covered with moderate-sized, reticulated scales. Toes lengthened and slender, with the lateral ones nearly equal, the inner one the strongest, and all strongly scutellated above; the hind toe as long as the inner, and strong; the claws moderate, strong, and curved, especially that of the inner toe. The front of the head, cheeks, and fore part of the throat denuded of feathers.

2. Neophron Percnopterus. (Linn.) Pl. Enl., 427, 429; *L'Ourigourap.*, Le V., Pl. 14; *Witte Kraai* of Colonists, lit. *White Crow*; *Percnopterus Ægyptiacus*, Steph.; *Vultur Ginginianus*, Daud., Lath. Hist., Pl. 5; Gould's B. of Eur., Pl. 3.

GENERAL colour dirty white; feathers of nape narrow, elongated, and pointed; quill feathers of wings black; secondaries greyish black; tail white; bill long, slender, horn-coloured at the tip, yellow at the base; space round the eyes, cheeks, ears, chin, and part of throat bare and yellow. Length, 2′ 3″; wing, 1′ 10″; tail, 1′ 11″. The young bird is wholly of a dirty dark-brown.

Said to be common in the Interior and about Namaqualand. Mr. Ayres (Ibis Vol., 1859, p. 236) states that it is rare in Natal. At this end of the Cape Colony it is decidedly a scarce species; but

B 2

few specimens have reached my hands.   I saw a single bird feeding on a dead horse, off which it drove three hungry white-necked crows (*Corvus Vulturinus*) not far from the Paarl.   It is very abundant at the Cape de Verde Islands, breeding in the rocky precipices in St. Vincents, at the back of the town, off which the steamers lie to coal.   I ascended to their nests, which were vast masses of sticks ; but was, I believe, too late (December) for their eggs. I have never heard of any instance of the bird breeding in South Africa.   It may do .so, however, and its nest should be sought in the mountain ranges about Beaufort West and the Zwartberg, where I saw it not unfrequently during my visit to those localities.

3. Neophron Pileatus. (Burch. Trav., II., p. 195.)

N. Carunculatus, Smith ; *Cathartes monachus*, Temm ; *Percnopterus Niger*, Less. Ibis, 1860, p. 236.

PLUMAGE brown, variegated with fulvous on the thighs ; tarsi black ; top of head, cheeks, and front of neck entirely bare ; lower part of neck and posterior portion, almost to the hind-head, covered with a close greyish down. Length, 2′ 2″; wing, 19″ ; tail, 7″.

Inhabits Kafirland (Dr. A. Smith) and Natal (Mr. Ayres).   I have not met with it, but have been informed that there are two Vultures of this genus inhabiting the country North of the Orange River and Damaraland   One of them is the common *N. percnopterus ;* the other probably the present species.

The Third Sub-Family,

VULTURINÆ, or Vultures,

have the bill long, strong, more elevated than broad ; the sides more or less compressed ; the base covered with a cere for nearly half the length of the bill ; the tip suddenly hooked over the lower mandible ; the nostrils placed in the cere, with the opening oblique and exposed ; the wings lengthened ; the tarsi covered with small scales, and the middle toe longer than the tarsus, while the lateral toes are much shorter.

Genus VULTUR, Linn.

Bill large, much compressed and flattened on the sides; the culmen elevated, and much arched from the cere to the tip, which is hooked and acute ; the nostrils placed in the cere, and rather ovate. Wings lengthened, pointed, with the first quill short, and the third and fourth the longest. Tail moderate and rounded, with the shafts of each feather strong, and projecting beyond the webs. Tarsi feathered below the knee, and the rest covered with small reticulated scales. The middle toe rather longer than the tarsi, and united to the outer by a membrane at the base; the hind toe as long as the inner one; and all covered with strong transverse scales, and armed with slightly-curved, strong, rather acute claws, especially those of the inner and hind toes. Head covered with scattered down; the occiput generally furnished with an elevated transverse crest of closely-set down; and the neck-ruff advancing upwards towards the hinder part of the head.

4. Vultur Occipitalis. Burch. Trav., S. Afr., II., p. 329; Rüpp Atlas, t. 22; *Vultur galericulatus*, Temm., Pl. Col. 13; Le V., Pl. 12.

GENERAL colour above, blackish brown; thighs, under part of body, and neck white; quill feathers of wing and tail black; bare part of neck, space round eyes, and base of bill white; bill and feet flesh-coloured. Top of head covered with white down, which at the back is elongated and reversed. Length, 3'; wing, 1' 11"; tail, 9".

I received two specimens of this vulture from the late Mr. R. Moffat, jr., which were killed at Kuruman, where Mr. James Chapman tells me they are plentiful, and that their habits assimilate to those of the common vulture. Mr. Ayres quotes it from Natal; but it does not appear to come within the boundaries of this colony.

Genus OTOGYPS, Gray.

Characters in common with Vultur; but the head and neck are bare of feathers or down; and the sides of the latter are furnished with lengthened wattles (or wrinkles) from each ear.

5. Otogyps Auricularis. Daud. (Sp.) *L'Oricou*, Le V., Pl. 9; *Zwarte Aasvogel* and *Black Vulture* of Colonists.

GENERAL colour above, brown; head and neck bare; upper part of breast covered with short brown feathers, surrounded by a ring of white down; feathers of abdomen elongated, dark-brown in the centre, light-brown at the edges; thighs covered with thick white down. Length, 4'; wing, 2' 7"; tail, 13".

The Black Vulture is pretty generally distributed (but is not seen in such large flocks as the fulvous species, from which it keeps aloof), sailing round in enormous circles, at a great altitude. It usually hunts in pairs, and seems to have dominion over the common species.

Le Vaillant states that the eggs are white; but those which I have obtained from the Frenchhoek mountains, about 50 miles from Cape Town, are of a dirty white ground, profusely blotched and speckled with deep red-brown (dried-blood colour), especially at the obtuse end, where the blotches become confluent. Axis, 3″ 9‴; diam., 2″ 9‴. Another specimen is nearly spotless, and throughout of a dirty white. In the Zwartberg mountains this species builds on dense, flat-topped bushes, forming a vast accumulation of sticks, so closely matted together that a single nest will hold and sustain the weight of several men.   This is resorted to for several years, until the lice and insects generated in the mass become unbearable to the parent birds.   I have never heard of more than one egg in a nest.

Genus GYPS, Savigny.

Bill lengthened, with the culmen gradually hooked to the tip and rounded ; the sides rather swollen ; the nostrils oblong-ovate and oblique. The other characters are like those of Vultur ; but the head and neck are clothed with short down, and the nape ornamented with a ruff of lanceolate feathers, or of lengthened downy feathers.

6. Gyps Fulvus, Gmel. *G. Vulgaris*, Savig. *Vultur Kolbii*, Daud et Lath. ; *Le Chassefiente*, Le Vail., Pl. 10 ; *Aasvogel* of Dutch Colonists ; Rüpp. Atlas, t. 32.

GENERAL colour fulvous ; head and neck covered with short dirty whitish hairs ; lower part of cervix bare and bluish ; lower part of throat and middle of breast covered with short grey-brown feathers ; whitish down on the rest of throat, sides of neck and upper part of cervix ; a ruff of short white feathers on the back and lower part of neck ; wing and tail feathers very dark-brown.   Length, 3′ 9″ ; wing, 2′ 4″ ; tail, 12″.

The common Fulvous Vulture is found throughout the colony and Natal, and still lingers even in the neighbourhood of Cape Town. They breed in the most inaccessible precipices, depositing a few sticks in some hollow of the rock, on which I am informed they lay two eggs. Le Vaillant says these are bluish white ; but one given to me, as the egg of this bird, is of a dirty white, profusely blotched and speckled with very light brown, particularly at the obtuse end. Axis, 3″ 9‴ ; diam., 2″ 10‴.

At Nel's Poort is a breeding place of these birds. It is situated half-way up a vast inaccessible precipice, and is their constant resort at all times of the year. It is frequented by great numbers of birds, and their dung whitens the cliff to such an extent, that the spot is visible many miles off.   They have bred in this place from time immemorial, and from it they range over the surrounding country.   On killing a springbuck or any animal, it is curious to see how they come trooping in a lengthened string from this place to their anticipated banquet. Though not one may be visible when the shot is fired, in seven or eight minutes hundreds will be gliding to the spot.

7. Gyps Vulgaris. (Sav.) Rüpp. Atlas, t. 32 ; *Vultur Kolbii*, Cretzschen, nec Daud. ; *V. Rüppeli*, Natt. Mus. Vidob ; *V. Fulvus Rüppeli*, Schleg., Strickl· Orn. Syn., p. 10 ; Gurney, Ibis., 1860, p. 206.

PLUMAGE more or less of a dark-brown, each feather largely bordered with white or light fulvous ; bill yellowish. Size of *G. Fulvus.*

This species is not known to me as an inhabitant of the colony, though doubtless it may be found to the Eastward. It is said to occur at Natal but rarely.

The Second Family,

FALCONIDÆ, or Falcons.

have the bill partly covered at the base by a cere, compressed on the sides, and the culmen curved from the cere to the tip, which is hooked and acute ; the lateral margins either toothed or festooned ; the wings long and pointed ; the tail moderate and broad ; the tarsi and toes of various lengths and strong, and the claws long, curved, and acute, especially those of the inner and hind toes. The head and neck are covered with feathers, and the eyes placed laterally, deeply sunk, and protected by a horny brow.

The Sub-Family,

BUTEONINÆ, or Buzzards,

have the bill short, broad at the base, with the culmen much curved to the tip, which is acute, and the lateral margins festooned ; the cere covering more than the basal half of the bill ; the wings long, with the third and fourth quills generally the longest ; the tarsi lengthened, generally naked, and covered both in front and behind with broad transverse scales ; and the toes, including the hind one, rather short.

Genus BUTEO, Cuvier.

Bill short, broad at the base, laterally compressed at the tip, with the culmen much curved from the base, which is broad and rather flattened above, the lateral margins festooned ; the nostrils large, oval, and the lores clothed with hairy feathers. Wings very long, with the third and fourth quills the longest, and the inner and outer webs of some of the quills obliquely notched. Tail moderate, and even at its end. Tarsi lengthened, naked, covered with transverse scales before and behind ; but at the apex and on the sides

with small scales. Toes rather short, with the three anterior
ones united at their base ; the hind toe equal in length with
the inner, and both strong, and armed with a long powerful
curved claw ; the base of the toes covered with small scales,
and the apex with broad transverse ones.

8. Buteo Jackal. Shaw, Vol. 7, p. 173 ; *Le Runoir,* Le Vaill., Pl. 16 ; *Jackal Vogel* of Colonists.

A very variable species. Its chief hues are red and brown-
black, the latter preponderating on the head, neck, and
mantle. The throat is enlivened by white, which gradually
deepens to rust-red on the breast. Under wing-feathers
nearly white ; under tail coverts black, mingled with red ;
tail feathers deep red, each with a black spot towards the
tip.   Bill, cere, and legs yellow, irides ochreous grey.
Length, 1' 8" ; wing, 15" 8''' ; tail, 9' 3'''.
   This Buzzard is very common throughout the colony, and
its well-marked colours are plainly distinguishable at a great
distance ; it is therefore easily recognized. It appears to
capture its prey more by surprise and stealth than by hunting, sitting motionless on a bush, or tree, until some unsuspecting *gerbilus* or sand-rat *(Bathyergus)* emerges from its
hole, when down swoops the buzzard, clutches the victim in
its claws and bears it away, to be devoured at leisure. It is
a heavy-flying bird, but sometimes ascends to great altitudes,
wheeling in vast circles, and uttering a shrill, stridulous
cry.   Builds on trees, or in thick, high bushes.   The nest is
composed of small sticks and moss, thickly lined with feathers and wool.   Eggs two or three, rarely four, of a dirty
white colour, coarse in texture, and very thickly spotted, at
the obtuse end in particular, with dry-blood-coloured
blotches and stains.   Axis, 2", 5''' ; diam., 1" 10''' ; specimens,
however, vary much both in size and colour.

9. Buteo Desertorum. (Vieil.) *Buteo tachar-dus,* Mod. Auth. ; *B. Capensis,* Schleg ; *Le Rougri,* Le Vail., pl. 17 ; *Falco vulpinus,* Licht. ; *F. Cirtensis,* Le Vail. (fils.)

UPPER parts brown ; each feather having pale edges and a
black shaft.   Head, pale fulvous, streaked with brown.
Wing feathers dark-brown.   Tail feathers fulvous, inclined
to rufous, and narrowly barred with brown ; the broadest bar

at the tip. Under parts, pale fulvous; almost white on the chin and throat; streaked on the two latter, and blotched on the former with brown. Thighs rufous, faintly blotched with fulvous. Vent feathers pale fulvous. Length 1' 8"; wing, 14"; tail, 7". Irids yellow.

Not common, but widely distributed. A specimen lived for some time in my taxidermist's work-room, and would eagerly answer to his name, when called, by day or night. He would come to the hand and take from our fingers the bodies of the birds skinned for mounting, or pounce from his perch upon any stray mouse that ventured near him.

Fully adult birds become throughout of a deep rufous-brown, blotched with dark markings. In this stage they constitute Le Vaillant's species, called *Le Rougri*, Ois d' Af., Pl. 17.

The note of this bird is a weak stridulous scream. It frequents open country dotted with jungle, and is found also in the forest about George and the Knysna.

The Sub-Family,

AQUILINÆ, or Eagles,

have the bill moderate, the culmen elevated and straight at the base, and then much arched to the tip, which is hooked and acute; the sides much compressed, and the lateral margins more or less festooned: the nostrils placed in front of the cere, large and generally oval; the wings lengthened and acute, with the third, fourth, and fifth quills usually the longest; the tail long, ample, and mostly rounded at the end; the tarsi long, clothed with feathers to the base of the toes, or naked and covered with variously-formed scales; the toes long, strong, united at the base, especially the outer, and the lateral toes unequal both in length and strength, the inner being the strongest; the claws long, strong, much curved and acute, that of the inner toe the strongest.

Genus AQUILA, Mœhring.

Bill strong, straight at the base, and with the apical portion of the culmen much curved to the tip, which is greatly hooked and acute; the sides much compressed, and the lateral margins festooned; the nostrils placed in the cere large, and rather oblique. Wings lengthened and acute, with the fourth and fifth quills equal and longest. Tail long and wedge-shaped, or rounded at the end. Tarsi rather larger than the middle toe, robust, and entirely clothed to the base of the toes with feathers. Toes moderate, strong, lateral ones unequal, and all armed with strong, curved, acute claws; the inner the strongest.

10. **Aquila Pennata**. (Gmel.) Cuv., Pl. C., 33 ;

Butaetes Buteo, Less. ; *Butaetes Lessonii,* Smith ; *Falco Pennatus*, Pen. ; *Le Buse Gantée*, Le V., P. 18 ; *A. Minula*, Brehm ; Gould's B. of Eur., Pl. 9.

ABOVE brown ; the shoulder feathers margined with white ; ears and space behind the base of the lower mandible brown-black, under parts white ; the breast and anterior part of belly streaked with brown-black ; tail nearly even, black-brown, variegated with narrow, irregular, wavy bars of a darker tint, and all tipt with obscure white ; legs feathered to the toes. Length, 20" ; wing, 14" ; tail, 8".

Le Vaillant found this species only in the forests of Outeniqualand, perching on the summits of high trees, and shy and difficult of approach. Dr. Smith cites it as from near "Heeren Logement," in Clanwilliam. The only specimen I have seen is a young bird, shot by Mr. Jackson, at Nel's Poort. It is entirely of a dark-brown colour, in some places inclining to black. He tells me he was attracted to it by its peculiar cry, and has never seen another.

11. **Aquila Senegalla**. (Cuv.) Regne an., 1, p. 327 ;

Aquila Nævioides, Cuv., ib.; *A. Choka*, Smith., S. A. Quarterly Journal ; *Falco Rapax*, Tem., Pl. Col., 455 ; *Chok* and *Coo Vogel* of Colonists.

GENERAL colour rufous-brown ; tail and wing feathers dark-brown ; the former tipt with rufous. Bill black ; cere and toes yellow ; irides the same. Legs feathered to the toes. Length, 2' 8" ; wing, 1' 11" ; tail, 12½".

This bird seems very common at Kuruman, Colesberg, Nel's Poort, Beaufort, and the Karroo generally.

I should fancy, from the behaviour of one which I kept for some time in confinement, and which is now in the Zoological Gardens, Regent's Park, that it would make a good hunting eagle. Mr. Arnot, of Colesberg, from whom it was received, tells me it became quite as tame with him as with me. I have been informed by the Messrs. Jackson, of Nel's Poort, that these birds constantly accompany persons in pursuit of game, and have been seen by them to carry off wounded vaal knoorhaans (*Otis Vigorsii*) and hares. Their depredations on the flocks cause them to be killed on all occasions ; but they are still very numerous in the Karroo. I found a nest, evidently inhabited by young birds, in the month of January. It was a large mass of sticks in the top of a high, scraggy, and to me inaccessible tree, on the banks of the Dwass River, near Mr. Jackson's residence. One that I killed near the same place had frogs and fish in his throat : he was sitting by the river, close to the water's edge, evidently fishing.

12. Aquila Bonellii. (Temm.) Pl. Col., 289 ; *Aquila*
Fasciata, Vieil. ; Degl. Orn. Europe, I., p., 28. ; Bonap.
Consp , I , p. 14 ; *A. Intermedia*, Bonell.

GENERAL colour above, dark-brown, approaching to black,
mottled throughout with white ; below white, blotched
longitudinally on the breast and belly with dark-brown,
edged with light chestnut ; vent white, tinged with chestnut.
Tail barred. Legs white, feathered to the toes. Length, 2′ ;
wing, 17″ ; tail, 11″. Irides yellow ; cere and base of mandibles greenish-yellow ; anterior portion dark horn-colour.

This eagle is subject to great variation. The above description is taken from a fine example in the possession of Mr. C.
J. Andersson, killed at Otjimbinque. One in the S. A.
Museum is deep rufous below, blotched with black, and the
upper parts are brown. This bird is probably in young
plumage.

Previous to the acquisition of this specimen, which was identified by
Mr. J. H. Gurney (*A. Bonellii*), had never been observed, so far as is
known, South of the Equator. It was shot at Wynberg, in Mr.
Trotter's garden, after having made considerable havoc among the
fowls in two previous visits. Mr. Andersson procured several specimens, and informs me, *in epistolâ*, that "its flight is heavy, but when
once risen to a certain height it soars powerfully. It perches on trees
or rocks ; but to the best of my belief roosts only on the latter : its
food consists of small quadrupeds."

13. Aquila Verreauxii, Less., Cent. Zool., t. 38 ;
A. Vulturina, Shaw ; *Le Caffre*, De Vaill., Pl. 6. ;
Dassie Vanger and *Berghaan* of Colonists.

JET black, with half the back and rump white; legs feathered to the toes, which are bright yellow; claws black, very
strong, and curved. Length, 2′ 9″ ; wing, 2′ 1″ ; tail, 13″.
The young bird is mottled brown, inclining here and there
to black, according to its age.

Verreaux's eagle is not uncommon throughout the colony, wherever
rocky precipitous mountains are to be found. In these it fixes its
eyrie, sallying out daily at early dawn in quest of food, and returning
in the evening. One pair had their haunt in some of the rocks of the
" Devil's Hill," near Cape Town, and sailed over the "Camp Ground "
to and fro, night and morning, with the regularity of clock-work, to
their hunting grounds somewhere on the Cape flats. I have never
ascended Table Mountain without encountering a pair near the gorge
which opens on to the top, but they never suffered me to get within
range.

It is called " *Dassie Vanger*" (coney-eater) and " *Berghaan* " (moun-
tain-cock by the colonists, from feeding principally on the coney, or

rock-rabbit (*Hyrax Capensis*). It also attacks the klipspringer antelope (*A. Oreotragus*), darting at them when perched on pinnacles of the rocks (their favourite position), and hurling them headlong into the abyss below, into which they descend to feed on them at their leisure. They are also accused of killing lambs and sickly sheep, and carrying off the smaller antelopes from the low country.

Mr. Atmore writes, July 21st, 1864: "Fancy my knowing of two nests of *A. Verreauxii*, but in such inaccessible places that no one can get at them. Large nests in rocks about 1000 feet high, just on a ledge 300 or 400 feet from the top."

An egg brought me by Mr. Hugo, of Fransch Hoek, is chalky white, mottled throughout, and especially at the obtuse end, with rust-coloured and light-coloured spots. Axis, 3″ 5‴; diam., 2″ 4‴.

Le Vaillant evidently founded his "Caffre" upon this species, and, with his usual falsehood, enters into a long description of habits which the bird does not follow. In fact, he never obtained a specimen : only saw it at a distance, and invented his account. Had he secured one, he never could have stated that its talons were not fitted to carry away its prey : a single glimpse of the powerful, curved, sharp claws is enough to dispel this illusion.

Genus SPIZAETUS, Vieillot.

Bill moderate, the culmen straight at the base, and much arched at the tip, which is hooked and acute ; the sides much compressed, and the lateral margins festooned ; the nostrils large and rather rounded. Wings moderate, reaching to half the length of the tail, with the fourth and fifth quills equal and longest. Tail long, and slightly rounded at the end. Tarsi slender, much larger than the middle toe, and plumed to the base of the toes. Toes long, strong, the inner one much longer than the outer, which is united to the middle one by a membrane ; all covered above with small scales, except at the apex, where there are a few transverse ones, and each toe armed with a long, strong, and acute claw.

14. Spizaetus Coronatus. (Linn.) *Aquila Albescens*, Daud. ; *A. Coronata*, Smith ; *Falco Coronatus*, Linn.; *Le Blanchard*, Le Vail., Pl. 3 ; *Crowned Eagle*, Edw., Pl. 224. ; *Imperial Eagle*, Griffith.

EASILY distinguished from *S. Bellicosus*, and the other more common species, by the comparative roundness and shortness of the wings, and great length of tail. The head, crest, neck, and under parts yellowish white ; thighs varied with blackish brown lines and spots. Feathers on back brown, deeply margined with white ; tail barred, black, and grey ; feathers on occiput very long, forming a crest ; legs powerful, feathered to the toes. Length, 2' 11"; wing, 1' 9½"; tail, 1' 3".

Le Vaillant says it preys on small quadrupeds, builds on lofty trees, and lays two eggs.

One specimen of this bird has fallen under my notice, shot by Mr. G. Rex, at the Knysna. I know nothing of its habits or range.

15. Spizaetus Bellicosus. (Daud.) *Falco Armiger*, Shaw; *Aquila Bellicosa*, A. Smith; S. A. Zool., p. 144.; and Illust. S. Af. Zool. Av., Pl. 42.; *Le Griffard*, Le V., Pl. 1.

GENERAL colour above chest and neck dark grey-brown; beneath white, spotted with very dark-brown; thighs barred with the same; large wing-feathers, dark-brown; the lesser ones and tail transversely barred with grey. Talons black, much curved and strong; legs feathered to the toes. Length, 2' 7"; wing, 1' 11"; tail, 12".

According to Le Vaillant, this species feeds on small antelopes, hares, and gallinaceous birds, frequents forests, builds in lofty trees or rocky peaks; nest very strong, large, and flat; lays two round eggs, perfectly white.

It is very scarce in the colony, only two specimens having fallen under my notice. One was procured at the Knysna by Mr. George Rex, the other at Colesberg by Mr. Arnot. It is of this species that Mr. Atmore writes, *in epistolā* : " Just as we were leaving the Knysna, we heard of an eagle's nest in the forest, and under the tree the person who found it counted 95 heads of the little ' Blue Buck' (*Cephalopus Cærula*)." Mr. Tom Atmore informs me the tree was an enormous " yellow-wood," quite inaccessible; and the nest a huge mass of sticks impervious to a bullet. The Hon'ble Mr. Vigne informs me that one of these birds attacked his sheep-kraal and killed the lambs. It was ultimately caught in a steel-trap placed near the kraal.

16. Spizaetus Occipitalis. (Daud.) Vol. II., P. 40; Shaw, Vol. 7., p. 59; *Falco Senegalensis*, Daud.; *Morphinus Occipitalis*, Cuv.; Smith, Af. Zool., p. 148.

UPPER parts glossy-brown, approaching to black; darkest on extremities of wings and tail, and lightest on the shoulders and cheeks. Tail barred more or less faintly with white on the upper side; wings with black. Head crested; crest-feathers very long, and nearly black; feathers of head minutely tipped with white. Under parts almost black; legs feathered to the toes, and pure white; inside of quill feathers of wings and tail silvery-grey; barred with deep-brown. Length, 25"; wing, 16"; tail, 9" 9"; length of crest, 5" 6".

Mr. Atmore, to whom I am indebted for the only specimen I have seen of this beautiful bird, a young ♀, describes it as not uncommon in the neighbourhood of Traka. He writes : "The crested eagle is a gentleman of low tastes! One was eating an owl (*S. Capensis*) the other day! Tom could easily have shot him, but mistook him for a young *Buteo Jackal.* He is not uncommon about George, and not at all difficult to get at: sits for hours on a naked pole. I saw several during my survey, and might have bagged at least four, but had no gun with me."

Le Vaillant says they build on lofty trees, and line their nests thickly with feathers and wool ; lay two round eggs, blotched with brownish-red.

17. Spizaetus Spilogaster. Bp. Rev. Zool., 1850, p. 487. *Sp. Zonurus,* Müller, Naum., 1851, Part IV., p. 27. *Sp. Ayresii,* Gurney, Ibis., 1862, p. 149, pl. 4. *Sp. Leucostigma,* Heugh.

ABOVE, dark ashy brown, variegated with white; rump inclining to brown; each feather tipt with white; tail dark ash, barred with very dark-brown, tipt with white. Under parts white; more or less streaked and blotched on the flanks with clear dark-brown. Thighs and legs white ; feathered to the toes. Legs and cere yellow; eyes greenish yellow. Length, 19″; wing, 13½″; tail, 8″.

This pretty little eagle here described is, doubtless, rightly identified by my friend, Mr. Sclater, Secretary of the Zoological Society, who writes :—"Your bird is probably a small male of *S. Spilogaster.* It agrees well with a specimen, ex Gulam, in the Derby Museum, except (1) in smaller dimensions, (2) shafts of primaries are *black*, not white, (3) white edging at the extremity of tail is more distinct."

S. A. Museum specimen ♂. Length, 17″; wing, 13″; tail, 7″.
Derby Museum do., ♀. Length, 22″; wing, 16″; tail, 10″.

My specimen was taken alive in the neighbourhood of Cape Town, probably wounded by a gun-shot. He lived a few days in my possession, and was remarkably bold and fearless. He eat everything thrown to him—raw meat, birds, rats, and even fish. I have never seen another specimen in any collection which I have examined, and conclude, therefore, that it must be very rare in South Africa.

Genus CIRCAETUS, Vieillot.

Bill robust, rather straight at its base ; convex above, compressed on the sides ; cutting margin of the upper mandible nearly straight, the tip hooked ; under mandible straight, the tip obtuse. Nostrils oval, transverse. Tarsi naked, thick, and lengthened, covered with reticulated scales. Toes rather short, the hinder and lateral nearly equal. Claws rather short, nearly of equal length, and but slightly curved ; the anterior and posterior the strongest. Wings long ; the third quill the longest.

18. Circaetus Thoracicus. (Cuv.) *C. Pectoralis,* A. Smith, S. Af. Q. Journ., p. 109 ; *C. Cinereus,* Vieill., Gal. des Ois., Pl. 12 ; *C. Funereus,* Rüpp. Neue Wirb. Abzss., t. 14., p. 35 ρ.

HEAD and neck blackish-brown, tinged with grey ; back and shoulders of the same colour, each feather tipt with white. Throat black and white ; breast brownish-black ; lower parts white ; tail grey, crossed with broad black bands, tipt with white. Length, 2′ 7″ ; wing, 1′ 9″ ; tail, 12¾″. Iris pale straw-yellow

Rather rare in the colony, frequenting mountain-ravines clothed with timber, and keeping to the same spot for many successive years. Shy and suspicious, it rarely falls before the gun, although no opportunity is lost of killing a bird so destructive to young lambs.

I saw several pairs on the East Coast of Africa, and shot two at Fazy, a native village within a degree and a half of the Line; here they seemed quite fearless and allowed an easy approach. Mr. H. Gird, M.L.A , obtained a noble pair on his farm, Oliphant's Fontein, which are now mounted in the Museum. Mr. Atmore writes from Blanco, probably of this species : he says : "I have heard of another eagle inhabiting the forest—head, neck, and back blueish, belly white ; but I have not seen him yet. He preys on poultry, monkeys, and *cats !* Last week one carried off my hostess's favourite " Tom," and she now hopes my Tom (his son) will shoot him,"—so do I !

Mr. Henry Jackson has sent an egg of this fine bird from Nel's Poort. He says they lay but one in a nest. It is pure white; Axis, 3″ 1‴ ; diameter, 2″ 4‴

19. Circaetus Bacha. (Daud.) Shaw, Vol. 7, p. 157 ; *Hæmatornis Cheela* and *H. Bacha,* Lath. ; *Le Bacha,* Le V., Pl. 15 ; *Falco Bido,* Horsf. ; *F. Albidus,* Cuv., Temm., Pl. Col. 19 ; *Buteo Melanotis,* Jerd.

UPPER parts deep brown ; shoulders spotted with white. Under parts a paler brown ; thickly spotted with white ; two bars extend across the tail, appearing ochreous-brown on the top, and white below ; occipital crest very broad, black, and banded with white. Tail tipt with white ; thighs barred with white. Length, 2′ 2″ ; wing, 1′ 6″ ; tail, 10″.

Le Vaillant affirms that he found this species in Namaqualand, frequenting rocky, desert places ; and preying on the dassie (*Hyrax Capensis*). He also states that it is fierce and solitary, building in deep rocky caverns ; the nest a heap of dry branches, on which is massed leaves and moss. Eggs two, rarely three ; and he gives a long account of its habits, painting a wonderfully graphic picture of its ferocity, and the terror it excites among the dassies. Unfortunately for M. Le Vaillant's character, it is too probable that the bird only existed in

this colony in his vivid French imagination. These word-pictures seem
to run away with our neighbours, and carry them into the land of
Fable, as witness the wondrous gorilla-tales not long since given to the
world. Sober truth obliges us to disbelieve the whole account, and to
relate what we have seen of them. They were very plentiful at Point
Pedro, in the North of Ceylon, and frequented the jungle-dotted plains
throughout the Northern Province. It was no uncommon thing to
see three or four on the wing at once, wheeling round in airy circles ;
and from these peculiar markings they could be recognised at a great
distance. They built in banian trees, usually a large, strong nest of
sticks, without any lining, and laid three eggs, of dull-white colour,
with a few dried-blood-coloured blotches at the obtuse end. Axis,
2" 7"' ; diameter, 2". They fed on snakes, lizards, and other reptiles
and insects. They were particularly partial to the large trees on the
banks of tanks, and from them swooped down on the frogs which
came up to sun themselves on the floating logs or weeds.

Genus PANDION. Savigny.

Bill short, culmen covered from the base to the tip, which
is hooked and acute ; the sides compressed, and the lateral
margins slightly festooned ; the nostrils moderate, sublinear,
and rather oblique. Wings reaching to the tip of the tail,
with the second and third quills equal and longest. Tail
moderate and rather even. Tarsi short, very strong, and
entirely covered with reticulated scales. Toes long, much
padded beneath, and united at the base to the middle toe,
mostly covered above with small scales ; the claws very long,
much curved and rounded beneath.

20. Pandion Haliaetus. (Linn). Cuv., Pl. Enl.
414. The Osprey.

GENERAL colour deepish-brown ; wing-feather approaching
to black ; tail indistinctly barred with white. Head with a
white horse-shoe mark, extending from the back of the eye,
more or less broken with brown. Chin, belly, thighs, and
vent white ; the latter blotched slightly with rufous. Chest
variegated with brown. Length, 2' 1" ; wing, 20" ; tail,
8¼". Iris light-yellow ; legs livid.

Quoted as from South Africa by M. Verreaux, and found at Natal by
Mr. Ayres, who says they frequent the salt-water lakes near the sea.
I have never seen it from the Cape Colony. My description is taken
from a European specimen.

Genus HALIAETUS, Savigny.

Bill large, culmen straight at the base, and curved to the
tip, which is hooked and acute, the sides compressed, the
lateral margins slightly festooned, and the lores naked ; the
nostrils moderate, linear, and oblique. Wings lengthened and

acute, with the third, fourth, and fifth quills nearly equal and longest. Tail moderate and rounded. Tarsi short, strong, covered in front with transverse narrow scales, and with small irregular ones posteriorly and on the sides. Toes long, mostly covered above with transverse scales; the claws long, curved, and acute.

21. Haliaetus Vocifer. *Falco Focifer*, Shaw; *Le Vocifer*, Le Vail., Pl. 4; *Groote-visch-vanger* or *Witte-visch-vanger* of Colonists.

HEAD, breast, and top of back and tail pure white; wings and back nearly black, the larger feathers edged with white; upper parts of the wings reddish-brown. Belly and thighs deep-reddish brown. Cere and legs yellow. Irides yellow. Length, 2′ 8″; tail, 11″.

Found generally in pairs, frequenting the embouchures of rivers, lakes, and other parts of the colony, where wood and water afford a congenial home. Feeds on fish, crabs, and reptiles, and will not refuse carrion sheep, &c. (Le Vaillant says he has found antelope bones in their nests), resorting usually to some bare rock or dead tree to devour its quarry.

Le Vaillant found it on the Orange River. Mr. Chapman brought it from the Zambesi, and I saw it as far North as Fazy. It is common at the Knysna, forming an enormous nest of sticks and rubbish. I did not see its eggs; but Le Vaillant says they are white, and shaped like those of a turkey, only larger.

22. Haliaetus Leucogaster. (Gmel.), Pl. Col. 49; *Blagrus Leucogaster*, Blyth; *Falco Blagrus*, Shaw; *Le Blagre*, Le Vail., Pl. 5.

HEAD, neck, breast, belly, and thighs shining white; wings light-greyish; tail the same, with broad white tip. Length, 2′ 6″.; wing, 2′; tail, 10″.

Le Vaillant states it is found in South Africa, near the mouths of rivers, but is doubtful if it inhabits this country. I have never met with it, and from knowing it well in Ceylon, should have recognised it on the wing. It is, however, fair to state that Mr. T. H. Bowker, a first-rate sportsman, on being shown a specimen in the Museum, instantly challenged it as a bird he had seen fishing on the coast of the Albany district, but was not able to obtain. Mr. A. V. Jackson also described to me an eagle which he once saw, and which from his description I should have thought to have been *P. Leucogaster*.

It has occurred to me, while thinking over Le Vaillant's faulty descriptions and errors, that many of them may be accounted for by supposing that he saw a bird and failed to obtain it; but thinking it

c

new, or that it was a species with which he was acquainted, he, in the one case, described it as it appeared to him in the casual glance ; in the other, from a foreign specimen. His inventions of habits, nests, and eggs, &c., are deliberate falsehoods.

Genus HELOTARSUS, Dr. A. Smith.

The characters in common with Haliaetus : but the tail is exceedingly short and truncated. Tarsi short, covered with small scales; those posteriorly the largest. Toes long, the basis of all covered with small scales, and the tips with transverse ones.

23. Helotarsus Ecaudatus. *Helotarsus Typicus*, Smith ; *Falco Ecaudatus*, Shaw ; *Terathopius Ecaudatus*, Less. ; *Le Bateleur*, Le Vail., Pl. 7 and 8 ; *Berghaan* (cock of the mountains) of the Colonists.

HEAD, neck, and nearly the whole of the upper and under parts of the body black ; back and tail deep-red ; lesser wing-coverts rufous ; head crested and frilled ; cere deep-orange ; feet crimson. Length, $2'$; wing, $16\frac{1}{2}''$; tail $5\frac{1}{2}''$.

This eagle is very uncommon in this colony ; a single pair frequent some high mountain ranges near Caledon, and I have heard of a few other spots where it is said to exist, or once did do so. I saw several along the East Coast of Africa, and obtained a living specimen from the Governor of Mozambique, which lived with me upwards of two years. It fed voraciously on carrion of all kinds, including fish, but never molested living things—indeed, I kept him for a long time in the fowl-house with the fowls, but had to turn him out, as he devoured the eggs as fast as they were laid by the hens.

I have seen a specimen from Lake N'Gami, procured by Mr. Chapman ; and Mr. Atmore killed another with a stick in a wood at the Knysna.

The Sub-Family, FALCONINÆ, or Falcons,

have the bill short, the culmen curved from the base to the tip, which is more or less furnished on the sides with teeth ; the cere covering the nostrils, sometimes rounded, and sometimes long and linear ; the wings lengthened and pointed, with the second and third quills generally the longest ; the tail lengthened, and more or less rounded ; the feet of various sizes ; and the toes usually long and slender.

Genus FALCO, Linn.

Bill short, strong, with the culmen much arched from the base to the tip, which is acute ; the sides compressed, the lateral margins strongly toothed near the tip ; the nostrils placed in a short cere, naked and rounded, with a central tubercle. Wings lengthened and acute, with the second and third quills the longest, and the first and second notched near the tip. Tail long and rounded. Tarsi short, strong, covered with small irregular scales, and the tibial feathers covering the knee. Toes lengthened and strong, the lateral ones unequal ; the hind toe long, armed, as well as the inner, with a strong hooked and acute claw.

24. Falco Peregrinus. (Linn.) Pl. Enl, 430, 421, 470, 459 ; *F. Barbarus*, L. ; *F. Communis*, Briss.

GENERAL colour above, deep bluish lead-colour, barred with black ; crown of the head and upper part of neck nearly black ; greater wing-feathers dusky, with oval white spots. Tail similar to the back, and much barred ; beneath each eye a patch of black. Under parts, from chin to bottom of the breast, yellowish white, a brownish streak down the shaft of each feather. Thighs and remainder of body dirty-white, barred with deep-brown. Bill blue ; the cere yellow. Length, 1' 9" ; wing, 14" ; tail, 8".

The above description and measurements are taken from a fine European female in the South African Museum. A young male is more rufous on the back, and the under parts more mottled with brown. It is likewise much smaller.

Mr. Sclater quotes a single specimen as having been received from Natal; but I have not seen it from within the limits of the Cape Colony, where the next species seems to take its place.

25. Falco Minor, Bp. ; *F. Peregrinoides*, Smith ; *Spervel* of Colonists.

THE description given of *F. Peregrinus* will suffice equally well for this species, with the exception that all the specimens that have fallen under my observation seem to be duller-coloured. The great distinction is, however, in the size, the present bird measuring as follows :

♂ Length, $16\frac{1}{2}"$; wing, 12" 3''' ; tail, 8".
♀ ,, $18\frac{1}{2}"$; wing, 13"; tail, $8\frac{3}{4}"$.

The little peregrine seems to have a pretty general range over the colony, several specimens having reached me from different localities. I have likewise seen it on the wing several times near Cape Town, and purchased one in the flesh which was being carried through the town by a shooter. Mr. Atmore writes from Swellendam : " It is not rare about here, but very difficult to get, except in the breeding time, when they come after the poultry."

26. Falco Frontalis, Daud.; *F. Galericulatus*,
Shaw., Vol. 7, p. 149 ; *Le Faucon Huppé*, Le V., p. 28.

ACCORDING to Le Vaillant, this bird closely resembles *Hypotriorchis Tibialis* in size and general colouring. The tail is equally banded in grey and black. The folded wings extend beyond the tail. The crest is very conspicuous, and frequently elevated, extending over and beyond the head when at rest. The lower mandible is deeply truncated at the end, as well as notched on either side. Claws very sharp and strong. Female one-fourth larger than the male, with a smaller crest.

It frequents beaches, lakes, and rivers, as it only feeds on fish, crabs, and small shell-fish, which it can break open with its powerful beak. It either builds on sea-side rocks, or else on trees, near rivers abounding with fish. The young and old birds remain together till the next pairing season.

Le Vaillant does not inform us in what parts of the colony he procured his specimens ; but from his description we are led to believe it must be widely distributed. Sundevall denies the existence of such a bird, but suspects it to be an example of *Falco Communis* with a false crest inserted. I have never met with this bird, nor has it been sent to me by any of my correspondents. I fear it is another of Le Vaillant's fabrications.*

27. Falco Biarmicus, Temm., Pl. Col. t., 324.

ADULT ♂, general colour above bluish ash ; transversely barred with dark-grey ; head clear rufous, marked on the forehead, side, and nape with black : under parts clear vinaceous, marked on the thighs with a few black spots. Tail underneath barred grey and ash. Cere and legs yellow ; eye orange. Length, 17″ ; wing, 12½″ ; tail, 7″.

A fine ♀ is more rufous-brown in the general appearance, and is much mottled on the breast and belly.

Scattered throughout the colony, but not common anywhere. It is very rapid on the wing, and a great scourge to poultry and game of all kinds. It will not hesitate to dash at a flock of pigeons feeding close to the door of a dwelling and in the midst of people working on the werf. I have never yet heard of a nest being discovered.

---

* While in London lately, Lord Walden informed me that, being engaged in studying the Drongo shrikes, he visited a museum on the Continent, in which was retained a type specimen of Le Vaillant's, for the purpose of inspecting the bird from which Le Vaillant had made his description. Certain peculiarities in the structure led his lordship to believe that the bird in question was an *old friend* in a *new coat*, and mentioning his suspicions to the curator of the museum, that gentleman had the specimen damped, as if for remounting, when the fact revealed itself that a false white breast had been carefully gummed upon the original skin, from which the natural black feathers had been carefully removed.

Genus HYPOTRIORCHIS, Boie.

The characters in common with Falco, but the tarsi more or less lengthened, somewhat slender, and covered in front with large hexagonal scales. Toes very long and slender.

28. Hypotriorchis Ruficollis. *Falco Ruficollis*, Swain, W. Af., Vol. 1, p. 107, Pl. 2; *F. Chicqueroides*, A. Smith, S. Af. Q. J., p. 233.

GENERAL colour above and below light bluish-ash; much barred with brown-black; tail tipt with white, and crossed near the end by a broad black band; head deep-rufous, with black eyebrows and moustache; chin white; throat and chest vinaceous; legs yellow; bill horn-coloured, yellow at base. Length, 14″; wing, 9¾″; tail, 7″.

The only specimen of this elegant little hawk that has fallen under my notice was shot by my friend, A. V. Jackson, Esq., at Nel's Poort, in the Beaufort division.

29. Hypotriorchis Subbuteo. (Boie.) *Falco Subbuteo*, Lath.; *F. Cuvierii*, Smith; *The Hobby.*

HEAD and upper parts of neck dark-blue grey; rest of upper parts pale-blue. Shafts of all the feathers black; over each eye a narrow rufous white stripe; below each eye a black crescent. Sides of neck, throat, breast, and belly tawny-white. On the two last parts are many black blotches. Under tail-coverts and thighs rufous. Outer vanes of wings hoary-blue; inner vanes dull-brown, crossed with white bars. All margined and tipped with white. Tail slightly rounded, the two central feathers blue-grey only; the others blue-grey, banded with pale rufous, and tipped with white. Length, 14″; wing, 10″ 9‴; tail, 6″ 2″.

Several specimens of this hawk have been received, viz. :—a fine female from Swellendam; a pair, ♂ and ♀, from Mr. Jackson, at Nel's Poort; one ♂ purchased in the flesh in Cape Town. A rich-coloured male also fell to my own gun on the Cape Flats; and Mr. Atmore has procured it near Blanco. Mr. Sclater writes: "Never before received from South of the Equator." *

---

* This observation of Mr. Sclater's opens up a curious subject of inquiry. Have this and other species only lately found their way down the continent? Or have they escaped the notice of observers? I incline to the former supposition, as I cannot conceive that some of our common species should have escaped the notice of such men as Dr. A. Smith and Le Vaillant. Look, for instance, at the extreme abundance of *Cypselus Apus* and *Hirundo Rustica* throughout the colony. How came Le Vaillant not to include these among his swifts and swallows? Surely not because they are European, as he enumerated and figured *Cypselus Melba* and other European species.

30. Hypotriorchis Tibialis. (Daud.) *Falco Tibialis*, Swain. B. of W. Af., Vol. II., p. 212; *Le Faucon à culotte noir*, Le Vaill., Pl. 29.

TOP of head and thighs black-brown; wing and tail-feathers the same, edged with dull white; back and wing-coverts greyish brown, with dark stripes down each centre. Lower part of body rufescent, with brown streaks. Cere of the bill and legs yellow. Bill lead-colour; claws black. Size about that of the peregrine.

Inhabits Namaqualand, according to Le Vaillant, but is extremely rare. Dr. Smith appears not to have met with it. Sundevall declares that Le Vaillant has described the Indian *F. Juggur*, and that it is not an African species. I believe that he had in view a specimen of *F. Minor*.

Genus TINNUNCULUS, Vieillot.

The forms agree with those of Falco; but the tarsi are more or less lengthened, strong, and covered in front with large transverse hexagonal scales. Toes moderate and strong.

31. Tinnunculus Cenchris. (Naum.) *F. Gracilis*, Lesson; *Falco Tinnunculoides*, Natt.; *F. Tinnuncularias*, Vieil.

HEAD, shoulders, and tail ash-coloured; back rufous; under parts vinaceous, more or less spotted with dark-brown; throat and chin white; wing-feathers brown-black; tail tipped with white, and crossed at the end with a broad bar of black; legs and cere yellow; iris yellow-brown. Length, $12''$; wing, $9\frac{1}{2}''$; tail, $6\frac{1}{2}''$.

This species appears periodically in countless thousands in the Damara country, and apparently occasionally strays into the colony. On one of these occasions the pair now in the Museum were obtained by Mr. Cairncross, of Swellendam, in 1860. He informed me that they were feeding on the locusts, and after gorging themselves, perched on the summits of high trees, from which they were easily shot. Mr. Sclater, to whom the pair was sent in 1860, wrote me that they were the first he had seen from the South of the Equator.

32. Tinnunculus Rupicolus. *Falco Rupicolis*, Daud.; *F. Capensis*, Shaw; *Le Montagnard*, Le Vail., p. 35; *Roode Valk* (red hawk), *Steen Valk* (rock-falcon) of Colonists.

HEAD, back, and sides of neck dull blue-grey; back, belly, and shoulders deep rufous, with irregular black spots. Breast rufous, with black lines. Primary wing-feathers black; secondaries blackish, crossed with irregular rufous bands. Tail bluish grey, banded with black, all its feathers broadly tipped with white, next to a broad black bar. Cere round eye bright-yellow; cere of bill and legs orange-yellow; irides brown. Length of fresh-killed male, 11″ (female, 14″); wing, 10″; tail, 7″.

This little hawk is common all over the country as far as I have been. It takes the place in this colony of the *Windhover* of Europe, which it resembles closely in all its habits; hovering in the air over small birds, mice, &c., and falling suddenly from a considerable height upon its quarry. It nests in trees (or the high bushes which do duty for trees in this country), and its eggs, 3—5, are similar in colour to those of the European birds, having a reddish-brown ground, profusely spotted with very dark marks of various sizes and shapes. They, however, vary extremely, even to being pale cream-colour, with the minutest possible spots of brown. In size they also vary, but the shape is pretty constant, being rather round. Axis, 1″ 7‴; diam. 1″ 4‴.

33. Tinnunculus Rupicoloides. (Smith.)

HEAD, neck, shoulders, and scapulars pale tawny, clouded with pale rufous; head and neck with black longitudinal stripes; back, shoulders, and scapulars have broad, brown transverse bars. Throat and under parts pale fawn; breast and belly striped brown; quill-feathers brown, inner vanes banded pale rufous; tail brown, with about six white transverse bands; tips of feathers white. Bill bluish black; base of lower mandible yellow. Length, 16″; wing, 12″; tail, 8″.

This kestrel is very rare near the colony, but becomes more common towards the Zambesi, where Mr. Chapman procured it in some abundance. Mr. Andersson got it in Damaraland.

Dr. Smith gives the West Coast of South Africa, near the Orange River, and the interior to the North of that, as the chief habitat of this species; and states that, like its congener, *T. Rupicolus*, it perches on rocks. Mr. Henry Jackson has just sent this species, with its eggs, from Nel's Poort. The eggs resemble those of the preceding, but are rather larger. The nest was found in a tree, and the parent bird shot from it.

The Sub-Family, MILVINÆ, or Kites,

have the bill short, weak, with the tip hooked and acute, and the lateral margins sinuated; the nostrils basal and lateral, with the opening mostly in the form of an oblique slit; the wings long and pointed; the tarsi hardly longer than the hind toe, and robust; the toes moderate, broad, and padded beneath.

Genus AVICIDA, Swainson.

Bill strong, the upper mandible furnished with two teeth, and the lower mandible emarginated. Wings lengthened, with the fourth quill the longest. Tarsi not longer than the hind toe, the basal half plumed and robust. Toes remarkably broad, the inner toe longer than the outer; claws strong and curved.

34. Avicida Verreauxii, Lafr. *Hyptiopus Caffer*, Sund.; *A. Cuculoides*, Swainson. Birds of W. Af., Vol. I., p. 104, Pl. I.

ABOVE cinereous, with the back and scapulars brown; the throat and breast pale cinereous; body whitish, crossed by broad brown bars; vent and under tail-coverts fulvous, immaculate; tail even, cinereous, with a broad terminal bar. Length, about 16″; wings, 13″; tail, 8″.

A species chiefly found in Western Africa, but sent from Natal by Mr. Ayres. It has not occurred to me.

Genus PERNIS, Cuvier.

Bill weak, gradually curved from the base to the tip, which is hooked, the lateral margins curved, the side compressed, the gonys gradually advancing upwards; the nostrils long, narrow, placed obliquely in the cere; the lores thickly clothed with scale-like feathers. Wings long, broad, with the third, fourth, and fifth quills the longest. Tail long, broad, and somewhat rounded. Tarsi short, with the basal half plumed, and the rest covered with reticulated scales; toes slender, with the last joint covered with transverse scales, the inner one the longest; the claws moderate and curved.

35. Pernis Apivorus. *Falco Apivorus*, Linn.; *Buteo Apivorus*, Jeny.; *Honey Buzzard; Le Tachard*, Le Vail., Pl. 19; *Falco Tachardus*, V.; *B. Madagascariensis*, Smith.

TAIL with four broad and numerous small dusky bands; wings with two similar bands. Adult male with the anterior parts of the head brownish-grey, the upper parts deepbrown; the throat white, with longitudinal dark lines; the rest of the lower parts white, with broad bands and spots of brown. Female with the forehead bluish-grey; the upper parts deep-brown; the lower pale, yellowish red, with large reddish-brown spots. Length, $24\frac{1}{2}''$; wing, $16''$ $9'''$; tail, $11\frac{1}{2}''$.

I have not seen any specimens of this bird from the Cape Colony; but two have been procured in Natal by Mr. Ayres. My description is quoted from Macgillivray's "British Birds." It is a very variable species.

It is now generally believed that Le Vaillant's "Tachard" is founded on a specimen of this bird.

Genus MILVUS, Cuvier.

Bill laterally compressed; the culmen straight at the base, and then much curved to the tip, which is acute, the lateral margins somewhat straight; the nostrils oval, and placed rather obliquely in the cere. Wings very long, with the third and fourth quills the longest. Tail very long and broad, rounded or more or less forked at the end. Tarsi very short, partly plumed from the base, and the rest covered with scales. Toes rather short, with the outer united at its base with the middle one; the claws long and curved.

36. Milvus Ater. *Falco Ater et Austriacus*, Gmel.; *M. Govinda; M. Œtoleus*, Lesson.; *M. Affinis*, Gould.

GENERAL colour deep hair-brown, tinged with rufous on the under parts, particularly on the flanks, thighs, and under tail-coverts; each feather has a black shaft. Tail much forked, and barred with grey-brown. Length, $21''$; wing, $18''$; tail, $10''$.

This kite was received from Mr. David Arnot, of Colesberg, who shot it in a street of that town. It does not appear to have been seen before this instance in Southern Africa. I have since had specimens from the Zambesi, brought down by Mr. J. C. Chapman. From his descriptions, it appears to resemble in habits those of India and Ceylon, frequenting by preference the neighbourhood of native towns, near water, and feeding upon offal, particularly that of fish. It wheels round in large and easy circles, rising and falling, without flapping its wings for an immense length of time. It catches up its food with its foot, and eats it in the air, bending the head down, and bringing up one or both feet to meet it. Some eggs in my possession, procured by me in Ceylon, are of a dull white, with here and there small dry-blood-coloured dots: axis, $2''$ $2'''$; diam., $1''$ $9'''$.

37. Milvus Parasiticus. *Le Parasite*, Le Vail., Pl. 20.

THE whole of the upper parts tan-brown, each feather with a paler edge; breast tan-brown; top of head, neck, cheeks,

D

and throat whitish or light brown, the centre of each feather being dark-brown, inclining to black, gives a striped appearance. Lower part of body, thighs, and under part of tail cinnamon-colour. Wings very long and pointed, tail deeply forked. Legs and cere yellow. Length, 21″; wing, 17″; tail, 10″.

Le Vaillant states that it inhabits Great Namaqualand, builds amongst trees or rocks, but prefers marshy ground, where it makes its nest on some raised spot amongst the reeds. Eggs four, speckled with red. It is a bold and hardy bird, pursuing its prey even in inhabited places.

In November, 1863, I observed a single specimen flying over the village of Ceres, in the Cold Bokkeveld. I have also received it from Kuruman, and observed it along the East Coast as far as Fazy. A pair frequent the homestead of the Hon'ble T. H. Vigne, at Tyger Hoek, River "Zonder End" (1865). The flight and habits of this species differ in no way from those of the common kite of India; indeed, until I shot one, I did not discriminate that any difference existed. It is rare within the colony.

Genus ELANUS, Savigny.

Bill small, much compressed, the cutting margins slightly sinuated; the tip of the upper mandible very much hooked and prolonged; the cere short; nostrils large, oval. Wings very long, acuminated, reaching beyond the tail; the second quill longest; the first emarginated near the tip of the inner web. Tail short, almost even. Feet short, thick: the tarsus half-feathered; the lower part covered with minute roundish sub-equal scales. Toes thick, free; the outer toe much shorter than the inner, and of equal length to the hinder toe. Claws large; the lateral and the hinder ones smooth and round; the middle claw sharply carinated on its inner side.

38. Elanus Melanopterus. (Daud.) Leach.;
Falco Melanopterus, Daud.; *Elanus Cœsius*, Sav.; *Le Blac*, Le Vail., Pl. 36 and 37; *Fal. Vocifer*, Lath.

UPPER parts bluish grey, under parts pure white. Forehead whitish. Tail white on the upper side, slightly tinged with grey. Eye-brows black; shoulders black. Feet bright yellow. Eyes carmine in adult; bright-yellow in the young bird. Length, 12″; wing, 10″ 9‴; tail, 5″ 6‴.

According to Le Vaillant, it builds in the fork of a tree; the nest large, lined with feathers and moss. Eggs white; four or five in number. It perches on the tops of trees or bushes, and utters a frequent and very piercing cry, especially when in flight. Its food consists of insects. I have also found it devouring small birds and reptiles. It is a bold and fearless bird, and generally allows of approach within gunshot. It is migratory, appearing about Cape Town in the month of May. I have never heard of its nesting in this country. To a certain extent this species is gregarious. I counted nine roosting in one tree at Eerste River, and several more were flying about close by. I fancy they had been attracted by an exodus of white-ants which had taken place that afternoon.

The Sub-Family, ACCIPITRINÆ, or Sparrow Hawks,

have the bill short, much arched from the base to the tip, which is acute ; the sides compressed, and the lateral margins greatly festooned ; the wings generally long ; the tail ample ; the tarsi lengthened, and shielded in front ; the toes of various lengths, and generally slender ; the claws long and very acute.

Genus ASTUR, Auct.

Size large ; form robust ; nostrils large, oval, obliquely longitudinal. Bill short, the festoon prominent, lores thickly clothed with minute feathers. Wings short, often not reaching beyond the middle of the tail ; the fourth and fifth quill nearly equal and longest. Feet strong, moderate in length and size. Tarsi, feathered beyond the knees, short, broad, transverse, and smooth ; the divisions prominent ; posterior scales the same ; but those towards the knee joint and the lateral scales are small and reticulate. Lateral toes unequal ; the posterior shortest. Hinder and inner claws half the length of the former, and nearly equal.

39. Astur Melanoleucos, Smith ; South African Quarterly. Journal, Vol. I., p. 229, and Zool. S. Af., Pl. 18 ; *Astur Smithii*, Kaup. Isis, 1847, p. 196.

THE head, neck, and breast dark brownish-black ; back pale-brown ; shoulders much diversified with white ; body, tail, and thighs white ; tail-feathers equal, usually dashed with red. Length, 20". Dr. A. Smith (loc. cit.)

A fine pair of these hawks, ♂ and ♀, shot by Mr. Atmore and his son near Blanco (George) differ materially from the phases of plumage described by Dr. Smith, the ♂ is throughout of a rich brown-black ; the chin and vent are white, with streaks and blotches of the prevailing colour. Length, 18" ; wing, 11" 5" ; tail, 9½".

The ♀ resembles the ♂ on the upper side, but below she is white, blotched with the prevailing colour on the sides of the chest ; two large patches of the same are over the head of thigh-bone, the thighs themselves being of the same colour, mottled with white. Her length is $21\frac{1}{2}''$; wing, 13" 9'''; tail, 12".

Among the many good things sent to the Museum by Mr. Jackson, of Nel's Poort, appeared a young individual (♀) of this rare bird. While staying with him during the month of December 1863, I saw, just out of gunshot, a bird which I feel sure was the adult specimen. Mr. J. informed me that he had frequently seen a similar bird, but could never get within range of it.

Mr. Atmore writes that his birds had proved very destructive to poultry in their neighbourhood. Their nest, in a large tree, contained three young birds, and an egg just hatching.

Genus ACCIPITER, Brisson.

Bill very short, with the culmen much arched to the tip, which is acute ; the sides much compressed, and the lateral margins festooned ; the nostrils placed anteriorly in the cere, large and sub-oval, partly concealed by the projecting hairs of the lores. Wings moderate; with the fourth and fifth quills nearly equal and longest. Tail long, ample, and nearly square at the end. Tarsi longer than the middle toe, covered in front with nearly obsolete scales. Toes more or less lengthened, and padded beneath the joints ; the lateral ones unequal ; and the inner and hind toes equal in length, and both with a strong curved claw.

40. Accipiter Tachiro. (Daud.) *Accipiter Polyzonus*, Lesson., Tr. d'Orn., p. 58 ; *Astur Zonarius*, Tem., Pl., Col. 377, 420 ; Hartl., O.W., p. 15 ; *Ac. Tachiro*, Auct. (the young) ; *Le Tachiro*, Le V., Pl. 24.

ABOVE uniform ashy-brown ; throat whitish, finely rayed transversely with brown ; anterior part of neck, breast, and belly dull white, regularly rayed with transverse brown, or pale rufous bars ; vent and under tail-coverts white, with some fine transverse brown lines ; wing-feathers light-brown, banded with dark-brown, inner vanes marked with white towards the quills. Tail long ; beneath ash-white, with transverse brown bands ; above brown, with darker bands ; tips white. Length, 15"; wing, 8" 3'''; tail, $7\frac{1}{2}''$.

YOUNG BIRD.—Above brown, each feather margined with rufous, and usually with white at the base ; beneath pale Isabella colour, with numerous large oval dark-brown blotches ; thigh transversely barred. Legs yellow.

I have received specimens of the young bird from several of my correspondents, who tell me it is not uncommon in the forest districts ;

but I have never yet seen one in adult plumage. Le Vaillant, who figures the young bird under the name of *Le Tachiro*, describes its eggs as white, blotched with red, and three in number. His testimony on this point must be taken *quantum valet.*

Mr. Andersson has presented specimens procured in Damaraland, in all stages of plumage, to the South African Museum.

41. Accipiter Polyzonoides, Smith, Zool. S. A., Pl. 11.

ABOVE pearl-grey, shaded with brown; throat speckled with light-brown; all under parts of body white, closely banded with small wavy, brown bars; under tail-coverts white. Back and upper sides of wings deep grey; inside of wings greyish-white, with small dark bars. Tail grey, with about six dark-brown bands; tips of feathers white. Tail long, slightly rounded. Cere of bill and legs yellow. Length, $11\frac{1}{2}''$; wing, $7''$ $9'''$; tail, $6''$.

According to Dr. Smith, who found this species about lat. 26° S., it feeds on small birds, and prefers the neighbourhood of rivers.

42. Accipiter Minullus. (Daud.) *Falco Minullus*, Daud; *Le Minulle*, Le Vail., Pl. 34; Smith's Af. Zool., p. 153.

UPPER parts dark-brown; throat and chin white; sparsely streaked with brown; under parts very much blotched with dark-brown; the markings here and there edged with rufous; thighs rufous-brown; tail above brown, faintly barred with a darker shade, but below these bars show plainly on a grey ground. Length, $10\frac{1}{3}''$; wing, $6''$; tail, $5''$.

According to Le Vaillant (who states that it inhabits the country between the Gamtoos River and Kaffraria), a bold and fearless little hawk, preying on small birds or insects, hunting in pairs, and breeding in trees. Eggs five, spotted with brown at the ends. My esteemed correspondents, the Messrs. Atmore, procured a single specimen near George.

43. Accipiter Rufiventris, Smith, Ill. Zool. S. A., pl. 93; *A. Exilis*, Temm., Pl. Col., 496.

UPPER parts brownish-blue, glossed with purple, chiefly on the head and neck; ear-coverts, and all the lower parts of the body and thighs, mottled with reddish orange and white. Tail above barred in shades of brown; all the feathers tipped with dull white; under tail-coverts white. When folded, the wings reach to the middle of the tail. Feet yellow, claws very long, slender, and curved. Female more rufous on the

shoulders and back; mottled on the breast. Length, 16¼";
wing, 10"; tail, 8". The male is smaller, being about 13" in
length.

Found generally, though sparingly, throughout the colony. Two
eggs, said to be those of the present species, were forwarded to me from
Tulbagh: they are of a dirty white colour, irregularly and obscurely
blotched here and there with pale blood-coloured marks: axis, 1" 9"';
diam., 1" 5"'.

I have shot this species in the act of hovering like a kestrel, and as it
preys much on birds and small quadrupeds, particularly field-mice
(*mus pumila*), I do not so much wonder at this habit. At other times
I have seen it glance like lightning through a copse, and whip off a
bird from a branch in passing. It will also eat coleoptera and white-
ants. I saw a pair constructing a nest of sticks in a thick fir-tree. It
was placed over an horizontal forked branch, cleverly supported by
two large sticks across the foundation. I was too early for the eggs.

Genus MELIERAX, Gray.

Bill moderate, gradually arched to the tip, broad at the
base, and with the sides compressed towards the tip, the cere
covering half of the bill, and the sides of the gape naked;
the nostrils placed in the cere, large, and rather oval. Wings
long, with the third, fourth, and fifth quills nearly equal and
longest. Tail long and ample. Tarsi nearly twice the length
of the middle toe, and covered in front with transverse scales.
Toes rather short, with the lateral ones unequal; the outer
toe shorter and weaker; the hind toe as long as the inner,
and equally strong.

44. Melierax Gabar. *Accipiter Gabar*, Daud.; *A. Erythrorhynchus*, Swain., B. of W. Af., Vol. 1, p. 121; Class. of Birds, Vol. 2, p. 215; *Le Gabar*, Le Vail., Pl. 33.

ALL the upper parts and head brownish-grey, darker on the
mantle and occiput. Throat and breast blue-grey; belly
white, barred with grey. Large wing-feathers brown, the
centre ones tipped with white. Upper and under tail-coverts
white. Upper tail-feathers clear-brown, barred with dark-
brown; lower feathers barred, black and white. Cere and
legs red. Iris, according to Mr. Atmore, bright crimson in
adult, yellow in young bird. Length, 14"; wing, 8"; tail, 7" 6".

Le Vaillant found one white egg, and three young birds in a nest, in
a mimosa tree, built of flexible twigs and thorns, lined with feathers.
General in the colony, though not near Cape Town, frequenting the
wooded banks of rivers and kloofs in mountains, preying on small
birds and reptiles. Mr. Atmore states that they whistle very much,
and better than *M. Musicus*.

45. Melierax Niger. *Accipiter Niger ; Sparvius Niger*, Vieill.

GENERAL colour black ; tail-feathers each with three white spots above and four below. Primary quill-feather grey-white, with small black and ashy-coloured spots.

Dr. Smith, in Illustrations of the Zoology of South Africa, gives this bird as an inhabitant of South Africa—*sed non vidi.* By some naturalists it is thought to be simply a melanism of the preceding species, *M. Gabar,* which it equals in dimensions.

46. Melierax Musicus. *Falco Musicus,* Daud. ; *Le Faucon Chanteur,* Le V., Pl. 27 ; *Blaauwe Valk* of Colonists, lit. "Blue Hawk."

UPPER parts and breast pearly-grey ; belly white, variegated by many brownish-blue lines ; shoulders light grey ; rump white. Larger wing-feathers black ; the tail dusky, tipped with white, and crossed by broad white bars, except upon the two middle feathers. Cere and legs red. Irides orange ; according to Mr. Atmore, dark-brown. Length, 24" ; wing, 13" 6''' ; tail, 12".

It devours hares, quails, partridges, and other small game, reptiles, and locusts ; builds either in the fork of a tree or a thick bush. The eggs are four, round, and pure white : axis, 2' 4''' ; diam., 1" 9'''.

Le Vaillant states that the male sings for hours together in the twilight of morning and evening, and sometimes through the night. This I never noticed at Nel's Poort, where it is very abundant, as it is throughout the Karroo. It will perch on the top of a high tree, utter its mellow-piping whistle, and fly off again. I have sometimes heard it call while on the wing.

The Sub-Family, CIRCINÆ, or Harriers,

have the bill moderate, short, with the culmen much curved to the tip, which is hooked, the sides compressed, and the lateral margins festooned ; the wings lengthened and pointed ; the tail long, broad, and even, or rounded at the end ; the tarsi usually lengthened and slender ; and the toes rather short.

Genus POLYBOROIDES, A. Smith.

Bill broad and elevated at its base, the sides much compressed, and the culmen much curved to the tip, which is hooked ; the nostrils placed in front of the cere, with the opening forming a longitudinal slit. Wings very long, with the third, fourth, and fifth quills longest. Tail very long, and rounded at the end. Tarsi the length of the middle toe, and entirely reticulated. Toes long and slender, with the outer toe the length of the hind one. The cheeks round ; the eyes and cere entirely denuded of plumes and hairs.

47. Polyboroides Typicus, Smith, Ill. S. Af.

Zoolog. Pl. 81, 82 ; *Falco Gymnogenys*, Tem. ; *Gymnogenys Madagascariensis*, Less.

GENERAL colour rusty pearl-grey, with a row of large black spots from each shoulder, head crested, and with throat and chest bluish. Flanks, thighs, belly, and vent profusely barred, black and white ; wing and tail feathers black, and tipt with white, the latter with a broad white bar across the centre ; bare space round the eye, cere and legs light-yellow. Iris dark-brown. Length, 2' 1" ; wing, 1' 6" ; tail, 12".

Young birds have much more of the rufous tinge, particularly on the head and crest. Specimens of this curious bird have been procured in various parts of the colony—by Mr. Gird, at the Paarl ; Mr. Cairncross, at Swellendam ; and Mr. Atmore, in the George district. Mr. Chapman also found it in the Interior. Mr. Gird observed his specimen beating over a marsh abounding in snipe. The stomach of one I opened contained the remains of lizards. Mr. Gird's bird was probably engaged in the pursuit of frogs.

Mr. Atmore writes :—" Meiring's Poort : Got a fine adult female of *P. Typicus*. She was full of frogs. This accounts for their sitting so long on stumps, &c., by pools of water. They are very ' getatable,' not at all shy, but scarce."

Genus SERPENTARIUS, Cuvier.

Bill moderate, broad and elevated at its base, the culmen much arched to the tip, which is hooked, and the sides much compressed ; the nostrils lateral, with the opening large and oblique. Wings long, with the third, fourth, and fifth quills nearly equal and longest ; armed on the shoulder with an obtuse spur. Tail very long, wedge-shaped, with the two middle feathers prolonged. Tarsi much lengthened, slender, covered in front with transverse scales. Toes very short, with the anterior ones united at their base by a membrane; the hind toe remarkably short, and rather elevated, and all covered above with transverse scales. The lores and space round the eyes naked.

48. Serpentarius Reptilivorus. (Daud.) *Gypogeranus Capensis ; Falco Serpentarius*, Gmel. ; *Vultur Serpentarius*, Lath., Pl. 2. ; *Ophiotheris Cristatus*, Vieill., Gal. Pl. 260 ; *Le Mangeur de Serpents*, Le Vail., Pl. 25 ; Secretary Bird ; *Slangvreter* of Colonists.

HEAD, neck, breast, and mantle greyish-blue, slightly shaded with reddish-brown on the wing-coverts. Larger quills black. Throat white, under part of tail white, tinged with russet. Thighs black, tinted brown. Tail-feathers very long, black at the base, then paling into grey, and tipped with white; two long central tail-feathers bluish-grey, tipped with black and white. Crest of ten feathers arranged in pairs, the lesser ones at the top black and grey, the others black. Legs very long, toes short, claws blunt. Length, 48″ ; wing, 27″ ; tail, 26″.

Widely distributed throughout the colony When a pair establish themselves in any locality, they speedily drive out all others of the same kind, and will breed in the same nest for a long period. The nest is a huge structure, added to yearly, placed sometimes in a low bush, sometimes in a thick thorny mimosa. I am told they never lay more than two eggs, which are of a dirty dull white, profusely dotted with light, brownish-red blotches at the obtuse end, and sparsely over the whole shell : axis, 3″ 1‴ ; diam., 2″ 4‴. The young utter a gutteral, rattling cry, precisely resembling the call of the Stanley Crane. They are a long while ere they can walk, as their legs seem unable to support the weight of their bodies, and snap with the least exertion. One which I was rearing for the Acclimatization Societies of Melbourne and Sydney trod in a small wooden bowl sunk in the ground, and instantly fell, breaking his leg* and wing. I spliced them both, but he died in ten days. I am informed that the texture of the nest is so loose, that the legs of the young hang through the interstices until they acquire sufficient strength to be bent under them.

Genus CIRCUS, Lacépède.

Bill moderate, elevated at the base of the culmen, and arched to the tip, which is hooked ; the sides compressed, and the lateral margins festooned ; the nostrils large, oval, and partly concealed by the curved hairs of the lores. Wings long, with the third and fourth quills nearly equal and longest. Tail long and rounded on the sides. Tarsi long, slender, and compressed, the outer side covered with transverse scales, and the inner with small scales. Toes

---

* All who have tried to rear these birds notice this brittleness in their bones. I have, since the above was written, lost several more from similar causes. I have known them snap a leg if suddenly startled into a quick run!!

E

moderate, with the outer one longer than the inner; the claws long, slender, and acute.

49. Circus Swainsonii, Smith, Af. Zool., p. 162; *C. Superciliaris*, Id., S. Af. Q. J., p. 385; *C. Palidus*, Sykes; Gould's B. of Eur., Pl. 34; *C. Dalmaticus*, Rüpp.

ABOVE ash-grey, clouded with yellowish-grey; front, upper tail-coverts, shoulders, and all the under parts white; throat and breast tinted pearl-grey, sometimes streaked with rufous. Tail-coverts barred with yellowish-brown. Upper side of tail faintly barred with brown; lower sides white, obscurely barred. Figure long and slender. The bill strongly hooked; the cere covered with bristles; tail long, slightly rounded. The young much resemble *C. Ranivorus*, but is more distinctly striped on the breast. Length, 17½"; wing, 10"; tail, 9½".

Its food consists of small quadrupeds, reptiles, birds, &c.; prefers a low damp locality, and has a wide range in South Africa. Smith says he has seen a few specimens near Cape Town, Natal, the Mouth of the Orange River, and the Tropic of Capricorn. Mr. Jackson sends it from Nel's Poort, and I have shot it near the Observatory, Cape Town, in company with *C. Ranivorus*.

50. Circus Cinerascens, Montagu; *Falco Cinerascens*, Temm.; *Buteo Cinerascens*, Jen.; *The Ash-coloured Harrier; Circus Montaguii*, Vieill.; Gould's B. of Eur., Pl. 35.

♂ GENERAL colour of upper parts, neck, and breast light bluish-grey; the quill-feathers of the wings inclining to black; belly and thighs white, the latter with the flanks longitudinally streaked with bright rufous. Tail pearl-grey above, white beneath, the inner webs obscurely barred. Two outermost feathers on each side barred with rufous, which is deepest on the basal part of the feathers. Irides, legs, and feet fine yellow. Length, 17"; wing, 14½"; tail, 9".

♀ General colour umber-brown above; below pale rufous-yellow, faintly streaked along the shaft of each feather with a darker tint; collar round the neck rufous. Outer tail-feathers white, the rest barred distinctly white and brown. A young female shot by myself on the Cape Flats is very dark-brown above, many of the feathers being edged with bright rufous; below bright rufous, streaked with black.

The ash-coloured harrier is rare in South Africa. Mr. Atmore procured a fine adult ♂ at Swellendam, and I shot an adult ♀ and the young bird mentioned above near the Observatory at Capo Town, below which is a marsh much frequented by harriers of several descriptions. I also killed a fine ♂ in the crater of an extinct volcano at Johanna, one of the Commoro Islands, where I saw many specimens.

It preys upon reptiles, mice, and small birds, remains of which I found in the stomach of those procured. Mr. Atmore writes : " This fellow hunts his ground like a pointer, and drops suddenly on its prey, which, from his minute inspection of fences, I suspect to be mice."

51. Circus Ranivorous. (Daud.) *C. Levaillantii ;* Smith (the young bird); *Le Grenouillard,* Le Vail., p. 23.

ALL the upper parts of the body earthy-brown, that is, all the visible parts of the feathers ; the hidden parts are white on either side of the quill. Throat and cheeks thinly sprinkled with whitish feathers, with a brown band ; lower parts clear-brown, slightly mingled with white. On the legs rusty red feathers, edged and spotted with white ; wings brown, with transverse white bars on the under sides. Tail rufous-brown, and banded. Length, 19″ ; wing, 15″ ; tail, 9″.

YOUNG PLUMAGE.—Neck, throat, and anterior part of breast deep-brown ; lower parts reddish-brown ; shoulders in two shades of brown. Fore part of neck and breast white, or reddish-white ; wings deep-brown, with pale tips. Tail deep-brown, with pale bands and tips. Iris brown-yellow.

Feeds chiefly on frogs and small aquatic birds, and sometimes fish. Abundant throughout the colony, frequenting marshy places. The nest is built amongst rushes, and is made with the stalks and heaped-up leaves of water-plants. Eggs three or four, of a dull white : axis, 1″ 10‴ ; diam., 1″ 6‴.

The marsh below the Observatory is a favourite breeding-place for these birds, and several nests are annually made among the rushes. I found it very abundant at Zoetendals Vley, and saw it even in the dry Karroo country about Beaufort West.

52. Circus Maurus, Temm. ; *C. Ater,* Gray nec Vieil.

GENERAL colour black, tinged with rufous ; rump white ; tail on the upper side with four broad bars of grey, which below show white ; inside of wing more or less white, upper side grey ; a few white spots on the belly and flanks ; cere of bill, legs, and irides bright orange-yellow ; claws black. Length, 20″ ; wing, 13″ 6‴ ; tail, 9″ 9‴. .

" By some writers, this species is thought to be a black variety of *C. Cinerascens,*" G. R. Gray.

This very handsome harrier is not at all uncommon in the neighbourhood of Cape Town, and is generally seen in pairs, beating the bushes for prey, and quartering the ground with the regularity of a pointer dog.

After a few heavy flaps with its wings, it sails along with its pinions elevated, swaying to and fro like a clock pendulum; suddenly it checks itself, lets fall a leg, clutches up a cowering lark or unsuspecting gerboa, and flies away with it to the nearest termite heap, on which it perches and commences its repast. If accompanied by its mate, a shrill stridulous cry soon brings it to its side, and the dainty morsel is shared between them.

It breeds among reeds, making a thick heavy nest cn any elevated root that may lift it above the water. I am told that the eggs are white, but have not seen them. This Mr. Atmore confirms.

53. Circus Melanoleucos, Vieil; *Falco Melanoleucos*, Gmel.; *Le Tchoug*, Le Vail., Pl. 32.

THE head, neck, mantle, and interscapulars very deep-brown, changing to clear-brown on the scapularies and part of wing-coverts. Rest of wings grey-white, mingled with feathers of equal white and brown. Larger wing-feathers black, the lesser pearl-grey. On the back of the head a mixture of black, white, and brown. The lower parts of the body, the very long feathers of the legs, and lower tail-feathers pure white. Rest of tail reddish grey-white, a large brown spot on the end of each of its central feathers.

Le Vaillant describes from a Bengal specimen, but states that he plainly saw the bird in the Long Kloof. I believe he has mistaken *Astur Melanoleucos* for this species.

54. Circus Acoli. *Fulco Acoli*, Shaw; *L'Acoli*, Le Vail., Pl. 31; *Witte Valk* and *Leeuwerkvanger*.

COLOUR on the head, neck, and mantle pale grey-blue; feathers on legs very long, hanging low on the tarsus. Lower part of wings very much darker; all the lower parts of the body are whitish, finely rayed with dark-grey; body long and slight; tail, legs, and tarsus long.

Generally distributed, according to Vaillant, who says it is most abundant in the districts of Zwartland, "Red Sand," and "Twenty-four Rivers."

It flies very swiftly, but always low. In the colony it frequents cultivated places; in the deserts it prefers sandy ones. Its chief food consists of moles, mice, and small birds. They build in thick bushes. Eggs four, oval, and dirty-white.

Professor Sundevall supposes this to be another of Le Vaillant's manufactured species. I believe he has seen, *at a distance*, the rare

Polyboroides Typicus, or the West-African *Astur Monogrammicus*, Swain., and has confounded its habits with those of *Circus Maurus*, as his account of them exactly agrees with what I have observed of the latter.

The Third Family, STRIGIDÆ, or Owls,

have the bill short, the sides compressed, the base concealed by projecting bristles, and the culmen much arched to the tip, which is hooked; the wings usually moderate and rounded; the tail broad and of various lengths; the tarsi generally short, and feathered to the toes; toes usually short, and either clothed with plumes or slightly covered with hairs; the claws long and acute. The eyes large, and encompassed with a radiated circle of slender, hairy feathers, forming a facial disk.

The Sub-Family, SURNINÆ, or Hawk Owls,

have the head small, without tufts, and the facial disk imperfect above the eyes.

Genus ATHENE, Boie.

Bill short, partly concealed by the projecting plumes, the sides compressed, the culmen much arched to the tip, which is hooked and acute; the nostrils basal, lateral, and hidden by the frontal plumes. Wings rather long and rounded, with the third and fourth quills the longest. Tail moderate and nearly even. Tarsi longer than the middle toe, and covered with plumes. Toes short, and covered with scattered hairs; the claws long, arched, and acute.

55. Athene Capensis. (Sav.) *Noctua Capensis,*

Smith, Zool. S. Af., p. 33.

HEAD and neck red chesnut-brown, with imperfect white bars; from neck to end of tail umber brown, glossed with purple; some feathers have large white blotches, others yellowish bars; wings similarly coloured and variegated. Breast, of all these colours, but each paler. Belly white, with large brown blotches; under part of tail pale-coloured; legs white, faintly marked in brown. Length, $8\frac{1}{2}''$; wing, $5''\ 9'''$; tail, $3\frac{1}{2}''$.

The few specimens known to Dr. A. Smith were all killed in forests in the Cape Colony. I have not met with it; neither does M. Victorin appear to have found it at the Knysna.

56. Athene Licua, Licht. Jard. Cont. to Orn., 1852, p. 142.

TAIL greatly exceeding the wings, with six rows of white spots ; toes slightly hairy.

The above curt description is the only one which I have been able to obtain of an owl said to exist in South Africa.*

The Sub-Family, BUBONINÆ, or Horned Owls,

have the head large, broad, flat on the top, and furnished with two prominent tufts ; the facial disk imperfect above the eyes.

Genus BUBO, Sibbald.

Bill moderate, broad, concealed at the base, and compressed at the sides ; the culmen curved to the tip, which is hooked and acute ; the lateral margins slightly sinuated ; the nostrils placed in the fore part of the cere, hidden by the projecting plumes, large, oval, and rounded. Wings rather long, with the second, third, and fourth quills nearly equal and longest. Tail moderate, rounded. Tarsi short, thick, and densely plumed. Toes moderate, and thickly plumed ; the top of each toe furnished with two or three transverse scales, which are more or less concealed by the plumes ; the lateral toes unequal ; the hind toe short ; the claws long, strong, and curved.

57. Bubo Verreauxii. Bp. Cons., Av., p. 49 ;
Bubo Lacteus, Temm., Pl., Col. 4.

GENERAL colour above and below brownish-grey, minutely variegated throughout. A white mark extends from the shoulder half-way down the back, and several white blotches appear on the wing-coverts. The facial disk is lightish-grey ; a semicircle of black extends from each horn, under the chin, but they do not unite by several inches. The horns are dark, rusty-brown. The stiff hairs, which project along the bill, are black along their shafts and at the tips. This species has been separated from *B. Lacteus* on account of its larger size. Iris, according to Mr. Atmore, black and red. Length, 26″ ; wing, 18½″ ; tail, 6½″.

Several specimens of this noble owl have come to me from my correspondents. Mr. Jackson says it is not uncommon in the mountains near his residence. Mr. Arnot has sent it from Colesberg, and Mr.

* Schlegel in his catalogue of the Museum of Natural History of Holland, which has but just reached my hands, while these sheets are passing through the press, gives this as identical with the preceding species.

Atmore procured it at Blanco. This latter gentleman's account I transcribe :—

"Meiring's Poort, June 19th, 1863.—I have at last obtained the large owl described to be as 'big as a vulture.' I send him herewith. Eyes black, with a narrow brick-red ring round the iris. He is very destructive to poultry, and judging from the capacity of its swallow, could take down a chicken at a gulp. The throat is of a peculiar construction, and I now regret I did not send it to you in spirits. The fat all round it was half an inch thick. His call-note is often mistaken for that of a leopard. One killed a dikkop *(Ædicnemus Capensis)*, and left his skeleton close to the house, so it appears he victimises them as well as poultry. He throws up no quids."

Mr. Jackson corroborates Mr. Atmore's statement of its being destructive to poultry. He had many hens sitting in boxes in a building with high walls, and closed doors, secure from every prowler but a winged one, for the window in the roof being left open, each night a hen disappeared. Mr. Jackson set a wire snare over each box, and soon had the satisfaction of noosing the robber, who, even with the wire round his leg, would not let go his prey, and was killed "red-handed."

58. Bubo Capensis, Smith, Zool. S. Af, Pl. 70 ;
Le Moyen-duc, Le Vail. The large Eared Owl.

GENERAL colour umber-brown, the under side being the lightest, variegated with irregular spots of dark-rufous on the head and shoulders, and with bars and spots of deep orange on the rest of the body. Wings umber-brown, much varied by pale buff bars of irregular width ; each bar is spotted, or speckled brown. Feathers of tail yellowish-white, with narrow dark bars. Legs and toes ochreous-yellow ; collar brown ; spot on the breast white. Facial disk yellow-brown, with white wiry feathers near the beak. Length, 23″ ; wing, 16″ ; tail, 9″.

Its food consists of moles, rats, and small birds. Le Vaillant says it makes a nest in a rocky locality, of branches, dried leaves, and moss. Eggs three.

I shot a specimen on the Cape Flats in the day-time on the 25th May, 1855 ; it appeared to see perfectly well. Since I made this old note in 1855, I have only obtained one other specimen ; the species must, therefore, be rare. The example alluded to above was shot in broad daylight in the bright sun, and gave me a long chase, after being wounded at the first shot. He rose out of some rushes in the first instance, and then perched in several different places, but always keeping me well in view, and flying off on my approach. Two eggs were lately given me as the "eggs of the large owl," which probably belong to this bird : they were taken from a nest in the rocky sides of the Winterhoek Pass, by Mr. Short, the Superintendent of Convicts, to whom I am indebted for many good specimens. Eggs pure white, smooth, but not shining: axis, 2″ ; diam., 1″ 9‴ ; ends both alike rounded.

59. Bubo Maculosus, Vieil. Gal., Pl. 23 ; *Strix Africana*, Temm. ; *S. Nisuella*, Lath. ; *Le Chou-couhou*, Le Vail., Pl. 39.

ABOVE grey-brown, more or less variegated with light ochreous (or white) spots and blotches ; below, the ground colour is more ochreous, and the markings, chiefly fine wavy bars, with a few large blotches, are of the same brown as the back. A dark circle surrounds the facial disk, broken under the chin with rufous and white, immediately under which is a broad white collar ; facial disk grey. Feet feathered to the toes, lightish-yellow or dirty-white, speckled with brown. Tail broadly barred with ochreous brown. Length, 19″ ; wing, 13″ ; tail, 7½″.

The commonest owl in South Africa. It is found throughout the colony, and extends to Natal, and as far as the Zambes'. Mr. Blyth writes me word that it was brought from the Somalie country by Capt. Burton. Frequents forests, open lands, bush—covered or cultivated places indiscriminately. I think it is subject to partial migrations, as on several occasions I have, while hunting for game, come upon little parties of eight or ten individuals, which kept together in their flight. Pointer-dogs will stand to them as staunchly as to game. My late lamented friend, the Rev. John Fry, of Rondebosch, a zealous observer of birds, informed me that they regularly frequented a grove of fir-trees in his garden at one season of the year. I have noted them in these trees in October. About twilight they begin to move, ascending to the highest branch, and uttering their loud "hŏo, hŏo" for a quarter of an hour together, and then sailing off to the Cape Flats in search of their prey, which consists of mice, rats, moles, and sometimes frogs. They lay two purely white eggs : axis, 2″ ; diam., 1″ 9‴, in holes of rocks, and sometimes in an old termite's heap, or even on the bare ground.

I sink Le Vaillant's *Chou-couhou* (*S. Nisuella*, Lath.) into a synonym of this species on the authority of Sundevall, who states that he has seen Le Vaillant's original specimen, which is nothing more than *B. Maculosus*. I should think this correct. And as regards habits, those described by Le Vaillant are precisely those of *B. Maculosus*, particularly the cracking noise made by the bill, which this bird constantly makes.

Genus EPHIALTES, Keyserling and Blasius.

Bill moderate, the base concealed and broad, the sides compressed, with the culmen flattened at the base and curved to the tip, which is hooked ; the lateral margins curved ; the nostrils rounded, placed in the fore part of the cere, and covered by the basal plumes. Wings long, with the second, third, and fourth quills nearly equal and longest. Tail short, and more or less even. Tarsi rather long, and covered with short plumes to the base of the toes. Toes long, covered with small scales at the base, and with three or four transverse scales at the tip; the claws moderate, strong, and curved.

60. Ephialtes Senegalensis. (Sav.) *Scops Senegalensis*, Swain., B. of W. Af., Vol. 1, p. 127; *Le petit Duc*, Le Vail., Vol. 1.

"VARIEGATED with grey, ferruginous, and black; ears grey, without any marginal band; marginal base of shoulders whitish; second quill shorter than the fifth." (Swain. loc. cit.)

This bird was mistaken by Le Vaillant for "Le petit Duc" of Europe (*Scops Europæus*). Mr. Swainson seems to have been the first to discriminate between them, and to point out the specific differences; his type specimen is in the Cambridge Museum. Le Vaillant states he saw it in the Camdeboo. I saw a bird in the possession of the Count de Castelnau, said to have been procured at the Knysna, which was either this species or the common Damaraland *Scotophilus Perlatus*. I could not, however, obtain a description of it at the time, and quote from memory.

61. Ephialtes Leucotis. *Strix Leucotis.* Tem. Pl. Col. Pl. 16. Swain, Birds of W. Af., Vol. 1, p. 124.

GENERAL colour brownish grey, longitudinally striped with black. Face white; margined with a black band. Egrets very long, more or less tipt with black. Under plumage with a deep black line down the centre of each feather. Legs white, faintly mottled. Length, 11"; wing, 8"; tail, 4".

This pretty little owl is apparently common in Damaraland, and I have received specimens from Mr. David Arnot, procured in Mahura's country. Mr. Ayres also obtained it at Natal. It does not come into the colony; I therefore know nothing of its habits, except that it roosts in trees and feeds on coleoptera.

The Sub-Family, SYRNINÆ, or Owlets,

have the head large, smooth, and furnished with two small tufts; the facial disk moderate and complete, or nearly so.

Genus SYRNIUM, Savigny.

Bill moderate, strong, with the base broad, and concealed by the frontal plumes; the sides compressed, and the culmen arched to the tip, which is hooked; the nostrils basal, lateral, and medial, with the opening large and oval. Wings long

F

and rounded, with the fourth and fifth quills equal and longest. Tail long, broad, and more or less rounded. Tarsi short, strong, thick, and densely clothed with plumes. Toes moderate, densely plumed to the end ; claws long, acute, and slightly curved.

62. Syrnium Woodfordii, Smith, Zool. S. Af., Pl. 71.

♂ ALL the upper parts very dark reddish umber, variegated with white spots, and sienna-coloured, and white bars. Wings generally yellowish brown, marked with dark bars, and yellowish, white blotches. Collar white, with brown bars. When lying smooth, the breast and belly appear alternately in bars of white and reddish pale-brown. Tail reddish-brown, barred with pale-brown. ♀ Everywhere of a much lighter colour. Length, 13¼"; wing, 9"; tail, 6".

Not common, but has a wide range in the colony. Food small quadrupeds, birds, insects, &c.

I have received two specimens from the Knysna, procured by the late Mr. Villet in that neighbourhood ; but know nothing of its habits.

63. Syrnium Choucou, Lath.; *Strix Africana*, Shaw ; *Le Choucou*, Le Vail., Pl. 38.

THROAT, front of neck, breast, the whole body, lower part of wings, legs, tarsi to the toes, covered with white, thick, silky feathers ; those on the leg are long enough to cover the toes ; the back is almost hidden by the same profusion of fine slender feathers. Top of head, back of neck, and mantle grey, russet-brown. The wing-coverts, of this last colour, are varied with white spots, and each feather is tipped with white. The tail has twelve feathers, two in the centre brown only ; the others are brown, barred white ; under part of tail wholly white.

Le Vaillant never found the nest or eggs ; but states that his Hotten-tots told him they nested in broken trees. Eggs white. In flight, it constantly utters a short sharp cry.

Inhabits Outeniqualand, according to Le Vaillant ; but according to Professor Sunderall, a fabulous species.

Genus OTUS, Cuvier.

Bill moderate, mostly concealed by the frontal plumes, broad at the base, and compressed at the sides ; the culmen curved towards the tip, which is acute ; the nostrils lateral, medial, and hidden by the plumes. Wings very long, with the second and third quills the longest. Tail moderate, even. Tarsi long, and covered with plumes. Toes moderate, more or less plumed, the lateral ones unequal, the outer one shortest; the claws long, curved, and acute.

64. Otus Capensis. *Phasmoptynx Capensis,* Smith, Ill. Zool. S. Af., Pl. 67, and S. A. Q. J., 1855, p. 306.

UPPER parts of head, neck, back, upper tail-coverts, breast, and lesser wing-coverts reddish-brown, varied with a more yellow shade. Body and legs pale yellow-ochre, barred with brown. Wings chiefly brown, barred with yellow; most of the feathers with a much paler margin. Tail barred brown and yellowish-white. Facial disk pale, dull-yellow, marked with brown; deep-brown round the eyes. Collar deep-brown.

Rare in South Africa, and only found in marshy places. Usually gregarious. It conceals itself during the day amongst long grass, or reeds, &c. They prey upon water-insects, mice, and lizards. Dr. Smith gives no locality for this bird, which must be very scarce, as no specimen has ever fallen under my observation from any part of the colony. Mr. Gurney includes it in his list of birds procured at Natal by Mr. Ayres, Ibis Vol., 1862, p. 26.

The Sub-Family, STRIGINÆ, or Owls.

have the head smooth, with the facial disk very large, sub-triangular, and complete.

Genus STRIX, Linn.

Bill long, mostly concealed at the base by the projecting plumes; the sides much compressed, and the culmen arched to the tip, which is acute and hooked; the nostrils large, partly covered by a membrane, with the opening oval and slightly hidden by the hairs. Wings very long, with the second quill the longest. Tail short and generally even. Tarsi much longer than the middle toe, slender, and covered with short soft plumes. Toes long, and covered with scattered hairs; the lateral ones unequal, the outer much shorter than the inner; the hind toe short and thick; the claws long, curved, and acute.

65. Strix Affinis. (Blyth.) *Strix Flammea,* Linn. var.; the common White Owl of S. Africa.

UPPER parts rather deep reddish-orange, profusely mottled with dark grey, and spotted with white. Tail with four bars of deep-brown, variegated with grey. Under parts ochreous-

yellow, spotted with small arrow-head-shaped marks of a
dark-brown ; on the feet, and half-way up the legs, short, stiff,
yellowish hairs.   Facial disk white, tinted with vinaceous,
darkest at the inner angle of the eye, and surrounded with a
circle of dark-brown, approaching to black, on the lower half.
Length, 16'' : wing, 12'' ; tail, 5'' 6''.

I cannot bring myself to consider this bird as a mere variety of the
European *S. Flammea.* I have shot reputed *flammea* in Ceylon, in
Egypt, Greece, England, and here, and have seen specimens—indeed,
have them before me while I write—from Europe, England, India,
two species from the Cape, *S. Flammea* and *J. Capensis,* and the
Australian *S. Personata.* I look upon them all as distinct, though
as closely related to each other as the *black crows.* There is as much
difference between them, and I therefore adopt Mr. Blyth's name
Affinis for our Cape bird.   *Strix Affinis* is common all over the colony,
breeding in the roof of nearly every farm-house that I have entered.
In Mr. Fry's fir-trees, a pair or two may be found at any time.   They
occupy the same perch for days together, the ground below being
covered with their pellets.   On examining these, they will be found to
consist almost entirely of the bones of mice—once only have I found
the skull of a bird.

Eggs, two to four in number, pure white, not shining : axis, 1'' 7''' ;
diam., 1'' 4'''.

66. Strix Capensis, Smith, Zool. S. Af., Pl. 45.

UPPER parts dark-brown ; on each feather a white spear-
shaped spot, larger and more distinctly marked as they near
the tail.   Shoulders and inside of wings orange, spotted
with brown.   Sides of neck and under parts pale, reddish-
orange, each feather with a light-brown, arrow-headed spot
on its tip.   Facial disk vinaceous near the eye, brownish-
black, surrounded by a ring of white feathers, tipped with
brown.   Tail paler above ; here and there barred with brown
below, nearly white.   Length, 17'' ; wing, 13½'' ; tail, 5''.

Found about Table Mountain (Smith), from which locality I have
also obtained one pair ; but the species must be very rare, as but few
other examples have occurred to me.

Mr. Atmore writes : " Blanco, May 25th, 1864.   A boy reports an
owl's nest in a vley near this.   It was the ♀ of this nest probably that
Tom saw in the clutches of the crested eagle.   26th : Have visited the
owl's nest and shot the ♂ *(Strix Capensis),* found three large young
ones in the nest just getting quill-feathers ; so they breed early.   It is
not a rare bird here, but difficult to raise.   It inhabits the wide palmiet
vleys.''   Mr. Ayres has procured it in Natal.—Ibis 1859, p. 28.

Order II. PASSERES.

THIS Order consists of a numerous series of Birds, whose feet are more or less constructed for perching. The toes are generally placed three before and one behind, though sometimes the posterior one is capable of being brought forwards, and sometimes the outer toe is carried backwards, which occasions the toes to be placed two before and two behind.

The First Tribe, FISSIROSTRES, or Fissirostral Birds,

have the bill very wide, and the gape extended beneath the eyes ; with the culmen short, and curved to the tip. The feet of these birds are weak, and of various conformatious.

The First Sub-Tribe, FISSIROSTRES NOCTUR-NÆ, or Nocturnal Fissirostral Birds,

are remarkable for the size of their eyes, which are very large, and easily dazzled by the light of day. Their plumage is generally soft, which enables them to fly without noise.

The First Family, CAPRIMULGIDÆ, or Goatsuckers,

have the bill short, more or less depressed, very broad, and the gape extending beneath each eye ; the tarsi very short, and more or less exposed ; the toes long, and rather strong ; the hind toe closely united to the base of the inner toe.

The Sub-Family, CAPRIMULGINÆ, or Goat-suckers,

have the bill short and weak, with the gape, extending under each eye, extremely broad, and furnished with more or less lengthened bristles ; the wings long and usually pointed ; the tail more or less lengthened and round, or graduated and sometimes forked ; the tarsi more or less short, and almost clothed throughout with short plumes, or entirely denuded and scaled ; the toes moderate, with the lateral ones shorter than the middle toe, and of equal length; the claws of the middle toe large, and pectinated on the sides, those of the other toes small.

Genus CAPRIMULGUS, Linn.

Bill very short, with the gape, extending under the eyes, very wide, and furnished with long strong bristles; the culmen curved, and the sides much compressed to the tip, which is hooked and emarginated; the nostrils lateral, tubular, with the opening large, and partly exposed. Wings very long and pointed, with the second quill the longest. Tail rather long, broad, and more or less rounded, or very long and strongly forked. Tarsi shorter than the middle toe, strong, and more or less clothed with plumes. Toes strong; with the lateral toes short and equal; the hind toe very short, the middle toe long, and armed with a broad serrated claw.

67. Caprimulgus Natalensis, Sm., Zool. S. A., Pl. 99.

GENERAL colour variegated rufous-white, yellowish, and reddish-brown; scapularies marked with oblong blackish-brown spots; on the throat in front a triangular white mark; the four external wing-feathers brownish-red, each with a white patch. Tail banded with brownish-black. Length, 9" 3'''; wing, 6" 3'''; tail, 4" 6'''.

Mr. Swainson, in alluding to the difficulty of describing the singularly variegated, dull-coloured plumage of this family of birds, makes the following remarks:—"We might almost say, that in regard to *colour*, if a person has seen one species of *Caprimulgus*, he has seen *all*. There are, it is true, trifling variations, but these are sometimes so slight that none but an acute ornithologist would detect them; while, from their very nature, they are so difficult to describe, that the most laborious descriptions fail to convey their differences to the mind of the reader; the different shapes of the marks, bands, and spots, the manner in which they are blended, and the diversity of tints under which the different colours of brown, grey, rufous, and whitish appear, often upon a single feather, in these parti-coloured birds, defy all clear description, and render it impossible by these indications only, to discriminate the species."

Fully coinciding in these views, which apply equally to most of the partridges, I purpose simply to indicate our species of goatsuckers by some particular point in their appearance, referring the student to those works in which more elaborate descriptions have been given.

Dr. Smith says that this species may be readily distinguished by a number of strongly-marked and well-defined oblong, liver-brown spots, each more or less distinctly edged with white, or light buff, which are arranged in somewhat longitudinal rows, on or about the middle of each scapular feather, contrasting with the other colours. Dr. Smith only procured two specimens, both near Port Natal.

68. Caprimulgus Rufigena, Sm., Zool. S. A., Pl. 100.

THIS goatsucker has the usual colours and markings of its tribe, but may be distinguished from all those inhabiting the colony by its rusty, buff-orange collar. It has a broad white bar across the inner vanes of the first and second wing-feathers, which in the third and fourth feathers extends across both vanes. Two outer tail-feathers of each side broadly tipped with white. Length, 9" 4'''; wing, 6" 2'''; tail, 5".

Dr. Smith procured this species from the Eastern districts of the colony. The specimens we have seen came from Damaraland, where it seems common.

69. Caprimulgus Smithii. (Bp.) *C. Europæus* of Smith.

MAY be recognised from all the other species by its lighter grey appearance, and from the absence of any white patches on the wing and tail feathers. The spot on the throat is also very indistinct. Length, 9"; wing, 7" 3'''; tail, 4" 6'''.

I believe this to be the bird figured by Dr. A. Smith as *C. Europæus* (Zool. S. Af., Pl. 102). It is smaller than that species, and, as far as I know, the ♂ equally with the ♀ wants the white markings on the tail.

I have received specimens from Colesberg, Swellendam, &c., and shot it myself in the neighbourhood of Cape Town.

I observe that goatsuckers are more abundant here about March than at any other time, though stray specimens may be found all the year round.

70. Caprimulgus Lentiginosus, Smith, Zool. S. A., Pl. 101.

THIS species may be distinguished by its greater size; by the head being generally speckled; and the ruddy colour of the wings; the tail also has no white about it. Length, 10" 6'''; wing, 7" 6'''; tail, 5" 4'''.

Dr. Smith procured two specimens in Great Namaqualand. Its habits appear to resemble those of the common species.

71. Caprimulgus Infuscatus, Rüpp., Atlas, 16 ; *C. Nubicus*, Licht. ; Ibis, Vol. 1859, p. 242.

GENERAL colour reddish-grey, finely mottled with black ; chin, throat, patch on wing, and two external tail-feathers white. Length, 8″ ; wing, 5″ 3‴ ; tail, 3¾″.

Natal: Ayres " Ibis "—loc.cit. non vidi.

72. Caprimulgus Atrovarius. (Sundevall.) Le Vaillant, Pl. 49 ; *C. Pectoralis*, Vieil ; *C. Africanus*, Steph.

MAY be distinguished by its much darker and distinct colouring, and broad rufous collar ; four outer wing and two outer tail feathers marked with white. Length, 10″ ; wing, 7″ ; tail, 5″ 9‴.

This is the bird that I identify with Le Vaillant's " Engoulevent à Collier." It is found about Cape Town, frequenting gardens and orchards. It is common at Rondebosch, and along the forest-clad Eastern face of Table Mountain.

Like all its congeners, it flies abroad as soon as the twilight begins to darken, and often perches on the rails of garden-fences, sitting lengthways along them. It has the churring note of the other goat-suckers, and lays two greyish, cream-coloured eggs, shaped much the same at each end, freckled and blotched with brown spots and patches : axis, 14‴ ; diam., 10‴.

The nest is a mere depression in the soil, under the shelter of a large stone or bush. At Mr. Vigne's farm, on the River Zonder End, a bird of this species has hatched her eggs, for several successive years, in a flower-bed close to a well-used path. I saw her sitting in the beginning of November : she allowed me to approach within two or three feet of her, and never once moved from her nest, though I visited her daily. Her eyes were always closed to within a mere thread-like crack, out of which she watched every movement, and she generally rested her head on a small dead branch that lay beside the nest.

73. Caprimulgus Forficatus, Vieil ; *C. Furcatus*, Cuv., Vol. 2., p. 78 ; *L'Engoulevent à queue fourchue*, Le Vail., Pl. 47 et 48.

GENERAL colour brown, variegated with black, red, and white, the latter chiefly on the belly, tail, greater wing-covers, scapularies, and under tail-covers. The black markings of the chest are larger than on other parts ; the wing and tail feathers are marbled brown ; the throat is reddish, transversely barred with black lines ; tail deeply forked. Length, 26″.

This species may, according to Le Vaillant, be distinguished from all our South Africau goatsuckers by its large size and deeply forked tail.

Le Vaillant states he found two specimens in a hollow tree, on the banks of the Lion's River, Namaqualand; but Sundevall declares that the bird is nothing more than *Nyctibius Grandis* of Guiana, with a false tail put into it, and that Le Vaillant's account of its capture is also false.

The Second Sub-Tribe, FISSIROSTRES DIURNÆ, or Diurnal Fissirostral Birds,

are distinguished from the nocturnal species by the closeness of their plumage.

The Second Family, HIRUNDINIDÆ, or Swallows,

have the bill short and weak, very broad at the base, and suddenly compressed to the tip; the wings lengthened, narrow, and acute; the tail more or less forked, the tarsi very short and weak.

The Sub-Family, CYPSELINÆ, or Swifts,

have the bill short, depressed, broad at the base, the sides suddenly compressed to the tip, and the margins inflected; the nostrils large and longitudinal on each side of the culmen; the wings extremely long and curved; the tarsi short and weak; the toes short, thick, and armed with short, strong, and much curved claws, the hind toe usually directed forward.

Genus CYPSELUS, Illiger.

Bill short and depressed, with the gape very wide, and the sides gradually compressed to the tip, which is curved; the nostrils basal, lateral, and large, with the opening longitudinal on each side of the culmen, and the margins beset with small feathers; wings lengthened, with the second quill longest; tail moderate, forked, or uneven; tarsi very short, and feathered to the base of the toes; toes all directed forwards, short, thick, and armed with short, curved, and compressed claws.

74. Cypselus Melba. (Linn.) *C. Alpinus*, Temm.; *Le Martinet à gorge blanche*, Le Vail., Pl. 243; *C. Gutturalis*, Vieil., Cuv., Vol. 2, p. 60.

GENERAL colour above light hair-brown; below the same, but with the throat, chin, and belly white; the sides of the

G

body mottled strongly with the same. The breast is also slightly mottled. The wings project considerably beyond the tail. Length, 8″; wing, 8″ 3‴; tail, 3″.

The great Alpine Swift is very abundant about Cape Town. I have seen it at the Knysna, and have received it from a valued correspondent at Swellendam, Wm. Cairncross, Esq., who has sent it with several other swifts and swallows killed in that neighbourhood.

C. Melba appears to reside principally in the mountains, only occasionally descending into the plains when some particular food is to be found. On the top of Table Mountain I have seen it in great abundance. Once, after lighting a fire there, I found myself surrounded by hundreds of them, hawking after the flies which the fire drove up. They dashed through the dense smoke and skimmed along the burning ground at the risk of singing their wings.

I never heard this species utter any note, even when in company with *C. Caffer*, screaming and chasing each other in a thousand airy evolutions.

Le Vaillant says that it breeds in rocks. I have never heard of their nests being found, though I have seen them flying into crevices of the rocky sides of Table Mountain. I also watched several pairs flying in and out of the rocky face of the " Ravine " at Simon's Town, but I had no means of ascertaining if the birds were in a breeding state, and the crevices were perfectly inaccessible.

75. Cypselus Apus, Linn., Pl., Enl. 242, f. 1;
C. Murarius, Temm.

ENTIRELY of a shot greenish-brown, with a whitish chin and throat. Length, 7″; wing, 7″; tail, 2″ 9‴.

The common European swift is extremely abundant about Cape Town during our summer; disappearing about the middle of April.

I have remarked that here they are not in the habit, as in Europe, of chasing each other round lofty buildings, uttering piercing cries and throwing up their wings. This evolution is performed by *C. Caffer*, and I am inclined to attribute it to some love-gambol, as *Caffer* breeds here, and *Apus* does not. *Apus*, however, constantly utters shrill screams when at a great altitude in the air; and my attention has thus often been called to flocks that would otherwise have escaped my notice.

I have this year (1861) kept a strict watch over their time of departure. The result has been as follows :—

April 9th. *Cyp. Apus* and *C. Caffer* about in abundance. *Hirundo Capensis* come into town; the other swallows having left us.

April 27th. Large flocks of *C. Apus* flying at a great altitude, and screaming; these were observed at night-fall—in fact just at dark.

May 3rd., morning. *C. Apus* and *C. Melba* in the upper regions, trending to the northward.

May 30th. Several specimens of *C. Apus* flying about the top of the " Lion's Rump," in company with a flock of *Cotyle fuligula*—three shot. After this they disappeared, with the exception of *Cotyle fuligula*, which remains with us all the year.

76. Cypselus Caffer, Licht.; *Cyp. Pygargus,*
Temm., Pl. Col. 460, 461; *Cypselus Leucothea*; *C. Leucorrhous,* Steph.; *Hirundo Nigra,* Vieil.; *Le Martinet à croupion blanc,* Le Vail., Pl. 244.

BACK bluish-black; wings and tail shining brown; head same colour as wings, lighter on the forehead. Throat and rump white; under side of body colour of the back. Length, 6″ 9‴; wing, 5″ 9‴; tail, 2″ 6‴.

This species can be easily distinguished from *C. Apus,* even on the wing, by its smaller size and conspicuous white rump. It is constantly in company with it and *C. Melba*; and indeed the former is rarely seen in any number without being mixed with it.

It is more domestic than the other swifts, breeding freely under the verandahs of houses in the city, and occupying almost every porch in the isolated farm-houses scattered over the country. It usually takes possession of the nests of *Hirundo Capensis,* driving out the rightful owners. Its eggs, generally four or five in number, are pure white, and singularly shaped, the small end rounding off very suddenly, as if truncated : axis, 1″; diam., 7‴.

I am convinced from the account given by Le Vaillant of the habits of his *Martinet à croupion blanc* (No. 244, Ois. d' Aff.), that this is the bird alluded to by him, though he only describes it as having the sides of the rump white. Probably he put off, like many others have done, obtaining a specimen of this common bird, thinking to do so at any time, came away without one, and subsequently described from memory.

77. Cypselus Velox, Cuv., Vol. 2, p. 59; *Le Martinet Velocifer,* Le Vail., 244, f. 2; *Hirundo Velox,* Vieil.

ENTIRELY of a blue black; the tail very forked, and the wings long, projecting nearly two inches beyond the tail when at rest.

This small swift, according to Le Vaillant, is found on the eastern side of the colony during the winter. He did not find its nest, and believes that it is only a migratory visitant. It frequents forests, and retires for the night into hollow trees. It never, however, perches on the branches, although he so figures it, and its flight is extremely rapid.

No collector has ever seen this little swift, and it is very doubtful if it ever existed, except in Le Vaillant's fervid imagination. He probably saw the forest-loving *Hirundo holomelas,* and failing to obtain it, manufactured from it " *Le martinet velocifere.*" Mr. Atmore writes : "I have looked in vain for *C. Velox,* and believe it to be a myth. Nothing like it is to be seen in the country traversed by Le Vaillant, but the common black swallow *(H. Holomelas),* which I found breeding in holes in banks. Had *C. Velox* been there I think I must have seen it."

Genus MACROPTERYX, Swainson.

Bill short, small, and depressed, with the culmen curved at the tip; the nostrils basal, lateral, with the openings placed on the sides of the culmen, longitudinal, and large. Wings lengthened, with the first and second quills nearly equal, and longest. Tail very long, and much forked. Tarsi shorter than the middle toe, thick, and naked. Toes long, slender, and compressed; armed with moderate, compressed, and curved claws.

78. Macropteryx Cristata. (Vieil.) Cuv., Vol. 2, p. 64; *L'Hirondelle Huppée*, Le Vail., Pl. 247.

GENERAL colour grey; darker on the wings and tail; lighter on the under sides. Wings and tail very long, the latter deeply forked; the two outermost feathers being very much elongated. Head crested.

Le Vaillant says he only met with a few of these birds in Namaqualand, and considers they were on their passage to some other part of the country. He gives no account of their habits; but from my remembrance of an allied species in Ceylon *(Macropteryx Coronata)*, I should judge that they were strictly arboreal, alighting habitually on trees, flying thence in search of food, and returning to their perch when tired. The introduction of this bird into the South African Fauna is supposed to be another of Le Vaillant's errors. Though a good species, it is not a native of South Africa, but of India.

The Sub-Family, HIRUNDININÆ, or Swallows,

have the bill short, more or less depressed, with the gape wide, and the sides gradually compressed towards the tip; the nostrils basal, lateral, and rounded; the wings lengthened, with the first quill the longest; the tail more or less forked; the tarsi short, generally naked and scutellated; the toes usually long and slender, and the claws moderate and slightly curved.

Genus HIRUNDO, Linn.

Bill short, much depressed, with the gape very wide, and the sides gradually compressed to the tip; the nostrils basal, small, oblong, and partly covered by a membrane. Wings lengthened, with the first quill longest. Tail more or less forked, the outer feather of each side sometimes lengthened much beyond the others. Tarsi shorter than the middle toe, and scutellated. Toes long and slender, with the lateral ones unequal; the claws moderate, curved, and acute.

79. Hirundo Rustica, Linn., Pl. Enl. 453, f. 1 ;
European Chimney-Swallow ; H. *Domestica,* Pall. Gould, B. of Euro., Pl. 54.

UPPER side steel-blue, with greenish reflections; quill-feathers of wing and tail rusty; throat and forehead deep rufous; under parts dull flesh-colour, with a broad collar of the same colour as the back, joining the red throat; tail deeply forked; the two outer feathers much prolonged; all the inner vanes with a patch of white, which together form a distinct white bar, most visible on the under side. Length, $6\frac{1}{2}''$; wing, $5''$; tail, $4''$.

The European swallow is a constant and common visitant to this colony, but, as far as I have ascertained, does not appear to breed with us, though I constantly see birds in young plumage. At the time I am writing (April 2nd) a few stragglers are still here; the main body is, I think, gone. They do not appear to congregate, as in Europe, previous to their departure; nor have I ever seen them together for any other purpose than hawking after flies, each on his own business. It is scattered throughout the country.

80. Hirundo Rufifrons. (Vieil.) Ency. Meth.,
L'Hirondelle a front roux, Le Vail., Pl. 245, f. 2 ; Cuv., Vol. 2 p. 64.

UPPER parts brilliant metallic blue; tail slightly inclining to green; forehead deep rufous; tail broadly barred with white, and forked; underparts white, with a very pale rufous tinge on the body; breast with a steel blue collar, more or less broken in the centre. Length, $6''\ 9'''$; wing, $5''\ 5'''$; tail, $3''\ 6'''$.

If any of our Cape swallows could be mistaken for the European species, this would in all probability be the deceiver. In fact, for a long time I *was* deceived by it, until one bright sunny morning, while watching the flight of some of these lovely aerial creatures, it struck me that the blue of the back and white of the breast looked brighter than in the old country bird. For some time they confined their course to the narrow river, on the bridge over which I stood, but at last one strayed for an instant over the bank, a well-directed shot laid him on the green sward, and I instantly recognised the true *H. Rufifrons* of Le Vaillant. They were breeding beneath the bridge, but I was unable at the time to get at the nests, which I have since visited, and found to resemble those of the European bird in shape and structure. Le Vaillant says he only found this species in the rainy season (our winter). In this he is most undoubtedly mistaken, so far as the Cape peninsula is concerned. No other swallow than *C. Fuligula* remains with us during this time, whatever they may do in the more inland districts; but, from all I can gather from my correspondents, I have no reason to think that even there they are to be found in the winter.

81. Hirundo Capensis, Gmel., Pl. En., 723 ; Cuv., 2, p. 63 ; *H. Cucullata*, Bodd. ; *L'Hirondelle Rouselline*, Le Vail., Pl. 245, f. 1.

HEAD deep rufous; back, wings, and tail black ; the former with a bluish tinge, the rest greenish ; the rump, nuchal collar, and all the under parts reddish-brown ; the shafts of the feathers being dark-brown, almost approaching to black. Tail deeply forked, and broadly barred with white ; throat pale ; eye dark-brown. Length to end of elongated tail-feathers, 8″ ; wing, 5″; tail, 4″ 6‴.

This is the household swallow of the colony, breeding freely about the houses, and in the country, often selecting the usual living-room of the family. In the city this familiarity is not permitted, on account of the dirt made by the birds ; but the Boer fixes up a board under the nest, to prevent the worst fouling, and considers that the rest is atoned for by the destruction of the myriads of flies, of which his little favourites rid him during the season of their stay. And only those who have sojourned in a Boer's house can estimate the plague of flies that infest it : they swarm on "bed and board;" they pollute the food and drink ; chairs, tables, walls, everything is blackened by them. No wonder, then, that the swallow is a welcome guest, and that to rob his nest is to get into the bad books of every member of the family.

As you sit at meals, the graceful bird hawks over the table, and snatches the flies from the walls and ceiling ; nor is this the only service he renders, for, sitting on the top of the window or door (always left open for his accommodation), he pours out a short but lively song, which enlivens the dreary solitude and silence of the lone homestead.

The nest of this species is always attached to the under side of the place chosen, and is composed of little pellets of mud, like that of the English swallow. In shape it resembles a gourd with a long neck, cut through longitudinally and glued up by the edges to the ceiling. It lays four or five pure white eggs : axis, 10‴ ; diam., 6‴.

I cannot see how Le Vaillant could have overlooked our European swallow, imagining that this bird had by some travellers been probably taken for it. Any one mistaking the two, except at a very great distance, could have had but small practice in discriminating between birds in their native wilds. But, singularly enough, Le Vaillant omits the two European species, *H. Rustica* and *C. Apus*, from his list.

The Rev. John Fry, of Rondebosch, lately related to me a singular instance of the reasoning powers of this swallow. The tube of a nest in his bath-room fell down, and was not replaced by the old birds, who brought up their young, till within a few days of their flight, when one more venturesome than his brothers and sisters crept to the edge of the nest and fell over. After vainly trying to replace their dead offspring, the disconsolate parents, although their nesting was nearly over, repaired the broken tube to prevent a recurrence of the catastrophe.

82. Hirundo Griseopyga. (Sunde.) Ofvers. Kongl.
Vet. Ak. Forhandl., p. 107.

BLUE-black, with white belly; rump grey; head fuscous; ear-covers black.

Inhabits Natal (Wahlberg)—non vidi.

83. Hirundo Gordoni, Jard.; *Hirundo Semi-rufa*, Sundevall, Ofvers. Kongl. Vet. Ak. Forhandl., p. 107.

ABOVE steel-blue black; rump and all the under parts light rufous; tail deeply forked; all the feathers, with the exception of the four middle ones, marked with white on the minor webs. The outermost much elongated and attenuated.

Inhabits Kaffraria. (Whalb. teste Hartlaub, Orn., W. Af.) I fancy it hardly comes within my range; Whalberg having invented his own limits for "Kaffraria."

84. Hirundo Atrocœrulea, Sundv.; Ofvers. Kongl. Vet. Ak. Forhandl., p. 107.

THROUGHOUT blue-black, occasionally an indistinct spot on the top of the external tail-feather.

Natal: Frequenting the open country, and said (teste Ayres) to be the most common of all their swallows—sed non vidi.

85. Hirundo Albigula, Bps.; *Hirundo Albigularis*, Strickland, 1848; *H. Rufifrons*, Less., nec. Vieil., Cont. Orn., 1849, t. 17.

FOREHEAD chesnut brown; back blue; beneath white; pectoral band black; wings scarcely exceeding the tail; inner webs of tail-feathers marked with quadrangular white spots. Length, 6½".

Natal: Ayres—non vidi.

86. Hirundo Lunifrons. (Sav.) Pr. Bonap., 128; Amer. Ornith., Pl. 7, f. 1.; and B. of Amer., Pl. 68.

UPPER parts deep brown, glossed with steel-blue; rump brick-red; under parts white, dashed with rufous; vent-feathers deep rufous; a rufous line extends from the base of the bill over the eye; wings and tail deep-brown, with a greenish lustre inside; the shoulder light rufous. Length, 5" 9'''; wing, 4" 6'''; tail, 2" 6'''.

Procured in the neighbourhood of Middelburg by Mr. A. V. Jackson. Builds in companies under rocks. I was first led to a knowledge of this species by observing an unusual appearance on an overhanging rock photographed during the journey of H.R.H. Prince Alfred through South Africa in 1860. On applying a strong magnifying power to the picture, I distinctly made out that the appearance consisted of a cluster of birds' nests. I at once concluded that they were constructed by some species of swallow unknown to me, and requested my zealous contributor, Mr. Jackson, to look well after them, if ever he found himself in the neighbourhood. This he has done, and tells me he counted about 20 nests, under a rock, clustered together ; he also obtained the only example which I have seen. Dr. Hartlaub, to whom this specimen was submitted, states that it is a young bird, of the American *H. Lunifrons ;* a species which has of late years been extending its migrations from its real habitation in a most remarkable manner.

87. Hirundo Dimidiata.

ABOVE entirely steel-blue ; the quill-feathers of the wing and tail dark brownish-black. This colour extends slightly, in the form of shoulder-knots, on to the collar. Under parts pearly-grey. Tail deeply forked. Total length, 6″ ; wing, $4\frac{1}{4}″$; tail, 3″.

This swallow is very rare in the neighbourhood of Cape Town, but becomes more common on the mainland. It will be as well to treat the peninsula bounded by False and Table Bays as apart from the continent. The vast tract of land called the "Cape Flats," together with Table Bay on the one hand, and False Bay on the other, quite shut it off from the mainland, so to speak ; and to aid in this isolation, the mainland ends in an abrupt precipitous wall of mountains, which are only to be passed in one or two places. Doubtless the peninsula of Table Mountain was once an island, entirely cut off from the mainland, and "Table" and "False Bays" part of a strait between. As soon as the hills are reached on the opposite side of Table Bay, this swallow commences ; and at Swellendam, Mr. Cairncross writes : "It is very common, and builds its nest generally under the thatch of an old mill or stable, where it is quieter than in a dwelling-house ; lays a small white egg, and tradition says it drives the sparrow and house-swallow (*H. Capensis ?*) from their nests, occupies them, and breeds therein. For this reason it receives no mercy from the farmer, but its eggs and young are destroyed whenever met with."

I found it abundant all along the route as far as Nel's Poort ; there it was breeding among the rocks, and under the eaves of Mr. Jackson's barn. They construct a nest of mud very similar to that of *H. Capensis,* but without the elongated neck. The eggs, three or four in number, are pure white : axis, 7‴ ; diam., 5‴.

Genus ATTICORA, Boie.

Bill short, weak, with the gape very wide, the sides gradually compressed, and the culmen curved to the tip ; the nostrils basal, lateral, and rounded. Wings lengthened, with the first quill the longest. Tail very long and much forked. Tarsi shorter than the middle toe and scutellated. Toes long and very slender, and the lateral toes unequal.

88. Atticora Holomelas, Sund.; Hartl. W. Af.;
A. Hamigera, Cass.; *Psalidoprocne Cypselina*, Cab.

OF a uniform black, with slight green tinge throughout. The inner vanes of the quill-feathers rusty towards their edges. Tail deeply forked. Total length, 5¼″, wing, barely 4″; tail, 2″ 9‴.

This little swallow first fell under my notice on the "Keurboom's River," Knysna district, where I saw it apparently breeding in holes in the banks, but was unable to investigate its doings more closely. I found it abundantly in the forest, hawking after flies over pools, frequently dipping into the water, and perching on the overhanging boughs in clusters of six or eight, to dry themselves.

Their habit of perching is noted by Mr. Cairncross, who writes: "This bird flies about very much like a bat (this resemblance also occurred to me when I saw it), amongst thick forests, and is generally more visible in rainy, heavy weather; but I have never seen or heard of their breeding here (Swellendam). They remain here after the winter has set in. Sometimes I have seen them roost on trees at the bottom of my garden, where I shot the specimen sent."

Genus COTYLE, Boie.

Bill moderate, very flat, extremely broad at the base, and gradually narrowed towards the tip; the nostrils very prominent and rounded. Wings very long, with the first quill the longest. Tail moderate, and emarginated, or nearly straight at the end. Tarsi rather shorter than the middle toe, slender and scutellated. Toes rather short, very slender, the lateral ones unequal; the claws moderate and slightly curved.

89. Cotyle Fuligula, Hartlaub.; *H. Fuligula*,
Licht.; *H. Hyemalis*, Forst; *L'Hirondelle Fauve*, Le Vail., Pl. 246, f. 1.; Cuv. Vol. 2, p. 61.

UPPER parts uniform hair-brown; throat and chest dull salmon-colour; belly and vent as on upper parts, but tinged with rufous; tail very slightly forked, and with a row of white spots, one on each feather: these are sometimes obsolete. Eye reddish-brown. Length, 6″; wing, 5″ 6‴; tail, 2″ 4‴.

This swallow remains with us all the year round, merely shifting its quarters from a town to a country residence. As soon as our summer visitants are gone, these crowd into the villages and take their places. I have counted upwards of one hundred sitting together on the cornices of the Dutch Church in Cape Town on a cold day, basking in the sun. They breed in the mountains, and seem generally distributed, as I have received specimens from Damaraland, Swellendam, Beaufort, and Colesberg; I have also found them nesting under the eaves of houses in November. They lay 3 or 4 eggs, of a creamy white, much spotted with brown, with here and there a grey patch;

H

iu some these spots seem to form a circle at the obtuse end: axis, 10'''; diameter, 6'''. The flight of this species is very slow and sailing, and they seek their prey usually near rocks, and are more crepuscular in their habits than any of our other species, often flying so far into the darkening twilight, that they can scarcely be distinguished from the bats with which they mingle.

90. Cotyle Cincta, Bodd.; *Hirundo Torquata*, Gml., Pl. Eul. 723; *Brown-collared Swallow*, Griffith, Cuv., Vol. 2, p. 67.

GENERAL colour above, brown; beneath white; pectoral band and thighs brown; spot in front of eye black; a white line extends from the nostrils over the centre of the eye; eye very dark-brown. Length, 5" 9'''; wing, 5"; tail, 2" 6'''.

The first specimens of this bank swallow reached me from Capt. Bulger, of H.M. 10th Regiment (2nd Batt.), who procured it at Windvogelberg, on the frontier. I subsequently, in October, 1865, discovered it about 14 miles from Cape Town, hawking about a small stream; it was there in some abundance. I again recognised a pair sitting on the telegraph wires near Somerset West; and on arriving at Mr. Vigne's farm found a pair breeding in the bank of the River Zonder End. The nest was about three yards deep, in a low bank. I did not obtain the eggs. The parent birds never seemed to fly far from the spot, but skimmed up and down the river. On my pointing them out, the Messrs. Vigne, who have paid some attention to the birds found in their neighbourhood, pronounced them strangers to them; and I do not think they have been in the vicinity of Cape Town till this year. My son tells me they have been abundant about Cape Town during the whole of this year. Several fine species added to the Museum shelves, the produce of his gun, attest the correctness of his eye in more senses than one.—Dec. 1866.

91. Cotyle Palustris, Steph., Bp. Consp., p. 342; *C. Paludicola*, Vieill.; *C. Paludibula*, Rüppell.; *L'Hirondelle de Marais*, Le Vail., Pl. 246, f. 2; *C. Paludicola*, Rüp.; Cuv., Vol. 2, p. 61, var.

GENERAL colour hair-brown; the edges of the feathers light rufous; belly and vent white; tail very slightly forked. Some specimens are throughout of a dull dark-brown. Length, 5"; wing, 4" 4'''; tail, 2" 5'''.

The Cape bank swallow is the earliest comer of all our migratory swallows and swifts. It is rarely seen far from water, and breeds in the banks of rivers or artificial dams, over which it continually hawks for flies. It lays three or four white eggs, of the same size and shape as those of our European *H. Riparia*, and the nest is often run to the depth of two or three feet into the soil, when it is loose and friable.

Mr. Cairncross, of Swellendam, informs us that, if the winter is mild, it remains about that part of the country during the whole year. Mr. Jackson has sent it from Nel's Poort. He also states that it stays with him all the year round.

92. Cotyle Ambrosiacus. (L.) *Cypselus Ambrosiacus*, Gmel. ; *Hirundo Riparia Senegalensis*, Briss ; *Hirundo Ambrosiaca*, L., Pl. Col. 460, f. 2 ; *C. Parvus*, Licht., Hart., Orn., W. Af.

GENERAL colour grey-brown ; head and wings darker ; throat whitish ; under part lightish ; tail furcated, outer feathers elongated. Length, 6″ ; wing, 4″ 10‴.

This little swallow may be at once distinguished from all our other species by its very lengthened tail. It is said by Temminck to be found at the Cape of Good Hope ; but it has not yet fallen under my notice.

The Third Family, CORACIADÆ, or Rollers,

have the bill more or less lengthened, and broad at the base, with the sides compressed towards the tip, which is hooked ; the nostrils basal, lateral, with the opening linear or oval ; the wings long and pointed ; the tail generally short and even ; the tarsi short ; the toes moderate, sometimes free, or the outer toe united at the base with the middle one.

The Sub-Family, CORACIANÆ, or Rollers,

have the bill moderate, higher than broad at the base, the sides much compressed towards the tip, which overhangs the lower mandible ; the nostrils basal, lateral, linear, and oblique ; the wings moderate ; tail moderate, and the side-feathers sometimes lengthened ; the tarsi short ; the toes moderate, and free at the base.

Genus CORACIAS, Linn.

Bill long, straight, more elevated at the base than broad, the sides much compressed, and the tip bent over that of the lower mandible ; the nostrils basal, linear, oblique, and partly covered by a feathered membrane ; wings moderate and pointed, with the second and third quills the longest ; tail long and rounded, with the lateral feathers sometimes lengthened ; tarsi much shorter than the middle toe ; toes moderate, free at the base, with the outer toe nearly as long as the middle one ; the hind toe long, the claws moderate, curved, and acute.

93. Coracias Nuchalis,[*] Sw., B. of W. Af., p. 110 ; *C. Crinita*, Shaw ; *C. Levaillantii*, Rupp. var. ; Shaw, Vol. 7, p. 391.

GENERAL colour rufous, tinged with green on the back, inclined to red on the under side, with the shaft of the feathers white ; wings dull rufous green, with purple reflections ; quill-feathers dark blue.   Tail blue, with the two centre feathers dull-green, even at the end, but rather long. Head more rufous than the body ; forehead and chin buff-coloured ; cheeks rusty red, with white shafts to many of the feathers.   Total length, 14" ; wing, $8\frac{1}{2}$" ; tail, $6\frac{1}{2}$".

Le Vaillant procured this bird in Kaffraria. I have received it from the neighbourhood of Otjimbinque, in Damaraland ; and one specimen was shot near Middelburg, on the Eastern side of the colony, and sent to me by the late Mr. J. O'Reilly, magistrate of that place.

94. Coracias Garrula, Linn., Pl. Enl., p. 486.

HEAD, neck, breast, and belly are of a light pea-green ; back and scapulars reddish-brown ; the ridge of the wings and upper coverts rich deep blue; the greater coverts pale green ; the quills dusky, inclining to black, and mixed with deep blue ; the rump blue ; tail somewhat forked ; the lower parts of the feathers are dusky green ; middle parts pale blue ; tips black.   Length, about 12".

Said by Hartlaub, B. W. Af., p. 29, to inhabit Natal. I have not a specimen to refer to, and the foregoing description is taken from Bewick's British Birds.

95. Coracias Abyssinica. (Lath.)   Shaw, Vol. 7, p. 393; *Albifrons*, Shaw ; *C. Senegalensis*, Gmel., Edw. B., Pl. 327, var. ; Cuvier, Vol. 2, p. 186.

HEAD, neck, wing-coverts, and under-parts, light bluish green ; forehead and chin, light buff; back, dark-brown buff; shoulders, azure-blue ; quill-feathers of wings and two middle tail-feathers black, the former reflecting blue, the latter green ; base of all tail-feathers, except the outermost, dark blue-black.   The outermost feather on each side very much prolonged, that part which exceeds the others being very narrow and black.   Rump blue.   Length, 18" ; wing, $6\frac{1}{2}$" ; tail, centre feathers, $6\frac{1}{2}$" ; exterior pair, 11".

I have received this species from one or two places along the northern border of the colony ; also from the neighbourhood of Springbok Fontein, in Namaqualand, and from Damaraland.   In the colony it is looked upon as a great rarity.

---

[*] Swainson's type specimen is in the Cambridge Museum.

96. C. Caudata, Linn.; *Coracias Angolensis*, Bris.; *C. Natalensis*, Licht.; Sh., Vol. 7, Pl. 1, p. 394.

UPPER part of head, neck, shoulders, back, and wing-coverts ferruginous-green, the rusty colour appearing strongly on the back; rump bright-blue; forehead and chin buff-coloured; under side of neck and breast ferruginous, glossed with purple, the feathers of the throat having white shafts; cheeks deep ferruginous; belly and vent pale blue-green, which is the colour of the tail-feathers, with the exception of the two centre, and the prolonged tips of the two outer-feathers, which are green, more or less inclining to black. Total length, 15″; wing, 7″; tail, to centre feather tip, $5\frac{1}{3}$″; to tip of exterior pair, $8\frac{1}{2}$″.

This species is found in the same localities as *C. Abyssinica*; neither of them have fallen under my own observation in their natural haunts. Mr. R. Moffat, jr., forwarded several specimens from Kuruman, and it has been procured in Natal.

The Fourth Family, TROGONIDÆ, or Trogons,

have the bill short, strong, sub-triangular, broader at the base than elevated, with the top margins generally toothed; the gape furnished with bristles; the nostrils basal, lateral, more or less concealed by projecting feathers; the wings moderate and rounded, with the fourth and fifth quills usually the longest; tail long and even, with the lateral feathers graduated; the tarsi short, and more or less covered with feathers; the toes placed two anteriorly and two posteriorly, each pair unequal.

Genus APALODERMA, Swainson.

Known from the others of this family by the lateral margins of the upper mandible being nearly smooth, those of the lower strongly sinuated; tarsi moderate, naked, and scutellated; toes long, slender, and the two anterior ones free at the base.

But one species of this family is found within the limits assigned to this list, namely—

97. Apaloderma Narina. *Trogon Narina*, Vieil., N. Dict. Hist. Nat. 8, p. 318; *Le Couroucou Narina*, Le Vail., No. 228--9; *Bushloorie* of Colonists. Cuv., Vol. 2, p. 476; Gould, Monog. Trogons, Pl. 26.

GENERAL colour above, brilliant, shining green; tail dull green; the three outer feathers on each side more or less white; wings grey, very delicately pencilled with white; under side, from vent to breast, brilliant carmine; from thence to chin green, as on the back; bill yellow. Length, 11″ 6‴; tail, 7″; wing, 5″ 3‴.

Found throughout forests and wooded kloofs in all the eastern portions of the colony. About the Knysna it is scarce; but wherever found it exhibits the same shy, creeping habits. When apprehensive of discovery, it sits motionless on its branch until alarmed at some act of the fowler, when it precipitates itself headlong into the bush, and is instantly lost to view. It feeds on fruits and insects, and utters a loud moaning note, or, perhaps, more correctly speaking, a hoot. Le Vaillant states that they nest in holes of trees, and lay four, nearly round, white eggs; but this requires confirmation. I have never met with any one who had seen their nests, though I made particular enquiries for them when at the Knysna.

The Fifth Family, ALCEDINIDÆ, or Kingfishers, .

have the bill more or less lengthened, generally straight, and broad at the base, with the tips of both mandibles usually acute. The nostrils lateral; the wings more or less long and rounded; the tail mostly short and rounded on the sides; the tarsi short and strong; the toes long, sometimes three before and one behind, and sometimes two and two, and of various lengths.

The Sub-Family, HALCYONINÆ, or Crab-hunters.

have the bill broad at the base, and gradually compressed towards the tip; the lateral margins straight, or turned upwards towards the end of the upper mandible; the gonys lengthened and gradually ascending, or curved upwards to the tip of the lower mandible.

Genus HALCYON, Swainson.

Bill long, straight, broad, nearly quadrangular, culmen slightly inclining towards the tip, near which the margin is slightly sinuated; gape smooth; wings broad, short, rounded; tail very short; feet syndactyle; scales of tarsus obsolete.

98. Halcyon Senegalensis. (Linn.) *Alcedo Senegalensis*, Auct. Pl. Elum., t. 594; Swain. Illus. of Zool., t. 27.

GENERAL colour of back, tail, and wings blue, with a black stripe extending from the shoulder down the wing; some of the outer quill-feathers also black. Head, neck, breast, and under parts cinereous, palest on the centre of the breast and belly ; a black patch between the eye and the bill, the upper mandible of which is red, the lower black. Length, about 8½" ; wing (apud Schlegel), 3" 9''' to 4" ; tail, 2" 6''' ; bill, 15''' to 18'''.

This bird appears to me to be identical with my *Cinereifrons.* My description is taken from the figure in the Planches Enluminée. Schlegel also says it resembles *Cinereifrons* and *Dryas*, but is much smaller. I am inclined to think it may be a small race.

Specimens have been procured at the Mouth of the Monocusi River, north-east of Natal, by Mr. Ayres. Ibis Vol., 1856, p. 265.

99. Halcyon Senegaloides, Smith S. `A. Z., Vol. 2, p. 144, Zool. S. A., Pl. 63.

UPPER and lateral parts of head and of neck grey-green ; interscapulars and scapulars green ; back a clear, dark blue-green ; shoulders and extremities of wing-feathers black, middle of wings deep blue-green, inner vanes towards quills white ; tail blue-green ; throat, centre of belly, and vent white ; breast and sides of belly greyish ; bill and legs deep red ; space between eye and base of bill black. Length, 9" 4''' ; wing, 4" 3''' ; tail, 2" 10'''.

Inhabits the country about Port Natal. "Resorts to thickets, and feeds upon insects, especially *crustacea* and *gryllæ*." Smith—loc. cit.

100. Halcyon Swainsonii, Smith, S. A. Z., Vol. 2, p. 143 ; *Halcyon Semicærulea*, Gmel. ; *Alcedo Cancrophaga*, Forst. ; Rüpp. Wirbeth. Ab. t. 24, f. 1.

HEAD, back, and sides of neck and interscapulars dull white, the first more or less tinted with grey-brown, the last margined behind with a broad transverse black band ; back, middle of wings, and tail blue, with a green gloss, particularly the two centre feathers of the latter ; shoulders, scapulars, and points of quills black, inner vanes of several of the outermost wing-feathers white towards quills ; throat and centre of breast pure white ; sides of breast, belly, and vent ferruginous ; bill, legs, and toes light red ; claws dark horn-coloured. Length, 7" 6''' ; wing, 3" 4''' ; tail, 2".

Inhabits the interior of South Africa. Smith—loc. cit. Hartlaub Orn. W. Af., p. 33—sed non vidi.

101. Halcyon Fuscicapilla, Lafr., Mag. Zool, 1863., Ois. t. 18 ; *Dacelo Vaillantii*, Temm.

HEAD, dark ashy-grey ; back and rump azure-blue ; wing-coverts black ; wings and tail blue ; nuchal collar ferruginous-grey; under parts the same, but clearer, and some of the breast-feathers have dark shafts; throat albescent; bill and feet red. Length, 9″; wing, 4″; tail, 3″

Extends from Kaffraria to Natal, and inland as far as Kuruman.

Mr. Atmore assures me that it is plentiful at Swellendam, and that when he is digging in his garden specimens may always be seen perched on the trees near him, from which they dart down on the worms, slugs, &c., which he turns up with his spade.

It is not found in the neighbourhood of Cape Town; but I received a pair from a Mr. Dixon, said to have been killed on the Eastern frontier.

102. Halcyon Striolata. (Licht.) *Alcedo Variegata*, Vieil. Encycl., p. 397; *Chelicutia Striolata*, Reich., Alc., p. 39, fig. 3155; Jard. Contrib., V., p. 57; *H. Pygmœa*, Rupp. Atlas, t. 28, b. •

GENERAL colour brown; head striped with black, and slightly crested; wing and tail feathers blue; rump bright-blue, beneath white; the breast striped with brown; bill red. Length, 5″; wing, 3″; tail, 1″ 5‴.

Said by Mr. J. Verreaux to inhabit Natal—sed non vidi.

103. Halcyon Cyanotis, Swainson, B. of West. Af., Vol. 2, p. 103; *Todus Cœruleus*, Auct.; *Alcyone Coronata*, Smith; *T. Pictus*, Bodd.; *Alcedo Nutans*, Vieil., Encycl., p. 396; *Ispidina Picta*, Kaup. Eisv., p. 12.

SIDES of head, ears, and neck, rufous, glossed with purple-violet; crown of head and nape black, with transverse blue spots; chin and throat white; breast and body beneath, orange-rufous; bill and feet red. Total length, 4½″; wing, 2″; tail, 1″.

Inhabits South Africa, Hartlaub., Orn., W. Af.—Natal: Ayres, Ibis Vol. 1859, Pl. 246—sed non vidi.

104. Halcyon Cinereifrons. (Vieill.) Hartlaub, Orn., W. Af., p. 32; *Alcedo Malimbica*, Shaw; *H. Torquatus*, Sw., B. of West. Afr., Vol. 2, p. 99; *H. Malimbica*, Cass.

HEAD above, brown; ears blue, margined by a black line; back wings and tail blue; shoulders and wing-coverts black; under plumage white; breast with a blue band; upper mandible and feet red ; lower mandible black. Total length, 10″ ; wing, 4″ 6‴ ; tail, 3″ 6‴.

Natal. Hartlaub, loc. cit. non vidi.

The Sub-Family, ALCIDININÆ, or Kingfishers,

have the bill more or less long, straight, and slender, with the culmen sloping to the tip, which is acute ; the sides much compressed ; and the gonys long and ascending.

Genus ALCEDO, Linn.

Bill more or less long, slender, and straight, with the culmen sloping to the tip, which is acute ; the sides compressed ; the lateral margins straight ; the gonys lengthened, and slightly sloping upwards ; the nostrils basal, with the opening oblique, and covered by a membraneous plumed scale ; wings short, with the first quill nearly as long as the second and third, which are equal and longest ; tail short, broad, and rounded ; tarsi very short and robust ; toes unequal ; the outer toe nearly as long as the middle one, and united nearly to the tip ; the inner toe short, and united at the base ; the hind toe rather short and broad ; the claws short, compressed, and curved.

105. Alcedo, Semitorquata, Swainson, Zool. Ill., Pl. 151 ; Cuv., Vol. 2, p. 410 ; Hartlaub, Orn., W. Af., p. 35.

GENERAL colour above, blue ; brightest on the rump, neck, and head ; in the last this colour assumes the shape of narrow bars ; throat whitish, inclining to buff, which is the colour of the rest of the under parts ; bill black ; legs light-red. Total length, 7½″ ; wing, 3¼″ ; tail, 1″ 6‴.

This species does not extend into the Table Mountain peninsula. I have never received it from any place nearer than Swellendam, and know nothing of its habits.*

106. Alcedo Cristata. (Linn.) Pl. Enl., 755 ;
Corythornis Cristata.

ADULT.—Head with large blue and black crest ; upper part mazarine blue ; under parts rufous ; chin and ear-tufts white ; bill and feet coral red. Length, 5½″ ; wing, 2½″ ; tail, 1½″.

YOUNG.—Crested as in adult ; upper parts much variegated

---

* 1865. One specimen since received from the "Salt River," near Cape Town, shot by Mr. John Reid, jun.

I

with black; under parts rufous, mingled with black; ear-tufts and chin white; spot near nostril rufous; bill and legs black.

This beautiful little kingfisher is abundant throughout the colony, wherever a stream or marsh exists which can supply it with its necessary food. It breeds in banks, and lays four or five glistening white eggs—so transparent that the yellow yolk shines plainly through the shell. I have not myself seen the nest; but have been assured by many who have, that it consists of nothing but the bones of the delicate little fish upon which the bird habitually feeds.

107. Alcedo Quadribrachys. (Temm.) Bp. Consp. I., p. 158. Strick. and Jard, Contri., 1851, p. 134, Pl. 79.

BODY above, cheeks, neck, and breast, blue-black; top of head barred with black; ears and throat fulvus; below reddish. Length, 6" 5'''; wings, 2" 10'''; tail, 1" 5'''.

Natal: Mr. Ayres, Ibis Vol., 1859, p. 245—sed non vidi.

108. Alcedo Natalensis, Smith, Contributions to Nat. His. of S. Africa.

FRONT, top of head, and nape, deep indigo-blue, variegated by pale blue, or violet transverse bands; sides of head and neck light reddish-brown, with tints of purple; back fine bright-blue; shoulders blackish, with a shade of blue, in addition to numerous spots of clear light-blue; chin and throat white; breast, belly, and vent, pale chesnut; tail dusky-blue. Length, 4".

Inhabits the banks of rivers, &c., to the eastward of Kafirland. Dr. Smith, loc. cit., non vidi. I cannot trace this species in any author, and fancy it must be identical with *Halcyon Cyanotis.*

Genus CERYLE, Boie.

Bill long, straight, and strong, with the culmen slightly advancing on the forehead, and sloping to the tip, which is acute; the sides much compressed; the lateral margins rather dilated at the base, and straight to the tip; the gonys long and ascending. Tail long, broad, and rounded. Tarsi remarkably short and stout. The remaining characters very similar to those given with Alcedo.

109. Ceryle Maxima, Pallas; *Ispida Gigantea,* Swain., B. of W. Af, Pl. 11, p. 93; *Alcedo Afra,* Shaw.

HEAD crested; the whole of the upper plumage and breast bluish-grey, spotted with white; belly and vent deep rufous; chin white, with a few black spots; bill and legs black. In some specimens the under parts differ in the distribution of the colours. Length, 18″; wing, 8″; tail, 4″ 9‴; bill, 4″.

This, the largest of our kingfishers, is not at all uncommon, and is very widely distributed. It is migratory, and I know nothing of its nidification. It frequents the "Salt River" and other small streams in the neighbourhood of Cape Town, perches on trees, and is wary and difficult of approach. Its flight resembles that of the American "Belted Kingfisher," and like it, whilst on the wing, it utters a loud chattering cry. Indeed, it so closely resembles *Al. Alcyon*, that when I first heard and saw it, I thought an old friend was before me, in a new country; a near approach, however, soon dispelled the illusion.

Habitat: Cape Colony and Natal, extending into Damaraland. Feeds on crabs, fish, frogs, and reptiles; very common at Zoetendals Vley and along the River Zonder End.

110. Ceryle Rudis. (Linn.) Pl. Enl. 716; *Isipida Bicincta*, Swain., B. of West Af., p. 95.

CRESTED; body above variegated with black and white; below white, with two black collars on the breast; nuchal collar white; head and crest not entirely black, but variegated with narrow white lines. A white coronet extends from the base of the bill over the eye and round the head. Length, 11½″; wing, 5½″; tail, 2″.

This species is not uncommon, and is widely distributed. I have seen it about "Salt River," near Cape Town, but in greater numbers about the rivers, lakes, and estuaries of the Knysna, and have received it from Kuruman, Colesberg, and Kaffraria. Mr. Chapman found it all the way to Lake N'Gami. It hovers over the water while fishing, and plunges under the surface in pursuit of its prey. It breeds in holes hollowed by itself in sandy banks. I took a nest (Nov. 10th, 1865,) in a bank of Zoetendals Vley, which was placed at the extremity of a small hole, more than two yards deep. It was composed entirely of fish-bones and scales; and the eggs, six in number, were of a lovely shining white, almost similar in shape at each end: axis, 14‴; diam., 11‴. At the time I took them, the young were nearly ready for exclusion.

If *I. Bicincta* is distinct from *I. Rudis*, Auct., we certainly possess both species. I have carefully compared specimens with Swainson's description, and they accord entirely, with the exception of the nuchal collar being "white and *immaculate*." I find, however, that it is more broken by the black line of the ears in some specimens than in others. I have a specimen from Singapore which exactly resembles *I. Bicincta!* *

---

* Since the above was written, I have examined Swainson's type specimen in the Cambridge Museum, and am satisfied that the species is not separable.

The Sixth Family, MEROPIDÆ,

have the bill longer than the head, with both mandibles curved and acutely pointed; the sides compressed and sloping from the culmen; the nostrils basal, lateral, rounded, and partly hidden by short bristles; the wings more or less long and pointed; the last long, broad, and the middle feathers sometimes prolonged beyond the lateral ones; the tarsi very short; the toes long, with the lateral ones more or less united to the middle toe; the hind toe long, and broadly padded beneath.

The only Sub-Family, MEROPINÆ, or Bee-Eaters.

have the same characters as previously given for the Family.

Genus MEROPS, Linn.,

Bill long, curved, and pointed, with the culmen curved, keeled, and the sides much compressed to the tip, which is very acute; the sides sloping from the culmen to the lateral margins for its entire length; the gonys very long and arched; the nostrils basal, lateral, rounded, and covered with a few short hairs. Wings reaching two-thirds the length of the tail, and pointed, with the first quill spurious, and the second the longest. Tail long, broad, more or less even, with the ends of the two middle feathers more or less prolonged beyond the others. Tarsi very short, robust, and covered in front with narrow transverse scales. Toes long, the lateral ones unequal, the outer the longest, united as far as the second joint, and the inner as far as the first joint of the middle toe; the hind toe long, strong, and broadly padded beneath; the claws moderate, curved, and acute.

111. Merops Apiaster. (Linn.) Tem. Manuel, Vol. 1, p. 420; Swain., B. of W. Af., p. 76; *M. Chrysocephalus*, Gmel.; *Berg Schawler* (mountain-swallow) of Colonists.

FOREHEAD light-buff, fading into green; top of head, neck, and shoulders rufous; back orange; wings and tail green; under parts bluish-green; chin and upper part of throat golden-orange, followed by a narrow black line; ear-tufts black; middle tail-feathers pointed, and slightly elongated beyond the others. Length, 10″; wings, $5\frac{3}{4}″$; tail, 5″.

The European bee-eater, during its period of visitation, extends all over the colony. I have generally found it come in with the quail about August; but from constantly living in the town, I am unable to state anything with certainty of the migration of our periodical visitants.

It hawks after flies, uttering its cheerful, chirruping cry, and alighting on the summit of the highest bush in its neighbourhood. It breeds in the neighbourhood of Nel's Poort; Mr. Henry Jackson having found several nests in holes in banks. The eggs are pure white : axis, 12'''; diam., 10'''.

112. Merops Savignii. (Swains.) Le Vaillant, Guêp., p. 6, 6.; *M. Superciliosus*, Auctorum; *M. Ruficollis*, Vieil., Le Vail., Guêp., t. 16. Swain., B. of W. Af., Pl. 7, Zool. Ill., Pl. 76.

GREEN; front whitish; ears black; stripe above and beneath the eye, sky-blue; chin yellow, passing into chesnut; crown green; wings and tail glossed with fulvous. Length (excluding two long tail-feathers), $10\frac{1}{2}''$; wings, $5\frac{1}{2}''$; tail, $3\frac{3}{4}''$.

Found in Natal by Mr. Ayres. A single specimen was also forwarded to the Museum by T. B. Bayley, Esq., of Wynberg, having been shot by that gentleman on the Cape Flats. Mr. Dumbleton, of Wynberg, assures me that these birds periodically visit a circumscribed portion of the Cape Flats in considerable numbers. February 15th, 1866, a specimen was this day sent to the Museum by Mr. Bishop, shot near Kuils River.

113. Merops Ægyptius, Forsk; *M. Persica*, Pall; *Viridis*, var. γ, et *Superciliosus*, var. β, Gmel; *M. Savignyi*, Cuv. nec Sw.; *M. Superciliosus*, Licht., Le Vail., Guêp. t. 10; *M. Lamarkii*, Vieil.; *M. Viridissimus*, Swain.

GENERAL colour dark-green; front yellowish-white; forehead and eyebrows light-green; eyeband black; anterior part of throat yellow, under part maroon; tail olive-green; centre-feathers prolonged, and green-black. Length, $10\frac{1}{2}''$.

"Inhabits S. Africa."—Dr. A. Smith, non vidi.

There seems to me great confusion between the names of this and the preceding species. They are probably identical; but as Mr. Gray divides them, I follow his example, not having any means of forming an opinion on the question.

114. Merops Nubicus, Gmel., Pl. Enl. 649; *M. Cœruleocephalus*, Shaw, Nat. Miscel., Pl. 78; Swain., B. of W. Af., Vol. 2, Pl. 9; *M. Superbus*, Penn.

HEAD and throat sea-green, the latter margined below with black; eye-band black; neck, back, breast, and belly, bright red; wings and tail a dull-red; quill-feathers black at the tips, clouded with green; tail-feathers pointed with obscure green, the two centre ones much longer than the others, narrow towards the points, and of a blue-black colour; rump and vent light-blue. Length, 11½".

"Inhabits Africa; rare in South Africa." Dr. A. Smith—non vidi.

115. Merops Erythropterus, Gmel.; *M. Minutus*, Vieil., Encl., p. 392; Swain., B. of West Af, Vol. 2, p. 88.

UPPER parts, excepting the primary and secondary quills, and lateral tail-feathers glossy, changeable green; ear-tufts black; chin and half the throat, clear, full yellow, with a narrow edging in the middle of the throat of a delicate greenish-blue, followed by a transverse spot of the deepest black, placed in the middle of a patch of rich chesnut extending across the breast. Total length, 9"; wing, 3"; tail, 2" 7‴.

Dr. Hartlaub includes this among the birds common to Western, North-Eastern, and Southern Africa.—Orn. W. Af., p. 40.

Mr. Ayres also procured it at Natal, where he says it is resident all the year round in certain localities. As far as I can ascertain, it does not extend its range into this colony.

116. Merops Bullockoides, Smith, Zool. S. A., ·Pl. 9; *Melittophagus Bullockoides*, Boie.

FRONT, chin, and moustache white, with a greenish lustre; hind-head rather rufous; rest of the upper parts, with the exception of the rump, green; rump blue. On opening the wings, the tips of some of the feathers will be found to be black; ear-tufts black; under parts rufous, tinged with green; throat brilliant scarlet; vent blue. Length, 9"; wing, 5½"; tail, 4".

Dr. Smith first obtained this species in 25° south latitude. I have received several specimens through the kind exertions of Mr. David Arnot, of Colesberg, who, mindful of the interests of the South African Museum, never loses an opportunity of contributing to it. His specimens were procured in the Orange Free State.

117. Merops Hirundinaceus. (Lath.) B. of W. Af., Vol. 2, p. 91; *M. Taiva*, Griff., Cuv., Vol. 2, p. 404; *M. Chrysolaimus*, Jard. and Selby Illus., p. 99; *M. Azuror*, Less. Tr., p. 239; *M. Furcatus*, Stanl.

GENERAL colour green; chin orange; collar, belly, and tail-coverts bright-blue; tail much forked, the feathers with whitish tips. Length, 8½"; wing, 3½"; tail, 4½".

Found by Le Vaillant. "in abundance on the banks of the Orange

River, in South Africa" (Swainson loc. cit.); by Mr. Andersson at Otjimbinque; and Mr. Chapman as far as the Zambesi.

The Second Tribe, TENUIROSTRES,

have the bill of various lengths and forms, and always slender, and with the tip of the upper mandible entire, and usually acute; the tarsi short; the toes generally long, especially the hind toe, which is strong, and armed with a more prominent claw.

The First Family, UPUPIDÆ, or Hoopoes.

have the bill lengthened, slender, curved, or straight for its entire length; the sides compressed to the tip, which is entire and acute; the nostrils basal, and rather small; the wings long and rounded; the tail more or less long and rounded, or even; the tarsi strong, and usually short; the toes long and strong; the outer toe more or less united at its base; the hind toe long; the claws strong, long, and curved.

The Sub-Family, UPUPINÆ, or Hoopoes,

have the bill long, curved, or straight from the base; the culmen keeled, and the sides generally compressed suddenly at the base towards the tip, which is entire and acute; the nostrils basal, small, and mostly covered by a membraneous scale; the wings long and rounded; the tail more or less long, graduated, or even; the tarsi generally short and strong; the toes long; the outer toe united at its base; the hind toe long and strong; claws long, strong, curved, and acute.

Genus UPUPA, Linn.

Bill very long, slender, slightly curved from the base to the tip, which is entire and acute; the culmen keeled; the sides compressed from the base, and the gonys lengthened and curved; the nostrils basal, small, with the opening covered by a membraneous scale. Wings long and rounded, with the fourth and fifth quills equal and longest. Tail long, broad, and even. Tarsi shorter than the middle toe, strong, and covered with broad scales. Toes moderate; the outer toe longer than the inner, and united at the base to the first joint; the hind toe long; the claws moderate, compressed, curved, and that of the hind toe long, and nearly straight.

118. Upupa Minor, Gmel., Vieil, Ois d'Or., t. 2 ; *U. Cristatella*, Vieil., Gal., Des. Ois., t. 184 ; Shaw, Vol. 8, p. 138 ; *U. Africana*, Brecht. ; *U. Capensis*, Sw. Pl., En. 697.

GENERAL colour of the head, back, and under parts, deep ferruginous ; the wings black, with white patches ; tail black, with a white bar near the base ; back barred with black ; the long crest-feathers of the head tipped with black. Length, 10″ ; wing, $5\frac{3}{4}″$; tail, 4″ 7‴.

The lesser hoopoe is not found to my knowledge within the Table Mountain peninsula ; the nearest spot from whence I have received it has been Swellendam.* It may readily be distinguished from the European bird by its deep rufous colour—the situation of the white bar on the tail, close to the base instead of the centre; and the absence of white below the black tips of the crest. It is very common in the Karroo towards Beaufort and Nel's Poort, frequenting the thickets of mimosa trees which border all the rivers of that district. Its monotonous cry of "hoop-poop" resounds the entire day. It is shy and difficult of approach, searching for its insect prey about the rugged limbs of trees or crevices of rocks.

Mr. Atmore writes : "They are migratory ; we did not see them at Meiring's Poort from February till June : they then came in, but were in bad plumage."

Genus IRRISOR, Lesson.

Bill long, more or less curved throughout its length ; the culmen keeled, and the sides suddenly compressed from the base towards the tip, which is entire and acute ; the gonys lengthened and curved ; the nostrils basal and small, with the opening oval and covered by a membraneous scale ; wings long and rounded, with the fourth and fifth quills equal and longest ; tail more or less lengthened, broad, and graduated ; tarsi much shorter than the middle toe, strong, and covered with broad scales ; toes long and strong ; the outer toe longer than the inner, and united at the base beyond the first joint ; the hind toe very long ; the claws long, strong, compressed, and much curved.

119. Irrisor Erythrorynchos, Lath. ; *Promerops Erythrorynchos*, Auct. ; *Irrisor Capensis*, Less. ; *Promerops Moqueur*, Le Vail. ; *Promerops*, 1, 2, 3 ; Vieil., Ois d'Or, t. 6.

GENERAL colour black, glossed with green on the head, back, and under parts; purple on the wings and tail, and bronzed

---

* April, 1861. I have just received one from the district of Clanwilliam, on the Western coast.

on the shoulders; tail lengthened and graduated; all tho feathers, except the two centre ones, spotted with white near the tips; wings barred with white; belly and vent dull black; bill more or less curved, bright coral-red, as are also the legs. The bill differs in length with age and sex. Length to base of tail, 7″; wing, 7″; tail, 10″.

This bird inhabits the forests of the Knysna, and extends as far as the Great Lake. It frequents high trees, about the topmost branches of which it hunts unceasingly for insects. Its voice is harsh and resounding, and has acquired for it the name of "*Kackela*" among the Dutch, which signifies "the chatterer." It is a difficult bird to shoot, being very shy and wary, and for ever whisking round on its perch, taking special care, however, of its long tail, by lifting it over the branches. Its motions at these times reminded me very much of the habits of the pheasant cuckoos (*Centropus*). I never saw or heard of one of these birds being observed on the ground. The residents in the Knysna district informed me that these birds breed in hollow trees, and that their eggs were pure white and round.

120. Irrisor Cyanomelas. *Rhinopomastes Cyanomelas*, Vieil.; *R. Smithii*, Jard.; *Promerops Namaquois*, Le Vail., Prom., t. 11, 12; *P. Purpuratus*, Burch.

GENERAL colour of male, deep black, richly glossed with purple-blue on the head, neck, breast, and back; less so on the wings and tail, and not at all on the body; feathers of the wings with a white spot near the quills, and the five first have a patch of greyish-white about an inch from the tips; bill curved, very thin, and dark horn-coloured, approaching to black; legs the same. The females are brown underneath, so also are young birds. In one example in my possession, the brown tint is mottled with blue-black, as if there was a change of plumage going on. Length, 10″; wing, $4\frac{1}{8}″$; tail, 5″.

Brought from Kuruman by Mr. R. Moffat, and from Damaraland by Mr. Andersson.

The Second Family, PROMEROPIDÆ, or Sunbirds,

have the bill more or less lengthened, slender, and usually curved throughout its length; the nostrils broad, and covered by a scale; the wings moderate; the tail of various lengths, even or graduated; the tarsi moderate, and covered with broad scales; the toes moderate, and armed with curved and acute claws.

J

The Sub-Family, PROMEROPINÆ, or Sunbirds,

have the bill more or less long, slender, curved, with the sides compressed to the tip, which is acute, and the lateral margins sometimes finely serrated; the nostrils basal, and the opening clothed by a membraneous scale; the wings moderate; the tail more or less long, with middle feathers sometimes prolonged beyond the others; the tarsi usually short; the toes moderate, the outer toe rather longer than the inner, and united at its base; the claws moderate, curved, and acute.

Genus PROMEROPS, Brisson.

Bill lengthened, slender, rather curved, with the base broad, and the sides compressed to the tip, which is acute and emarginated; the gonys lengthened and slightly curved; the nostrils lateral, basal, placed in a lengthened groove, with the opening linear, and closed by a membraneous scale; wings moderate and rounded, with the fourth quill the longest; the inner webs from the second to the fifth are broad and emarginated near the middle, especially the latter one; tail lengthened, graduated, and composed of narrow feathers, the two middle much longer than the others; tarsi as long as the middle toe, strong, and covered with broad scales; toes moderate, the outer toe longer than the inner, and united at its base, the hinder toe long and strong; the claws moderate, curved, and compressed.

121. Promerops Caffer, Cuv., Vol. 2, p. 460;

Upupa Promerops, Linn.; *Meliphaga Caffra*, Ed.; *P. Capensis*, Less.; *Cinnyris Longicaudatus*, Vieil., Ois d'Or, t. 4; Le Vail., Pl. 287 and 288.

ABOVE, uniform cinerous-brown; greyish on the forehead, and rufous on the sides of the neck; chin whitish, with a brown moustache; under parts mottled rufous, grey, and white; vent a fine yellow. Length, 7''; wing, $3\frac{3}{4}''$; tail of male often extending 14''.

This honey-eater differs considerably in all its habits from the sun-birds with which it is associated. It is generally found in small parties, the members of which, though acting much in concert, still maintain their individuality. They follow one another to favourite bushes, chase each other, or feed amicably side by side, and if one is alarmed and flies off, the rest generally follow; yet they cannot be called gregarious in the strict sense of the word. They are fond of perching on the summit of high bushes, particularly *Proteaceæ*, from the flowers of which they extract their food, consisting of insects, and the limpid, saccharine juice which these plants supply plentifully, and which will often run in a copious stream from the bill of a fresh-killed specimen. Le Vaillant states that it constructs an hemispherical nest in the largest protea-bushes, covered externally with moss or lichen, and lined with hair and wool. The eggs he describes as olivaceous; but my son procured a nest, which he assured me belonged to this bird, cup-shaped, about three inches in diameter *within*, composed of rootlets, grasses, and the spicular leaves of firs, mingled with the silky, dark-orange seeds of the protea. The single egg contained in it instantly reminded me of that of the Bunting of England, being a pale dirty cream-colour, irregularly though sparingly marked with wavy tracery and hair-streaks, now and then inclining to blotches of a dark purple-brown, or a pale, indistinct purple: axis, 11″; diam., 8½″.

Genus NECTARINIA, Illiger.

Bill more or less long, curved, and acute; the base broad and rather depressed, with the culmen rounded and keeled between the nostrils; the gonys lengthened and curved, and the lateral margins finely serrated; the nostrils basal, lateral, and placed in a short, broad groove, with the opening concealed by a membraneous scale; wings moderate and rounded, with the third and fourth quills, or the fourth only, longest; tail moderate, broad, and slightly rounded, with the two middle feathers lengthened and narrowed; tarsi as long as the middle toe, and covered in front with very broad scales; toes moderate, the outer toe rather larger than the inner, the hind toe long and strong; the claws moderate, compressed, and curved.

122. Nectarinia Chalybea. (Linn.) Pl. Enl., 24–62; *Certhia Capensis*, Linn.; *Lesser Double-Collared Sunbird*, Cuv., Vol. 2, p. 350; Nat. Lib., Vol. 8, p. 132.

HEAD, back, breast, and throat, all shining, metallic-green; rump blue; wings and tail brown. To the green of the breast succeeds a narrow blue collar, followed by a red one, about half an inch broad, not extending below the yellow side-tufts. Length, 4½″; wing, 2″; tail, 1″ 10‴; bill, 1″, curved.

This is a very common species about Cape Town, and as far as Swellendam: there it becomes mixed with the greater double-collared species, *C. Afra*, L., which altogether ousts it in the George and Knysna districts, and supplies its place.

It is one of the boldest and most familiar of all our sunbirds, frequenting the flower-gardens in the midst of Cape Town, and even venturing into open windows to visit potted plants.

Nests, reported to be of this species, have been brought to me—pendant, domed, and porticoed structures, like those of others of the family that I have seen. Eggs, 3—5, minutely mottled grey-brown : axis, 8′′′; diam., 5½′′′. I can confirm the statements of my correspondents, having myself taken nests of this species containing eggs and young birds. They are not, however, always pendant, being sometimes supported by twigs interwoven with their structure. They are usually composed of cob-web, stuck over with bits of dead leaves, or chips of thin bark, always placed on the *outside* of a bush, never *among* the branches. I have, however, seen one placed on the side of a bush close to a rock, so that the bird had to fly round the bush to get at it. In appearance they exactly resemble the masses made and collected by one of our commonest spiders; and I have more than once seen an inhabited spider's-web forming part and parcel of the nest. Whether the nest was built in the web, or whether the spider found it a convenient place and selected it herself, or was brought with a bit of web by the birds, and then took up her abode and enlarged it, I cannot tell; but there the incongruous allies lived, and each brought up her own brood, or would have done so, had not I harried them both.

123. Nectarinia Afra. (Linn.) Ed. Birds, Pl. 347;
Cinnyris Smaragdinus, Vieil.; *Certhia Scarlatina*, Sparm.; *Certhia Erythrogaster*, Shaw; *C. Pectoralis*, *Le Sucrier à plastron rouge*, Le Vail., Pl. 300, f. 1, 2.

HEAD, back, and breast metallic-green, blue, and bronze-shot; rump blue; wings and tail brown; below the green of the breast is a narrow blue collar, succeeded by a broad red one, which extends nearly all over the stomach; vent brown; bright yellow tufts on each side. Length, 5½″; wing, 2¾″ nearly; tail, 2″ 3′′′; bill, 1¼″, curved.

This bird never visits the neighbourhood of Cape Town. It replaces *N. Chalybea* in the forest districts, but about Swellendam it is mingled with it, and keeps to the wooded river banks. At the Knysna it is abundant, and has the same habits as *N. Chalybea*, being only perhaps rather more shy. Le Vaillant states that it builds in the fork of a tree, and lays four or five eggs, which are dotted with fawn-colour on a bluish-white ground.

124. Nectarinia Verroxii. (Smith.) *Cinnyris Verroxii*, Smith, Zool. S. A., Pl. 57; Jardine's Sunbirds, Pl. 9.

ABOVE brownish; head, upper, and lateral parts of neck, back, and shoulders, dark bluish-green, with a strong metallic lustre; wings and tail brown; under parts yellowish-grey; tufts under the wings pale, brilliant scarlet. Total length, nearly 6″; wing, 2½″.

Rare. Inhabits ".Kafirland and the country eastward of it, towards Port Natal." "Like the other species of the group, it feeds upon small insects, and these it collects partly from the branches and leaves of brushwood and dwarf trees, and partly from flowers."—A. Smith, loc. cit.

125. Nectarinia Collaris, Vieil., N. Dict. d'Hist. Nat. XXXI., p. 502 ; *Le Sucrier Gamtocin*, Le Vail., Pl. 299 ; Jard. Sunbirds, Pl. 6.

THE head, back of neck, rump, upper wing-coverts, and tail, golden-green ; all the lower parts of the body dullish-yellow ; neck golden ; on the breast a collar of brilliant blue.

The female has no collar ; otherwise, though less bright, her plumage is similar to that of the male. Each has the bill and feet blackish ; the eyes brown.

Le Vaillant states that they keep together for some time in families, consisting of the parent birds, and seven or eight young ones ; also, that they are abundant near the "Gamtoos" River. Sundevall has it from Kaffraria. I have not seen it.

126. Nectarinia Bifasciata, Shaw, Vol. 8, p. 198 ; *C. Nitens*, Vieil., Jard. Sunbirds, Pl. 4. ; *C. Mariquensis*, A. Smith.

GREEN and gold, with black-brown quill-feathers ; belly and tail, and red pectoral bar, bounded above by a steel-blue one. Length, 4" 9''' ; wing, 2" 9''' ; tail, 2".

Dr. Smith found this species at Kurrichain ; and Mr. Verreaux cites it from Natal. Mr. Andersson and Mr. Kisch procured it in Damaraland.

127. Nectarinia Famosa. (Linn.) Pl. En., 83, f. 1 ; Cuv., Vol. 2, p. 364 ; Vieil., Ois d'Or, t. 37, 38 ; *Le Sucrier Malachitte*, Le Vail., Pl. 289 and 290.

GENERAL colour, rich shining green ; wings and tail black ; side-tufts brilliant yellow ; two centre tail-feathers prolonged three inches beyond the rest. Total length, 9" ; wing, 3" ; tail, 5" 5'''.

Common throughout the colony. It is very partial to the blossoms of the aloe, among which it finds an abundance of its insect food. I have found it plentifully up the ravines of Table Mountain, extending even to the top. It has a shrill, not unpleasing, but short song, and when pursuing a rival utters a piercing scream. It is very combative, and if two males meet about the same bush, a fight is sure to ensue, to the great detriment of their beautiful tail-feathers. The males lose their beauty in the winter season ; and the young birds are just like the females.

Eggs, reported to be those of this species, have been given me—they are of a dull, greyish-brown colour, minutely mottled all over. They

were said to have been taken from a domed nest, suspended from the
end of a twig. Since this was written (1865), I have found several
nests containing eggs, and can confirm the account given me by my
correspondent. The eggs are—axis, 10′′′; diam., 6′′′.

It abounds on the Cape Flats during the flowering of the protea,
from which it extracts the saccharine juice by means of its long brush-
tipped tongue.

128. Nectarinia Cinerea, Vieil.; *Certhia Cinerea,* Linn.; Lesson, Vol. 2, p. 28.

HEAD, neck, top of the back, and breast, greyish-brown; a
yellow line on each cheek; the lower part of the back, wing-
coverts, and rump, brilliant green; throat pale-yellow, mixed
with golden-green down the centre and on the breast; belly
white; tail brown. Length, 8½ inches, *French.*

M. Vieillot thinks that this is a young bird. I agree with him; and
should say it was the young of *C. Famosa!* As a species, I have never
met with it.

129. Nectarinia Olivacea, Smith, Ill., S. Af. Zool., Pl. 57.

THE colour of this species, above, is intermediate between
grass and olive-green, the head being strongly tinged with
blue; below it is light, yellowish-green, with an orange tint
on the throat, and on each axilla there is a small tuft of
brilliant yellow feathers, Length, from the base of the bill
to the point of the tail, 5′′; bill, 1′′ 3′′′.

Of this bird, Dr. Smith says: "In the same country in which we
found *C. Verroxii,* we discovered another species of the genus
Cinnyris, which appears to us undescribed, and which we shall hereafter figure under the name of *C. Olivaceus.*"

Mr. Ayres has procured it at Natal. (Ibis 1866, p. 347.)

130. Nectarinia Violacea. (L.) Pl. Enl., p. 670, f. 2.; *Certhia Crocata,* Shaw; Jard. Sunbirds, Pl. 16; Cuv., Vol. 2, p. 364.; *Cinnyris Aurantia,* Lath.; *Soui Manga Orangé,* Le Vail., Pl. 292.

HEAD, neck, throat, and shoulders, shining green; breast
shining violet; back, wings, and tail, olive-green; belly and
vent, deep orange and yellow; side-tufts yellow; centre
feather of tail prolonged about 1¼ inch beyond the rest, and
darker. Total length, 6½′′; wing, 2¼′′; tail, 3′′ 2′′′.

This sunbird appears to affect wild, uncultivated country and mountains, more than the other species, *N. Chalybea, Afra,* or *Famosa;*
indeed, I never observed it near habitations. It is plentiful on the

top and about the sides of Table Mountain. At the Kuysna, I found it in abundance among the uncultivated hill-sides, away from the timber. I saw a pair building a pendant, domed nest, with a projecting portico over the entrance, at Cape Delgado, on the East Coast of Africa. The nest was hung at the extreme end of a drooping branch of a *Casuarina*, close to the sea-beach; not far off was the nest of *N. Senegalensis*.

131. Nectarinia Fusca. Vieil.; *Le Sucrier Nama-quois*, Le Vail., Pl. 296.

Top of head, back of neck, and wing-coverts, brown, shot with dull violet. Throat, changing violet and olive-green; wings and tail brown-black, slightly shaded with violet; eyes, bill, and feet, brown. Female: uniform, greyish-brown on the upper parts; below, dirty-white.

The ♀ builds in a hole of a tree, lays four or five greyish eggs, and sits 18 days. Inhabits Namaqualand (Le Vaillant).
Professor Wahlberg found it in Damaraland.

132. Nectarinia Amethystina. (Shaw.) *C. Auratifrons*, Vieill., Pl. En., p. 590; Cuv., Vol. 2, p. 260; Nat. Lib., Vol. 8, p. 134; Jard. Sunbirds, Pl. 13; *Le Sucrier Velours*, Le Vail., No. 294; *Nec. Aurifrons*, Licht.

Top of head and forehead, shining green; throat and rump cupreus; shoulders the same, but with a tinge of blue; all the other parts deep, rich, velvety purple-brown. Length, 6″; wing, 3″; tail, 1″ 10‴.

This sugar-bird is not very uncommon in the forest districts of the colony; and I have also received it from the neighbourhood of Swellendam, through Mr. Cairncross, who informs me that the specimens sent by him were procured in the mimosa trees which line the banks of the rivers flowing through that district. I saw it at the Knysna, and its habits appeared similar to those of the common *N. Chalybea*. Le Vaillant found their nests in the thickest bushes and in holes of trees. Eggs five in number, spotted with olive-green, on a greyish ground.

Mr. Atmore writes: "Oudtshoorn, 24th January, 1863. I enclose you a couple of eggs of the black sugar-bird. The nest is a curious structure, hanging on the branch of an apple-tree, very rough outside, composed of short bits of stick, grass, and spider's-web—arched, as are the nests of all the tribe. The number of eggs appears to be two, as we did not take these till they were incubated; before they were blown they were of a soft, creamy-yellow colour. I have not seen this species west of the Gouritz River: on the other side they are plentiful, especially when the 'wilde dagga' is in flower.

The eggs sent by Mr. Atmore are of a beautiful creamy grey, with indistinct, confused blotches, spots, and streaks, chiefly at the obtuse end: axis, 9‴; diam., 6‴.

133. Nectarinia Senegalensis. (Linn.) Briss., Orn., Vol. 3, t. 34, f. 2 ; *Cinnyris Discolor*, Vieil., Ois d'Or, t. 8, 9 ; *Le Sucrier Protée*, Le Vail., Pl. 285, f. 2.

In repose, the front of the neck and breast is bright red ; when excited, the slightly raised feathers show spots of most brilliant golden-green and changing blue ; forehead and chin golden-green ; top of head, back of neck, the whole upper body and belly deep-brown, with a violet tinge ; wings and tail purple-maroon ; bill and feet black ; eyes red-brown. The female is of a uniform greyish-brown, deepening on the wings and back ; bill and feet brown.

Le Vaillant only found this bird in Kaffraria ; but it inhabits Senegal, and is found both on the East and West Coasts. He supposes they build in hollow trees, as the plumage of those he shot smelt strongly of decayed wood. I met with this bird plentifully at Zanzibar, and at Cape Delgado, on the East Coast. It frequented the mango trees, perching on the highest branches, and uttering a loud, shrill note. In the young males the breast is mottled black and yellow. Andersson seems to have found them plentiful in Damaraland ; his collection contained many specimens. I have also seen it from Otjimbinque, not far from Walwich Bay ; it must, therefore, be a widely-distributed species. I killed a specimen at Cape Delgado, clinging to, and apparently building, a nest similar in shape, and hung in like manner to those of the *N. Famosa*. I also shot down the nest from the elevated pendant branch of a *Casuarina* tree. It was composed chiefly of cob-web and bits of lichen, and had the usual covered porch. It was incomplete, and contained no eggs.

134. Nectarinia Natalensis, Jardine, Sunbirds, Pl. 12.

Coronal patch (confined to the forehead and crown) is of a rich bluish-green ; gular patch, of a golden-green, confined to the chin and throat, and is bordered on each side with a narrow maxillary stripe of the same colour as the crown. Upper parts, cheeks, and sides of the neck, very deep, soft, and velvety umber-brown, paler on the wings and tail, the bend of the wing having a violet patch. On the under parts, the fore part of the neck and breast are of a brilliant scarlet, appearing in different lights, waved with violet, from the structure of the feathers being the same as those of *N. Senegalensis* ; the simple apical tips only being scarlet, and producing all the brilliant effect. The remaining under parts are very deep blackish-brown, appearing in some lights almost black. Length, almost 6″.

I have quoted the full description given by Swainson, that there may be no mistake in the discrimination of this species ; though, for my part, I consider it identical with *N. Senegalensis*, but have not had an opportunity of comparing a specimen from Natal with one from Damaraland.

135. Nectarinia Œnea. (Vieil.) Encyc. Meth., p. 599 ; *Soui Manga Bronzé*, Le Vail., pl. 297.

UPPER parts, chesnut-brown, shading into blue, green, or violet ; wings and tail, bronzed-black ; lower parts black, shading olive ; bill and feet bright black ; eyes red ; tuft under wing, bright yellow ; bill very strong and curved. The female has a smaller and less curved bill ; wings and tail olivaceous ; bill and feet brown.

Builds in holes of trees ; lays 4 to 6 eggs, rosy white, spotted with red. According to Le Vaillant, inhabits Zwartland, and the vicinity of Sunday's River ; but I believe it to be the Ceylon species, *N. Lotenia*, and not an inhabitant of South Africa.

136. Nectarinia Pusilla. (Vieil.) *Le Sucrier Sucrion*, Le Vail., Pl. 298.

HEAD and neck of male, chesnut, brightened with glimmering shades of blue, purple, and green ; lateral and middle tail feathers bronze ; the rest of the upper parts dead, purple marroon ; rump and upper tail-coverts, brilliant purple ; under parts of the body, orange-red ; bill and feet blackish ; eyes maroon. The female much less. All the upper parts of body and wings, olive, yellowish-green ; rest of plumage pale-yellow ; bill and feet brown-black.

According to Le Vaillant, who knows nothing of its nest nor economy, it inhabits the neighbourhood of Van Staden's River, and is only found in the winter season. Sundevall declares it is a manufactured species ; *Nec. Zeylanica* forming the principal portion of it. I agree with him.

137. Nectarinia Cardinalis. (Vieil.) *Le Sucrier Cardinal*, Le Vail., Pl. 291 ; Cuv., Vol. 2, p. 364.

THE male bird has all the upper parts, including the tail-coverts, the two long tail-feathers, and also the breast, of a rich green, shot with gold, most brilliant on the top of the head ; from the breast, all the lower parts are of a soft carmine colour ; wing and lesser tail-feathers shot green and gold, upon a black ground ; bill and feet black ; eyes brown. The female is smaller than the male, of less brilliant, though similar colours above ; but the body and lower parts are yellow, and she wants the two long feathers.

Le Vaillant says he met with this bird only in the mountain

K

regions of Great Namaqualand : he neither found the nest nor the eggs. Le Vaillant's original specimen is in the Leyden Museum, and is composed of portions of *N. Chalybea*, with a red stomach inserted. I should think likewise the long tail-feathers had been stolen from *N. Famosa*.

138. Nectarinia Splendida. (Shaw.) Cuv., Vol. 2, Pl. 360 ; *Certhia Coccinigastra*, Lath. ; *Sucrier Eboulissant*, Le Vail., Pl. 295, Fig. 1 ; *Cin. Bombysinus*, Vieil. ; Jard. Sunbirds, Pl. 5 ; *Cinnyris Lucidus*, Less.

MALE.—Head and neck, brilliant shades of purple, violet, and blue ; the breast and body darker, spotted with glowing red, golden-yellow, and green ; the back, shoulders, rump, and upper tail-coverts, vivid green and gold ; wing and tail-feathers velvet-black ; bill and feet black ; eyes maroon.

FEMALE.—Above uniform brown ; tail and wings shaded with olive-green ; lateral tail-feathers edged with dull white ; lower parts greyish ; bill and feet brown-black. Length, $3\frac{1}{2}''$; wings, $2\frac{1}{2}''$.

Inhabits Great Namaqualand, towards Fish River, according to Le Vaillant, and builds in mimosas. Eggs white, four or five in number. I have never met with it in this country, nor has any one since Le Vaillant's time. It is a native of Senegambia, Guinea, and Congo.

139. Nectarinia Pulchella. (Linn.), Pl. En. 670, f. 1. ; Cuv., Vol. 2. p. 364 ; *Cinnyris Caudatus*, Vieil. ; Swain., Nat. Lib., Vol. 8, p. 123 ; *C. Pulchellus*, Vieil., Ois. dO'r. t. 41 ; *Sucrier Cossu*, Le Vail., Pl. 293, Fig. 1.

HEAD, neck, shoulders, and upper wing-coverts, changeable green, everywhere shot with gold, toned red on the scapularies ; breast crimson, margined with yellow ; the rump and upper tail-coverts, brilliant violet, glancing purple, or steel-blue ; wing and shorter tail-feathers, blackish-brown ; the two long tail-feathers red-gold ; bill brown, short, and straight. The female somewhat smaller ; body also yellow ; but the head, neck, shoulders, and rump are greyish-red, shaded olive, and slightly shot with gold ; wing and tail feathers reddish-olive ; bill and feet brown ; tongue long.

Inhabits Namaqualand, and lives chiefly on jasmines. Nest unknown. A Senegal species not found in South Africa by any one since Le Vaillant's time ; and it is doubtful if he did so.

140. Nectarinia Platura. (Vieil.) *N. Sylviella,*
Temm., Pl. Col. 347 ; *N. Cyanopygos,* Licht. ; Jard. Sunbirds, Pl. 19 ; *Le Sucrier Figuier,* Le Vail., Pl. 293, Fig. 2.

HEAD, neck, back, and wing-coverts, shining golden-green ; rump violet ; wing and tail feathers black ; the centre pair of the latter prolonged, and gilt with a reddish gold ; belly and vent fine yellow.

Le Vaillant says he discovered this species only in the forests of Great Namaqualand, feeding principally on an inodorous jasmine, climbing over the mimosa trees. Sundevall states it is only known from Western Africa.

141. Nectarinia Melanura. (Sparm.) Mus.
Carls., t. 5 ; Lesson, Vol. 2, p. 51 ; *Certhia Melanura,* Lath. ; *Anthornis Melanura.*

HEAD and back violet ; chest and belly inclined to green ; wing-coverts brown, and bordered with olive ; tail black, long, and forked. Length, 6″ 2‴.

This bird was first described by Sparmann as a native of the Cape of Good Hope. It is, however, a New Zealand species ; not a *Nectarinia,* but an *Anthornis ! !*
The two next birds have also been wrongly described as native of the Cape of Good Hope.

142. Sitta Chloris. (Sparm.) Lesson, Vol. 1, p.
361 ; *Acanthositta Chloris,* Gray ; *A. Tenuirostris,* Lafr.

BODY, green above ; white below ; tail black, yellow on the end ; a yellowish spot in the centre of the wing.

· Cape of Good Hope, Lesson (loc. cit.) ; but a New Zealand species, according to Sundevall.

143. Sitta Caffra. (Sparm.) Lesson, Vol. 1, p.
361 ; *Tatare Otaitiensis.*

BODY above, variegated with yellow and black ; below yellow ; feet black ; claws yellow.

Kaffraria : Lesson (loc. cit.) ; but a South Sea Island species, according to Sundevall.

Genus DICÆUM, Cuvier.
Bill short, curved, broad, and rather depressed at the base, with the culmen curved to the tip, which is acute, and finely serrated on the lateral margins ; the sides compressed, and

the gonys long and ascending; the nostrils lateral, and placed in a broad groove, with the opening linear, and closed by a membranous scale; wings moderate, and rather pointed, with the first quill wanting; the second nearly as long as the third and fourth, which are equal and longest; tail short and even; tarsi longer than the middle toe, strong; toes moderate; the outer one rather longer than the inner, and united at the base; the hind toe long and strong; the claws moderate, compressed, and curved.

144. Diceum Rufescens. (Vieil.) *Sylvietta Rufescens*, Vieil.; Lesson, Vol. 2, p. 19; *Le Crombec*, Le Vail., Pl. 135; *Nectarinia Rubracana*, Temm. *Certhia Erythropygia*, Lath.

UPPER parts, brownish-grey; all the lower parts reddish, deepest on the vent; bill long and curved, clear-brown; legs rather more red.

Le Vaillant found this bird about the banks of the Orange River, in Great Namaqualand, among the mimosa trees, amid the branches of which it hops in search of insects.

I have received specimens from Graaff-Reinet, Colesberg, Hope Town, Damaraland, and Swellendam. At this latter place, Mr. Atmore informs me: "It is not uncommon, creeping about the decaying fences, like our English hedge-sparrow.

The Third Tribe, DENTIROSTRES, or Tooth-billed Birds,

embraces a numerous series of birds, that have the tip of the upper mandible more or less emarginated and hooked; the tarsi vary in length, but are mostly slender, and covered with broad scales; the toes generally long, with the outer toe more or less united to the middle one at the base.

The First Family, LUSCINIDÆ, or Warblers,

have the bill subulate, more or less slender and straight, with the tip of the upper mandible curved and emarginated; the base sometimes broad, but the sides always compressed towards the tip; the nostrils basal, and placed in a membranous groove, with the opening exposed; the wings more or less long, sometimes rounded, and sometimes pointed; the tail of various lengths, truncated at the end, or rounded, and sometimes graduated; the tarsi more or less long, always slender ; the toes varying in length, with the outer one more or less united to the middle toe ; claws curved and acute.

The Sub-Family, MALURINÆ, or Soft-tailed Warblers,

have the bill moderate, more or less slender and straight, with the tip of the upper mandible curved, and sometimes emarginate ; the sides compressed ; nostrils basal, and placed in a membranous groove, with the opening exposed ; the wings short, and rounded ; the tail more or less lengthened, and rounded ; the tarsi generally long and slender ; the toes more or less long, and always slender, with the outer toe united to the middle one at the base ; the third toe long, and armed with a strong claw.

Genus DRYMOICA, Swain.

Bill entire, short ; rictus bristled ; wings very short, and rounded, the three first quills equally graduated, the fourth and fifth longest, the primaries hardly longer than the other quills ; legs pale ; the lateral toes equal ; tail graduated ; the feathers obtuse.

145. Drymoica Textrix, Smith, Z. S. A., Pl. 74, f. 1 ; *Sylvia Textrix*, Vieil. ; *Le Pinc-Pinc*, Le Vail., Pl. 131.

UPPER parts of head and neck, interscapulars, back, and shoulders, umber-brown, variegated with white and clear yellowish-brown ; sides of head and neck, dirty yellowish-brown, with small umber blotches ; wing-feathers brownish-red, edged with pale wood-brown ; chin and throat greyish white, indistinctly mottled with light umber-brown ; breast and belly pale senna-yellow, mottled with umber-brown spots ; flanks and vent, pale yellowish-brown, also spotted. Tail very short, and slightly graduated, two centre-feathers brownish-red, the rest umber-brown, all edged with wood-brown, and the three outermost broadly tipped with white. Length, 4" ; wing, 1" 10½''' ; tail, 1" 1'''.

This bird is abundant throughout the western end of the colony ; but not having any friend collecting for me in the eastern, I cannot speak as to that division. Near Cape Town it is very common, frequenting open plains covered with low scrub, or marshy places in which reeds grow. It climbs about these in a very clever manner, apparently sliding up and down ; when pursued, it drops to the tangled herbage at the foot of some bush, and cannot be induced to quit its hiding-place. It often hovers in the air at a moderate height over the bushes, uttering its ringing, metallic cry of "pinc-pinc-pinc," jerking about with rapid strokes of the wing, suddenly dropping into the bushes, and then remaining mute.

Le Vaillant attributes to this little bird a nest which is very common about the country, and is well known under the name of the "nest of the Kapokvogel," but he is quite in error. The fabricator of this nest, the true "Kapokvogel" (or cotton-bird), being the little *Ægithalus Minutus*. The nest of *D. Textrix* is a domed structure, generally supported between stems of grasses. The eggs, 4—5, are white, spotted with minute reddish spots: axis, 7'''; diam., 6''.

146. Drymoica Pectoralis, Smith, Zool. S. A., Pl. 75, f. 2.

Top of head, ear-coverts, back of neck, interscapulars, and back, brown; eyebrows whitish; under parts dirty-white; the belly and vent more or less ochreous; breast with a broad, dark-brown band; tail long, and much graduated, wood-brown, lightest at the tips, with a faint dark bar across each feather, except the two centre ones, which also want the light tips. Length, 5''; wing, 2'' 1'''.

Inhabits dry situations abounding in brushwood, and was found principally at and around Latakoo. It feeds upon insects, procured upon the shrubs and dwarf trees which it frequents. Dr. A. Smith, loc. cit.

147. Drymoica Pallida, Smith, Zool. S. A., Pl. 72, f. 2.

Upper parts, pale broccoli-brown, lightly washed with yellowish-brown; the head and upper parts of neck lightest; the tail darkest; ear-coverts rusty-white; chin, throat, and breast white, faintly tinged with wood-brown; belly and vent, pale senna-yellow; tail long and much graduated. Length, 5'' 11'''; wing, 2'' 1½''.

The only specimens procured by Dr. Smith were obtained about 300 miles north of Cape Town, in Namaqualand. "It occasionally occurs in thickets, but more generally upon dwarf trees in the vicinity of streams; and through these it passes rapidly from branch to branch in pursuit of insects, which form its chief food." Dr. A. Smith, loc. cit.

148. Drymoica Fasciolata, Smith Zool. S. A., Pl. 111.

Top of head, neck, and interscapulars, between orange-brown and oil-green; back and rump between hyacinth-red and yellowish-brown; across each wing a bar of pale cream-yellow; chin and throat pure white, with short narrow transverse umbre-brown bars; breast and belly senna-yellow, with curved bars; tail fan-shaped, rather silky, with semi-circular points, light reddish-brown, with a tint of green. Length, 5″; wing, 2″ 6‴; tail, 2″.

Inhabits districts covered thinly with small underwood, and in such places is found moving from bush to bush in search of its food, which it appears to take partly from the top of the bushes and partly from the branches, among which it passes rapidly. Open flat plains to the north-east of Latakoo. Dr. A. Smith, loc. cit.

149. Drymoica Natalensis, Smith, Zool. S. A., Pl. 80.

UPPER surface of head, back, and sides of neck, back, and rump, between reddish and buff-orange; each of the feathers, except those of the rump, marked along the shaft with a broad longitudinal dark umber-brown stripe; tail graduated, reddish-brown, edged with reddish-orange, tipped with pale buff-orange; behind each tip a broad transverse bar of deep umber-brown, indistinct on the centre feathers; wing-feathers brownish-red, margined with dull reddish-orange; chin and throat white; rest of the under parts senna-yellow. Length, 7″ 3‴; wings, 3″; tail, 3″ 4‴.

"Inhabits the neighbourhood of Port Natal, and the specimen described was shot upon reeds, among which it was flitting to and fro in search of insects, which, from the ingesta found in its stomach, appeared to have been its sole food." Dr. A. Smith, loc. cit.

150. Drymoica Ocularius, Smith, Zool. S. A., Pl. 75, f. 1.

TOP of head, back, sides of neck, interscapulars, back and lesser wing-coverts, umber-brown, striped with yellowish-brown; sides of head and ear-coverts reddish; chin and throat white; breast dirty-white, crossed by a narrow brown bar; under parts dirty-white; tail long, and much graduated, umber-brown, the two middle feathers being broccoli-brown. Length, 5″ 3‴; wings, 2″; tail, 3″ 3‴.

Dr. A. Smith states, loc. cit: "This bird has a rather extensive range in South Africa, being found, though sparingly, among the brushwood in the northern districts of the Cape Colony, and between those and the Tropic of Capricorn. It feeds upon insects; and, in quest of them, it is to be seen flitting from branch to branch in the most arid and barren situations." I have received several specimens from Kuruman, and found it myself in one particular locality near Mr. Jackson's place at Nel's Poort, Beaufort. In its method of flight and search after prey, it exactly resembled the common species, D. Capensis. Its nest also resembles that of the last named, and its eggs are pale blue, immaculate: axis, 6‴; diam., 5‴

151. Drymoica Subcinnamomea, Smith, Zool. S. A., Pl. 111.

UPPER surface of head, neck, back, and shoulders, between oil-green and orange-brown; forehead tinted cinnamon-red; wing-feathers, light reddish-brown, with a tinge of green; the primaries edged narrowly towards their base with cinnamon-red; rump ruddy; tail rather long, and slightly rounded, deep brownish-red; chin and neck inferiorly, liver-brown, variegated with narrow white transverse bars; breast and fore part of belly cinnamon-red; belly and vent coloured as the back. Length, 5" 8'''; wing, 1" 11'''; tail, 2" 8'''.

Dr. Smith only procured one specimen of this bird, which was killed on the top of one of the mountains of the Kamiesberg, in Little Namaqualand. "For some time before it was shot, it was beheld flitting from bush to bush, occasionally perching on their summits, at other times hopping rapidly to and fro among their branches, as if engaged in quest of insects, which were found to constitute its food." Dr. A. Smith, loc. cit.

152. Drymoica Substriata, Smith, Zool. S. A., Pl. 72, f. 1.

TOP of head, sides of neck, and shoulders, light broccoli-brown; back and upper parts of neck, interscapulars, back and tail-coverts, uniform rusty yellowish-brown; sides of head and ear-coverts pale wood-brown, the latter faintly streaked with dull umber-brown; eyebrows white, prolonged behind the eyes; chin, throat, centre of breast, and belly, ochrey white, the breast variegated with several narrow longitudinal umber-brown stripes; flanks and vent light yellowish-brown; wing-feathers pale brownish red, faintly margined with wood-brown; tail long, graduated, of a rusty broccoli-brown, the middle feathers darkest. Length, 5" 8'''; wing, 2" 1½'''; tail, 2" 19'''; tarsus, 7'''; bill, 6'''.

Dr. Smith met with but few of these birds, and those only on the banks of the Olifant's River, about one hundred miles north of Cape Town. Usually found in thickets composed of high brushwood or dwarf trees, in which it is seen rapidly flitting from branch to branch, apparently in quest of insects, which constitute its food.

Mr. Atmore procured this species, with its nest and eggs, at Traka; the latter are white, with the faintest tinge of green, marked with large blotches of dark and light-brown : axis. 8'''; diam., 5½'''.

153. Drymoica Aberrans, Smith, Z. S. A., Pl. 78.

TOP of head and back of neck, pale chesnut-brown, deepened with red; interscapulars and back, yellowish-brown; under parts, pale rusty olive-yellow, the throat and chin lightest;

sides of breast and flanks tinged with hair-brown; wings light browish-red; tail lengthened and graduated, coloured like the back, lightest at the points, with a faint dark mark behind the light colour. Length, 5″ 10½‴; wing, 2″ 4‴; tail, 3″ 2‴.

"The only specimens of this species seen by Dr. A. Smith" were shot near Port Natal, while perched "upon some brushwood which skirted a tract of marshy ground. Before they resorted to the brushwood, they were observed flitting to and fro among strong bushes, as if in quest of their food, which was found to have consisted of insects."—Dr. A. Smith, loc. cit.

154. Drymoica Affinis, Smith, Z. S. A., Pl. 77, f. 1.

Top of head, back and sides of neck, broccoli-brown, faintly tinged with yellowish-brown; interscapulars and back, intermediate between broccoli and yellowish brown, glossed with oil-green; rump pale yellowish-brown; chin, throat, centre of breast and belly, yellowish-white; flanks and vent, rusty senna-yellow; wing-feathers brownish-red, margined with light yellow-brown; tail elongated, and much graduated, light reddish-brown, edged with yellowish-brown, tipped with white, with a broad transverse brownish bar behind the white. Length, 4″ 7½‴; wing, 2″; tail, 2″ 6‴.

"Inhabits dry flats in the interior of South Africa, and flits to and fro, in search of insects, amongst the shrubs with which they are more or less coated."—Dr. A. Smith, loc. cit.

155. Drymoica Cherina, Smith, Z. S. A., Pl. 77, f. 2.

Top of head, and back and sides of neck, deep reddish-brown; interscapulars, rump, lesser wing-coverts, quill-coverts, umber-brown, the feathers edged and tipped with yellowish-brown; back yellowish-brown; tail very short, and rounded, umber-brown, the two centre feathers tipped narrowly, the others broadly, with white, and each, with the exception of the two centre ones, crossed by an umber-brown bar immediately behind the white tip; chin, throat, centre of breast and belly, yellowish-white; sides of breast and flanks tinted with broccoli-brown. Length, 3″ 11‴; wing, 1″ 11‴; tail, 1″ 7½‴.

"Inhabits situations covered with strong grass, and seeks its food, which consists of small insects, either on the ground or upon the grass itself, along the stalks of which it runs with facility. It also perches upon small shrubs." Dr. A. Smith, loc. cit.

L

156. Drymoica Ruficapilla, Sm., Z. S. Af., Pl. 73, f. 1 ; *La Rousse-Tête*, Le Vail., Pl. 124.

TOP of head and upper parts of neck, pale chesnut-brown ; interscapulars, back, and tail-coverts, pale broccoli-brown, washed with yellowish-brown ; tail the same, but brighter, and margined with light yellowish-brown ; quill-feathers of wing, pale brownish-red, with pale, yellow-brown edges ; chin, throat, centre of breast and belly, pale straw-yellow ; sides of head, neck, breast, belly, and vent, pale greyish-brown ; tail short, slightly graduated. Length, 4″ 5‴ ; wing, 2″ 1½‴ ; tail, 1″ 11‴.

"This bird occurs in various situations in the interior of the Cape Colony, and is usually found among brushwood or upon dwarf trees. It feeds upon small insects, and in quest of those it is to be seen flitting from branch to branch with great rapidity. As compared with many other species, it may be considered as rather a rare bird, and it seldom happens that more than one or two specimens are procured in any given locality which it is known to inhabit."—A. Smith, loc. cit. Dr. Smith believes this bird to be Le Vaillant's "Rousse-Tête." I cannot say I agree with him. The eggs resemble that of *D. Terrestris* in form and colour, but the spots are coarser.

157. Drymoica Levaillantii, Sm., Z. S. Af., Pl. 73, f. 2.

UPPER parts of head, upper and lateral parts of neck, bright chesnut-brown, the two first striped with umber-brown ; back and rump liver-brown, the feathers edged with pale wood-brown ; lesser wing-coverts, umber-brown, broadly edged and tipped with pale broccoli-brown ; under parts, pale wood-brown ; throat inclined to white ; tail light umber-brown, margined externally with pale chesnut-brown, and broadly tipped with dull wood-brown ; behind the light points each feather is marked with a circumscribed liver-brown blotch, plainly seen underneath ; eyebrows and sides of head, wood-brown ; tail moderately long, and graduated. Length, 5″ 1‴ ; wing, 2″ 3‴ ; tail, 2″ 6‴.

Common near Cape Town, among reeds, feeding on insects. Dr. Smith procured it some distance north of the Orange River ; and I have likewise received it from Swellendam, Colesberg, and Damara-land. The nest is oval, very lightly made, but supported firmly between the thin closely-set twigs of the "Rhenoster Bush." It is composed of wool, the down of plants, and thin bents of grass, very skilfully woven together ; a good-sized hole, near the top, gives access and egress to the birds, and the eggs are of a blue colour, marked at the obtuse ends with various-sized brown blotches : axis, 7‴ ; diam., 5‴.

158. Drymoica Chiniana, Sm., Z. S. Af, Pl. 79.

Top of head, back, and upper parts of neck, cinnamon-coloured; feathers narrowly edged and tipped with light yellowish-brown; feathers of the back, brownish-red, edged and tipped with wood-brown; rump pale-brown; wing-feathers brownish-red, edged with lightish-brown; tail rather long, and moderately graduated, two middle-feathers yellowish-brown; the other feathers light brownish-red, tips dull wood-brown, and behind them a broad bar of umber-brown; chin and centre of belly white, the rest ochreous. Length, 5″ 9‴; wing, 2″ 9‴; tail, 2″ 10½‴.

"Only one specimen of this species was obtained, and that was killed while perched upon some brushwood growing near the edge of a small stream to the northward of Kurrichain. In its habits it resembles the species already described; and at the time it was shot it was moving quickly about among the branches, apparently in quest of its food, which was found to consist of insects."—Dr. A. Smith, loc. cit.

159. Drymoica Terrestris, Sm., Z. S. Af., Pl. 74, f. 2.

Top of head and interscapulars, umber-brown, variegated with yellowish-brown; back of neck, back, and shoulders, clear yellowish-brown, with umber-brown streaks; rump umber-brown; chin and throat whitish; breast, belly, and vent, sienna-yellow; tail moderately long, and slightly graduated; two middle-feathers broccoli-brown, broadly margined and tipped with wood-brown; the other feathers brownish-red, broadly tipped with white, with a large umber-brown blotch just before the white, seen, as in all the species, most plainly on the under side; irids light-brown. Length, 4″ 3‴; wing, 1″ 11‴; tail, 1″ 11‴.

Dr. Smith (loc. cit.) first discovered this species upon the grassy plains lying between Latakoo and Kurrichain, feeding on the ground. It does not seem to affect bushes, dwelling only amongst the long grass.

Mr. Atmore found this species at Traka. He forwarded a couple of eggs, which are of a light verditer, unevenly blotched and spotted with dark and light dry-blood-coloured spots, and streaked at the obtuse end with delicate wavy brown lines: axis, 7‴; diam., 5‴.

160. Drymoica Subruficapilla, Smith, Zool. S. A., Pl. 76, f. 2.

Top of head, back, and sides of neck, light chesnut-brown, striped with umber-brown; lower parts of back and sides of neck, interscapulars, and back, umber-brown, variegated with

ashy-grey; chin, throat, middle of breast and belly, pale
yellowish-grey, passing into white; breast faintly speckled
with pale-brown; sides of breast, flanks, and vent, pale hair-
brown; tail graduated, dull reddish-brown, tipped with wood-
brown, with a broad transverse umber-brown bar immediately
behind the light tip. Length, 5″ 4‴; wing, 2″; tail 2″ 4‴.

"This bird occurs in various districts of the Cape Colony, and is
either found upon brushwood or among rushes or reeds. It feeds
upon insects, and, like others of the genus, flits generally from branch
to branch or from reed to reed in quest of its food."—Dr. A. Smith,
loc. cit.

I have received it from Swellendam, from Mr. Atmore, and saw it
along the River Zonder End. Dr. Smith's account of its habits is
perfectly correct.

161. Drymoica Capensis, Smith, Zool. S. A., Pl. 76, f. 1.

TOP of head, neck, interscapulars, back, rump, and tail,
between broccoli and yellowish brown; the feathers of the
latter, with the exception of the two middle ones, narrowly
margined with wood-brown; wings dull umber-brown; chin,
throat, and under parts of body white, with a yellowish tinge,
more or less marked with short subovate umber-brown
stripes; eye stripe whitish; eye, light hair-brown. Length,
4″ 11‴; wing, 2″ 1‴; tail 3‴.

"Inhabits various districts in the western division of the Cape
Colony, and is found in situations which abound with shrubs. It
feeds upon insects, and searches for these both upon underwood and
on the ground—generally, however, on the former. It is also occasionally observed among coarse grass or reeds, busily engaged climbing
or flitting about in quest of food."—Dr. A. Smith, loc. cit.

I have received it from Colesberg and Swellendam, and have procured a few specimens on the Cape Flats. I cannot help thinking that
this is the bird described by Le Vaillant as "Le Capocier," Pl.
130, Fig. 1, the fabricator of the beautiful nest figured on Pl. 129,
and called by Cuvier, at page 391, Vol. 6, *Motacilla Macroura*, and at
page 467, *Sylvia Macroura*, Gmel.

The description given by Le Vaillant of the nests of these birds is
too long for quotation: suffice it to say, it is a most beautiful structure
of cotton, gathered from the wild cotton-tree. The eggs are from five
to eight in number, of a pale verditer ground, spotted, blotched, and
striped, chiefly at the obtuse end, and sometimes in the form of a ring,
with brown (more or less dark) markings: axis., 7‴; diam., 5‴.

162. Drymoica Melanorhyncha. (Jardine.)
Contrib. to Ornith., 1852, p. 60.

ABOVE, pale brownish-grey; stripe between the nostrils and
the eye, white; under parts white, tinted with pale-brown;

vent a marked brown; tail-feathers, with the exception of the two centre, marked with an obscure band before their white tips; bill black. Length, 4″ 2‴; wing, 1″ 9‴.

"Builds among stalks of high weeds in Natal." Ibis, Vol. 2, p. 208. I have not seen it from the Cape Colony.

163. Drymoica Obscura, Sunde.; Ofvers. Kongl.
Vet. Ak. Forhandl., p. 103.

BACK, dusky-grey, much marked with brown; head rufous; forehead considerably, and back of head obscurely, marked with brown; belly dirty white; sides grey. Length, 5½″.

Kaffraria : (Wahlberg) non vidi.

164. Drymoica Curvirostris, Sund.; Ofvers.
Kongl. Vet. Ak. Forhandl., p. 104.

HEAD and back yellowish-grey, with ·black markings; belly yellowish-white; middle of belly and throat almost white; tail-feathers fuscous, tips yellowish-grey. Length, 6″.

Kaffraria (Wahlberg). Natal (Ayres) Ibis, Vol. 1863, p. 323.

165. Drymoica Thoracica, Gray; *Sylvia Gutturalis*, Boie; *Saxicola Thoracica*, Licht.; *Motacilla, Thoracica*, Cuvier, Vol. 1, p. 438; *Apalis Thoracica*, Grill., l. c., p. 31; *Sylvia Thoracica*, Shaw, Vol. 10, p. 562; *Le Plastron Noir*, Le Vail., Pl. 123.

GENERAL colour above, ashy-grey, tinted with olive-green, except on the tail; chin, throat, and under parts whitish, tinted with rufous-brown, deepest on the vent; a black collar extends across the chest, and a black mark from the bill to the eye; tail graduated, the outermost feather nearly all white, the second white at the tip, the third at the tip only; two tufts of black hair—like feathers, from the basal half of the back of the head; eye light-yellow; tarsus flesh-coloured; claws brown; bill black. Length, 5″ 3‴; wing, 2″; tail, 2″ 4‴.

Le Vaillant found this bird in abundance after crossing the "River of Elephants" as far as the tropics; also on the bank of the Orange River, and in Kafirland. He states that they always went in couples, male and female, breeding in November and December. The nest is placed among grass or low bushes; the eggs are six in number, and of a reddish white. The male has an agreeable song. I procured specimens in considerable numbers at Plettenberg's Bay (on the southern side of the colony). in the wooded ravines and deserted gardens in which the brushwood had sprung up. They hunted incessantly after insects, gliding about among the branches, peering up at the undersides of the leaves or thick branches, and darting up at the small insects which sought concealment in such situations.

Swellendam has furnished me with several specimens; it is also abundant at Nel's Poort, and all along the rivers in that part of the Karroo; and I likewise saw a pair which evidently were nesting in the rank herbage and scrub which line the crater of the minute volcano from which issue the hot springs of Caledon.

166. Drymoica Oxyura, G. R. Gray; *Sylvia Oxyura*, Shaw; *Le Figuier à Cuti-Pennes*, Le Vail., Pl. 133; Nat. Misc., Pl. 957.

UPPER parts red; under parts citron-yellow, lightish towards the vent, which is white; the first two feathers of the wing brown, as are also the ends of the next seven or eight; tailfeathers ending in sharp points like bristles.

Le Vaillant says he discovered this species on the banks of the Groote River, among the mimosa trees. Sundevall doubts this being an African bird, and thinks that Le Vaillant has figured an American species of *Synallaxis*.

167. Drymoica Ayresii, Hartl., Ibis. Vol. 1863, p. 325.

RESEMBLES *Cisticola Europœa*, but much brighter in colour; rump rufous; feet considerably larger; the tarsi smaller. Length of wing, 1".

This new *Drymoica*, figured in the "Ibis" for 1863, p. 325, was discovered by Mr. Ayres at Natal, frequenting open grassy country.

168. Drymoica Fulvicapilla, Bp.; *Sylvia Fulvicapilla*, Vieil.; *Le Rousse-tête*, Le Vail., Pl. 124.

UPPER parts, wings, and tail brownish; ash-coloured below; whitish on the vent; top of the head red-brown, or tancoloured; tail square; eyes red-brown; feet yellowish. Length, 4" 4'''; wing, 2" 9'''; tail, 1" 9'''.

The nest is constructed among low bushes amid which the bird lives. The eggs, from four to six in number, are white, dotted with very small vinaceous spots, like fly-blows. Inhabits, according to Le Vaillant, the districts of Cambedoo, Kafirland, and Namaqualand.

Mr. Atmore has sent two specimens from the neighbourhood of George.

169. Drymoica Subflava; *Drymoica Flavicans,*

Sunde.; *Sylvia Flavicans,* Vieil.; *Sylvia Limonella, Motacilla Subflava,* Gm., Pl. Enl., t. 584, f. 2.; *Le Citrin,* Le Vail., Pl. 127; Cuv., Vol. 6, p. 391 et 469.

RED-BROWN; beneath grey; rump pale; sides of body reddish; tail wedge-shaped. Length, 4"; 6'''.

Le Vaillant found this bird in Namaqualand, living in small families, feeding on caterpillars, spiders, and small insects. Their nest is oval, formed of the down of plants, and entirely closed, with the exception of a small hole for entrance. It is firmly fixed in a low bush, and the eggs, five or six in number, are reddish-white, with brown markings.

170. Drymoica Diophrys, Gray; *Sylvia Diophrys,* Vieil.; *Malurus Supercilious,* Cuv., Vol. 1, p. 469; *Le Double Sourcil,* Le Vail., Pl. 128.

ABOVE, reddish-brown; below white, tinted with russet; tip of the head ruddy, with a black eyebrow and moustache. Female, less bright than the male, and without the black stripes.

A doubtful species. Said by Le Vaillant to have been found in the Karroo, but very sparingly.

171. Drymoica Brachyura; *Eremomela Brachyura* (Vieil.); *Sylvietta Chloris,* Boie; *Sylvia Brachyura,* Vieil.; *Eremomela Flaviventris,* Sunde., Ofv., 1860, p. 102; *L'Olivert,* Le Vail., Pl. 125; *Sylvia Flaviventris,* Burch.

ABOVE, yellowish-green; under parts dirty white; tail very short, the wings reaching almost to the end of it; eyes light-hazel. Length, 4"; wing, 2"; tail, 1" 4'''.

Outeniqualand, Le Vaillant; Karroo, Victorin. Mr. Atmore has sent this species from Blanco, near George; Mr. Ayres from Natal; and Mr. Andersson from Damaraland.

172. Drymoica ˜Africana. (Gmel.) *Sylvia Africana,* Shaw, Vol. 10, p. 615; *Musicapa Dubia,* For.; *Turdus Tibicen,* Vieil., Cuv., Vol. 6, p. 377; *Dasyornis Africanus,* Smith; *Synallaxis Cantor,* Less.; *Malurus Africanus,* Swain. (MSS. in his own handwriting on Museum copy of Le Vaillant); *Motacilla Africana,* Gmel.; *Sphenurus Tibicen,* Licht.; *Le Fluteur,* Le Vail., Pl. 112, f. 2.

UPPER parts rufous, deepest on the head and rump; inclining to grey on the fore part of the back; and everywhere marked with very dark-brown, broad streaks down the centre of the feathers: these are nearly obsolete on the head, but extend along the centre of the tail-feathers.

Under parts, uniform light isabella-coloured, everywhere mottled with dark black-brown streaks, most evident on the flanks. Length, 8″ 9‴; wings, 3″; tail, 4″.

The plumage of this curious bird is of a singularly lax texture; the tail-feathers are long, pointed, and graduated, and furnished with webs, so thin that they are transparent; the shafts project beyond the webs, which are fined down to a point. Wings small and weak, hardly enabling the bird to fly fifty yards. If flushed more than once, it betakes itself to a clump of grass, or bush, and will suffer itself to be taken with the hand rather than rise again; for this reason it has acquired the name of "*Idle Jack*" and "*Lazy Dick*."

173. Drymoica Apicalis; *Catriscus Apicalis,*
Caban, Mus. Hein. I, p. 43; Ibis Vol. 1863, p. 323; *Bradypterus Brevirostris*,[*] Sundev.; Ofvers. Kongl. Vet. Ak. Forhandl., p. 403.

GENERAL colour, lightish-brown; back, rump, and quill-feathers light yellow-brown; beneath white, as is also an inconspicuous eyebrow; flanks the colour of the back, but paler; tail fuscus, the lateral feathers terminated by a semi-lunate white spot; upper mandible fuscous; the lower and legs yellow. Length, 7″; wing, 2″ 4‴; tail, 3½″.

Found in Natal by Mr. Ayres, who describes their habits as very similar to those of the preceding species.

The Sub-Family, LUSCININÆ, or Warblers,

have the bill more or less long, slender, and straight, with the culmen curved at the tip, which is slightly emarginated; the sides compressed, and the gonys long and ascending; the nostrils basal, and placed in a membranous groove, with the opening usually exposed; the wings moderate and sometimes rounded; the tail moderate and rounded at the end; the tarsi more or less lengthened, slender, and covered with broad scales, the divisions of which are sometimes obliterated; the toes more or less long and slender, the outer toe generally longer than the inner, and united at its base; the claws long, curved, and acute.

Genus CALAMODYTA, Meyer and Wolf.

Bill rather small and straight, with the culmen very slightly curved, and the sides compressed to the tip, which is scarcely emarginated; the lateral margins straight and slightly inflexed; the gonys long and ascending; the gape sometimes furnished with very short, weak bristles; the

[*] According to Dr. Hartlaub, see Ibis 1866, p. 140.

nostrils basal, placed in a membranous groove, with the opening oval and exposed; wings rather short, with the first quill very short, the second rather shorter than the third and fourth, which are equal and longest; tail moderate, broad, and rounded; tarsi slender, longer than the middle toe, and covered in front with broad scales, sometimes the divisions between them are scarcely visible; toes rather long and slender, the outer toe longer than the inner one, and slightly united at the base, the hind toe long and strong; the claws moderate, curved, and acute.

174. Calamodyta Babæcula. (Vieil.) *La Caqueteuse*, Le Vail., Pl. 121, f. 1; *Calamoherpe Loquax*, Boie.

UPPER parts, head, wings, and tail, sombre-brown, lightly glossed with olive; under parts light-brown; throat speckled with brown. Length about 5″.

Le Vaillant says he procured this species in the marshes of Verloren Vley, on the West Coast, and in Outeniqualand. The nest is constructed among reeds. The eggs are white, spotted with brown, and five or six in number.

The male utters a note which Le Vaillant represents by the words " *gri-gri-gra-gra*."

" Avis nobis ignota ; quam vero existere nos dubitamus."—Sundev., p. 37.

I have not succeeded in identifying this species, but am inclined to think Le Vaillant was alluding to *Calamodyta Rufescens* when describing the note and habits above given.

175. Sylvia Bœticula, Vieil.; *Cettia Bœticula,* Sund.; *Calamodyta Bœticula*, Vieil.; *Sylvia Isabella,* Boie.; *L'Isabelle*, Le Vail., Pl. 121, fig. 2.

" IN general appearance, this species resembles *La Caqueteuse*, and might be mistaken for the female of that bird. It is found in the same localities, constructs its nest in the same manner, and lays five or six pure white eggs."—Le Vaillant.

176. Calamodyta Rufescens. (Keys and Bl) Grill., l. c., p. 28.

UPPER parts isabella-coloured; throat, front of neck, belly, and all the under parts, with a slight reddish tinge. Length, 6″; wing, 2″ 8‴; tail, 2″ 5‴.

Received from Mr. Atmore, Swellendam, and is very common in reed-beds along the banks of rivers in the Karroo, and about Nel's Poort; also at Zoetendal's Vley. It is a noisy bird, continually utter-

M

ing a stridulous cry of cur-cur-crak-crak. I sought in vain for its nest in the months of November, December, and January.

I identify this bird with "L'Isabelle" of La Vaillant. Its habits accord well with those described by him as belonging to that bird.

177. Calamodyta Gracilirostris, Hartl., Ibis 1864, p. 348.

ABOVE, pale-brown; crown greyish; rump and upper tail-covers rufescent; tail brown; eyebrow and lower part of body white; vent pale-yellowish. Length, 6″ 5‴; wing, 3″; tail, 3″.

My son procured several specimens of this shy and retiring bird in rushes on the banks of the "Diep River," near the Observatory, Cape Town. He informs me he was attracted to them by their babbling, which he at once detected as new to him. Their stomachs contained the remains of minute *coleoptera* and other insects. Mr. Ayres has observed it in Natal.

178. Calamodyta Arundinacea, Linn; *Sylvia Turdoides*, Meyer, Tem., Man., 1, 191; *Turdus Arundinaceus*, L., Pl., Enl, 513; *Agrobates Brunnescens*, Jard., Gould's B. of Eur., Pl. 106.

ABOVE, pale olive-brown; rump paler; throat white, shaded with grey; chest, belly, and vent pale fulvescent; tail and wing brown, the feathers margined with paler colour. Length, 7″ 4‴; wing, 5″ 9‴; tail, 3″ 2‴.

South Africa: Hartlaub, Orn., W. Af., p. 61. Mr. Andersson brought specimens from Damaraland, apparently identical with a European bird in the South African Museum.

179. Calamodyta Natalensis. *Camaroptera Natalensis*, Hartl., Ibis Vol. 1863, p. 326.

ABOVE, reddish-ash; below, pale ash-coloured; under side of wings and tail whitish; top of head rufous. Length, 3″ 6‴; wing, 1″ 9‴; tail, 1″ 5‴.

" These birds are plentiful, frequenting rough weeds and grass; their flight is tolerably strong; their food consists of minute insects, eggs of moths, &c."—Ayres, Ibis, loc. cit.

180. Calamodyta Olivacea; *Camaroptera Olivacea*, Sunde., Ofver. Kongl. Vet. Forhald., p. 103.

DULL olive-green; below dirty white; thighs fulvous-yellow; sides of head and forehead cinereous.

Inhabits Natal: Ayres.

Genus ÆDON, Boie.

Bill moderate and straight, with the culmen curved, and much compressed to the tip, which is entire ; the gonys long and slightly ascending ; the gape not furnished with bristles ; the nostrils basal, pierced in a short broad groove, with the opening exposed and oval. Wings moderate, with the first quill short, and the second nearly as long as the third and fourth, which are equal and longest ; tail lengthened and rounded ; tarsi long, and covered in front with broad scales ; toes short, with the inner toe shorter than the outer one ; the latter is united at its base ; claws small, compressed, and curved.

181. Ædon Pœna ; *Erythropygia Pœna*, Sm., Z. S. A., Pl. 30.

HEAD brown ; neck rusty-grey ; anterior half of back yellowish-brown ; hinder part of back, rump, upper tail-coverts, and the basal two-thirds of tail, clear reddish-orange ; last third of tail-feathers, liver-brown, edges rusty, four outer ones of each side broadly tipped with white ; chin and throat white ; breast, belly, and vent, pale cream-yellow. Length, 6″ ; wing, 2″ 9″ ; tail, 2″ 9″.

"Inhabits arid districts, thinly covered with dwarf trees, among the branches of which it seeks its food. Its movements, while flitting from branch to branch, are rapid, and when resting or hunting for insects, it is constantly watchful, always ready to take alarm and to retreat when its haunt has once been discovered."—Dr A. Smith, loc. cit.

Dr. Smith does not indicate the locality where this bird was procured ; but I have received it from Mr. Moffat at Kuruman. Mr. Andersson procured specimens in Damaraland.

182. Ædon Leucophrys. (Vieillot.) *Turdus Pipiens*, Steph. ; *Thamnobia Leucophrys*, Cab. ; *Erythropygia Pectoralis*, Smith, Zool. S. Af., Pl. 49 ; *Le Grievetin*, Le Vail., No. 118, Pl. 49.

BROWN above ; posterior part of back and rump, deep reddish-orange ; chin, middle of abdomen, and vent, pale cream-colour ; breast light sienna-yellow, streaked with umber-brown ; wing feathers edged with greyish-white ; eyebrows pale cream-colour ; tail liver-brown ; the feathers rusty at their edges, and white at the tips, most visible on the outside feathers ; all but obsolete on the two centre ones. Length, 6″ 3‴ ; wing, 2″ 9‴ ; tail, 3″ 1‴.

Inhabits arid districts thinly covered with dwarf trees, among the branches of which it seeks its food, which consist of insects.—(Dr. A.

Smith loc. cit.) Le Vaillant states that he found it on the borders of the Gamtoos, Sunday, and Swartkop Rivers, among the mimosa bushes that fringe these streams. The male has a moderate power of song, and the female deposits four or five light-green eggs, spotted with brown, at the obtuse end, in a nest which is placed in the midst of a thick bush. I have received it with *E. Pœna* from Kuruman.

Genus BRADYPTERUS,* Sw.

Bill distinctly notched ; wings short ; the three first quills equally graduated ; tail rounded ; the feathers remarkably broad and soft ; feet large, strong, and robust ; the lateral toes equal ; claws slender, slightly curved.

183. Bradypterus Coriphœus ; *Sylvia Coriphœus*, Vieil. ; *Drymoica Coriphœus*, Bp. ; *Bradypterus Coriphœus*, Swain. (MSS. note on plate) ; *Le Coriphé*, Le Vail., Plate 120 ; the *Boschcreeper* and *Katlachter* of Colonists.

GENERAL colour, cinereous, darkest on the crown of the head, and tinged with rufous on the lower part of the back, wings, and two centre tail-feathers ; beneath paler, with belly, vent, and a board stripe down the throat, white ; a narrow white stripe extends over the eye ; tail-feathers black, broadly tipt with white ; bill and legs black. Length, 6″ ; wing, 2″ 9‴ ; tail, 2″ 6‴.

This bird is common throughout the country from Cape Town to Beaufort. It frequents bush-covered land, and well deserves its trivial colonial name, being always found running about the lower parts of the shrubs, very rarely flying over the tops of them. It nests also at the foot of a bush, forming a cup-shaped, rather flat structure, composed of hair, wool, and bents of dry grass. The eggs are of a lovely verditer.

184. Bradypterus Layardi ; *Phlexis Layardi*, Hartlaub., Ibis Vol. 1866, p. 139.

ABOVE, very dark-brown, slightly tinged with rufous ; under parts, from chin to vent, deep rufous-brown ; a slight spot of the same colour between the eye and the bill ; flanks brown ; plumage very lax and worn in appearance. Length, 6″ ; wing, 2″ 3‴ ; tail, 3″ 1‴.

This curious little bird, on which Dr. Hartlaub has formed a new genus (*Phlexis*) was procured by my friend Mr. Atmore in the neighbourhood of George.

I confess I do not like to separate it from *Bradypterus*, with which,

---

* Mr. Gray (Genera of Birds, p. 163,) makes this genus equal with *Drymoica*.

in my humble opinion, it has such close affinity. Dr. Hartlaub gives the characters of his new genus in the "Ibis," p. 1866, loc. cit.

185. Bradypterus Victorini, · Sundevall ; Zool.
Anteku Victorin, p. 29.

♂ ABOVE, ruddy fuscus; head cinereous; below, with the throat, fulvous grey; the sides, especially of the chest, more ruddy; side of the head beneath the eyes, fuscous white; wings and back coloured alike; the quill-feathers rather black; tail-feathers the colour of the back; plumage long and lax, especially on the rump.

This bird, named after M. Victorin, was found by him at the Knysna, and appears to have much the same habits as the rest of the genus, frequenting vleys.

186. Bradypterus Platyurus, Swain. ; *Sylvia Brachyptera*, Vieil. ; *Calamoherpe Saltator*, Boie ; *Le Pavaneur*, Le Vail., No. 122, f. 1 and 2.

♂ GENERAL colour above, dark dull-brown, below paler; upper mandible black; lower yellow; feet brown. ♀ generally lighter, with some dark lines on the throat and chest. Size about that of *B. Layardi*.

This bird was found by Le Vaillant in the marshes which occur near Plettenberg's Bay, and throughout the whole of the Outeniqua country.

187. Bradypterus Sylvaticus, Sundev. in
Grill ; Aut., l. c. p. 30 ; Ibis., Vol. 1862, p. 146.

GENERAL colour above, dark reddish-brown ; under parts whitish, and yellowish-brown on the flanks. In certain lights the tail appears barred with dark-brown ; inner surface of the wings, towards the shoulder, white, mottled with dark-brown ; legs and lower mandible, light horn-colour ; upper mandible, dark horn-colour ; iris, light hazel. Length, 5″ 7‴ ; wing, 2″ 4‴ ; tail, 3‴ ; bill, 6‴ ; tarsus, 10‴.

Received from Kuruman, from Mr. R. Moffat, jr. Common in the vleys about Worcester and Swellendam. Its notes and habits resemble those of the *Drymoicæ*. It is fond of perching on the tops of the reeds when alarmed, and sliding thence down their shafts to the tangled herbage at their roots, amid which it lies concealed from pursuit. It feeds on insects and the little mollusc (*Succinea Lalandii*) which is found abundantly in marshy spots.

Genus SYLVIA, Latham.

Bill more or less long, rather strong and straight, with the culmen gradually curved, and the sides compressed to the

tip, which is scarcely emarginated; the gonys long and
ascending; the gape furnished with a few short weak bristles;
the nostrils basal, and placed in a short broad groove, with
the opening exposed and sublunate; wings moderate, with the
first quill very short; and the second shorter than the third,
or third and fourth, which are the longest; tail moderate,
broad, and rounded on the sides; tarsi rather shorter than,
or as long as the middle toe, rather strong, and covered in
front with broad scales; toes moderate, with the inner toe
shorter than the outer; the latter united at its base; hind-
toe long, and armed with a long strong claw; the claws of the
fore toes curved, compressed, and acute.

188. Sylvia Natalensis, Smith; *Chloropeta Na-talensis*, Smith, Zool. S. Af, Pl. 112, Fig. 2.

UPPER parts, between broccoli-brown and oil-green; eyebrows and ear-coverts, straw-yellow; under parts, pale
lemon-yellow; shoulders, pale brownish-red, the feathers
edged with white; tail-feathers between brownish-red and
yellowish-grey, narrowly edged with white. Length, 3" 10'";
wing, 3"; tail, 2" 6'".

Dr. A. Smith obtained one specimen of this bird near Port Natal.
Its stomach contained insects.

189. Sylvia Trochilus. (Linn.) Pl. Enl. 581, f. 2; *Phyllopneuste Trochilus; British Willow Warbler*, McGill, Brit. Birds, Vol. 2, p. 371; *Sylvia Flaviven-tris*, Vieil., Gould's B. of Eur., Pl. 131, f. 1.

ABOVE, dull olive-green; chin and throat yellowish-green;
belly and vent dirty greenish white. Length, 5"; wing,
2" 8'"; tail, 2" 2'".

Natal, Ayres; and Damaraland, Andersson.

190. Sylvia Obscura, Smith, Zool. S. A., Pl. 112, Fig. 1.

UPPER surface greenish; the feathers of the wings and tail
edged with yellow; eyebrows, chin, throat, breast, belly, and
vent, yellow. Length, 5" 3'"; wing, 2" 3'"; tail, 2" 6'".

A single specimen of this bird was obtained by Dr. Smith near
Latakoo. Its stomach contained the debris of insects. I have
received it from Damaraland through Mr. Andersson.

The Sub-Family, ERYTHACINÆ, or Robins,

have the bill moderate, more or less slender, and rather depressed at the base, with the culmen slightly curved, and the sides gradually compressed to the tip, which is entire ; the lateral margins straight, and sometimes inflexed, the gape more or less furnished with bristles ; the wings generally short and rounded, though sometimes long and pointed ; the tail usually short and broad, sometimes even or rounded at the end ; the tarsi lengthened, slender, and covered with an entire scale ; the toes moderate, the lateral ones unequal ; the claws moderate, curved, and acute.

A bird belonging to the first genus of this sub-family *(Copsychus)* is here included, in furtherance of the plan I proposed to myself, of including *all* birds said to have been found within my limits. It is the

Copsychus Saularis, Linn. ; *Gryllivora Intermedia*, Swain. ; *Le Cadran*, Le Vail., Pl. 109.

GENERAL colour, steel-blue-black, with bar on wing, belly, vent, and outer tail-feathers white. Female brown, instead of black. Length, 7″ 6‴ ; wing, 4″ ; tail, 3″ 6‴.

This species was procured—so says Le Vaillant—in Great Namaqualand ; but this is another of that author's falsehoods. I have never seen it in any of the collections formed in that country ; and I should not fail to recognise it instantly, as it is a common bird in Ceylon, where I resided some years. It is now generally believed that Le Vaillant never found this species in South Africa, and I only admit it into this list to correct the error.

Genus SAXICOLA, Bechstein.

Bill moderate, rather depressed and broad at the base, with the culmen slightly curved, and the sides gradually compressed to the tip, which is slightly emarginated ; the lateral margins straight and inflexed ; the nostrils basal, lateral, and placed in a membranous groove, with opening rounded ; wings long, reaching to middle of tail, with the second quill nearly as long as the third and fourth, which are the longest ; tail moderate and nearly even ; tarsi longer than the middle toe, and covered in front with an entire scale ; toes moderate, with outer toe longer than the inner ; the hind toe long ; the claws rather short, slightly curved, and acute.

192. Saxicola Pileata ; *F. Imitatrix*, Vieil. ; *Turdus Minus*, Forst., Descrip. Anim., p. 12 ; *Traquet Imitateur*, Le Vail., Pl. 181 ; *Schaapwachter* of Colonists (lit. the Shepherd).

GENERAL colour above, rufous-brown ; feathers of the wings dark-brown, edged with the colour of the back ; forehead white: this colour extending in a line over the eye; top of the head black; a stripe of the same colour extends from the corner of the bill down the sides of the neck, and forms a broad collar across the breast; chin, throat, and belly white, the latter tinted with rufous, which becomes stronger on the flanks and vent; centre tail-feathers all dark-brown, the basal half of the rest white. Length, 6" 9'''; wing, 3" 9'''; tail, 2" 10'''.

The above description was taken from a fresh-killed specimen "still in the flesh," on the 19th of June.

This is one of the most favoured and favourite birds of the colony, over the whole of which it extends. He is protected and petted on account of his own natural sweet notes, and for his great powers of imitation. Perched on a white-ant's nest, he pours out a flood of song, chaunting long into the darkening twilight, when other songsters are gone to rest; and the morning light scarcely suffices to enable you to see the musician, ere he again commences his mellow notes or imitations.

He is a favourite with the farmer and the shepherd—the master and the man: the good-wife also casts a kindly eye on him, as he flirts his tail and wings on the mud wall of her little garden, and thumps her "young hopeful" who, unable to resist the tempting shot, is about to "shy a stone at him."

He breeds in the rat-holes, close to the houses, or out in the fields, and if wounded retreats into the first he comes to. He runs along the ground with great rapidity, and seldom alights on bushes—never, perhaps, on trees. If frightened from one stone, he flits to another, and in alighting opens his tail, so as to show the white patch on his rump. He is usually perched on an ant-hill, in most cases sees you before you see him, and acts accordingly!

193. Saxicola Bifasciata, Tem., Pl. Col. Pl. 472, Fig. 2; *Sax. Spectabilis*, Hart. Ibis.

MALE.—Top of head dark-brown, mottled with black; back and neck reddish-brown, mottled with black; rump clear yellowish-brown: the same colour prevails over the whole of the lower parts, except the throat, and a little way on the chest, where is a rich black patch extending to the head, and including the eye and ear; a yellowish-brown stripe passes from the nape of the neck on the one side, over the eyes and nostrils, to the nape of the neck on the other side; tail and secondary wing-feathers black; quill-feathers brown. Length, 7"; wing, 3" 6'''; tail, 2" 6'''.

FEMALE.—Wants the black markings, and is altogether less brilliant, the yellows being rufous, and the tail and wings dull-brown.

This handsome species much resembles *S. Pileata*, but is a stronger-looking bird. It was first sent by Capt. Bulger from Windvogel-berg, but is also common near Graham's Town. Mrs. Barber writes:

"Highlands, June 22, 1865.—The contents of the Museum box are as follows : *Saxicola Spectabilis*— ♂, ♀. These birds are dwellers amongst rocks, and frequent rocky mountains and hills, old stone-kraals, &c. They build their nest under shelving rocks, near or upon the ground, and sheltered by spreading ferns or long grass; and their eggs are usually three in number. The ♂ is fond of placing himself on some high projecting rock, and of making himself conspicuous by chirping away in a cheerful voice, either to annoy a rival or amuse his mate : he is also fond of opening and shutting his wings, "bowing and scraping," &c., and I have no doubt "thinks no end of himself." The ♀ is a very quiet, retiring little body. These birds (the ♂'s) possess the power of mocking other birds and animals. When we were living in the district of Graaff-Reinet, I had a beautiful spotted *Merecat*, a tame one: it was very fond of me, and was my companion in all my walks. These little animals have a peculiar bark, and produce a great variety of sounds with their pretty little voices : and these mocking-birds would imitate all the sounds that my pet made very accurately. I have also heard them mocking robins and sugar-birds, but they seldom use this power. We will get you some more specimens. I was surprised to hear that they were new—very much so, for they are common all over the Eastern Province, excepting near the coast. They prefer high situations. There are always five or six of them here : they sit upon the garden wall and upon the top of our house. A few days ago there was one sitting upon the gable-end of our stable, making a terrible row : a quiet old rock-thrush that belongs to the place, feeling disgusted with his noise, went up to him and gave him a smart peck, which sent him flying!! We never shoot birds that take up their abode with us, so we see all their odd ways."

194. Saxicola Cinerea. (Vieil.) Smith, Zool. S. A.; *Le Traquet Tractac*, Le Vail., Pl. 184 ; *S. Levaillantii*, Smith.

GENERAL colour, ashy grey; lighter on the lower part of the back, and becoming quite white on the rump; breast and throat grey; rest of the under side white ; tail with a white patch, which extends from near the point of the outside feather, across each of the others, to the base of the centre feathers, in the shape of a triangle ; some of the wing-feathers are bordered with white ; iris black. Length, about 6″; wing, 3″ 9‴ ; tail, 2″ 10‴.

Le Vaillant found this wheat-ear in the province of Outeniqua, perching on bushes, always on the move from one to another, and very wary. In flying they expanded the tail, exposing the white mark ; they also had the habit of opening and closing the wing, so peculiar to the wheat-ears.

They build at the foot of bushes on the ground. The female lays four bluish-green eggs, minutely speckled with brown, the specks sometimes forming a ring at the obtuse end : axis, 11‴; diam., 8‴. I found them abundantly at Nel's Poort, nesting in November. Mr. Atmore writes of them as follows :—" Blauco, Sept. 10th, 1864. The rock-

N

saxicola *(C. Cinerca)* is abundant here ;—and, by the way, how well this class of birds obeys the geology of the country : wherever there is karroo soil you find them. The same also with the ' Kalkoentje ' *Anthus (Certhilauda) Capensis*, which is found in every patch of grass country, but never in karroo soil ; for instance, they are plentiful here, and proceeding northwards they do not occur in the fifty miles of karroo you pass over on the way to Cango ; but in that narrow valley they are again plentiful."

195. Saxicola Monticola, Vieill. (Smith, Zool., S. A., No. 185) ; *S. Alpina ; Vitiflora Rupicola*, Boie ; *Le Traquet Montagnard*, Le Vail., No. 184, Fig. 2, and Cuvier, Vol. 7, p. 437 ; *Motacilla Maderaspatana*, Gmel. apud Cuv., Vol. 7, p, 475, nec Shaw, Vol. 10, p. 548.

ADULT, entirely black, except belly, shoulders, and the edges of tail-feathers, which are white. When young, nearly all the feathers, which when adult are black, are cinerous. Length, 7" ; wing, 4" ; tail, 3".

Le Vaillant found this bird inhabiting the mountains of Namaqualand, and never descending into the plains, except compelled by great drought. He describes them as very shy and difficult of approach, hiding themselves in holes or inaccessible precipices.

Mr. Andersson brought specimens from Damaraland ; I have received a few from Kuruman, and from Mr. Atmore, who procured them about Traka ; and also found it myself at Nel's Poort. Le Vaillant's account of their habits is substantially correct. I only found them among the rocky mountain sides, breeding in the holes and crevices.

196. Saxicola Albiscapulata. (Briss.) Neue Wirbelth (1835), p. 72, I. 26, fig. 1 ; *Thamnolacea Albiscapulata*, Bp. Consp., p. 302 ; *Thamnolæa Cinnamomeiventris*, Lafr.

♂ GENERAL plumage throughout glossy-black ; rump, belly, and vent deep rufous ; shoulders white. Length, 8⅓" ; wing, 4½" ; tail, 4" 7‴.

♀ Wants the white patch on the shoulder, and is altogether much duller in colour and smaller in size.

Sent from " Highlands," near Graham's Town, by Mrs. Barber.

197. Saxicola Rufiventer, Swain., An. in Menag., p. 293.

GREY-BLACK ; rump, tail-coverts, and body beneath, chesnut ; tail rounded, and somewhat lengthened. Length, 7" 3‴ ; wing, 4" 3‴.

" Inhabits South Africa."—Dr. Burchell's Coll." Swainson loc. cit.

I suspect this is nothing more than the ♀ of *S. Albiscapulata*, Briss., the preceding species.

198. Saxicola Infuscata, Sm., Z. S. A., Pl. 28.

GENERAL colour, rusty-brown; chin and throat of a dirty white; under parts grey, more or less tinted with yellowish-brown; wing and tail-feathers with a dash of umber-brown about them; and they and the wing-coverts are more or less edged with dirty white; tail square; iris black. Length, 7″ 6‴; wing, 4″ 9‴; tail, 3″ 6‴.

"This is the largest and rarest of the South African *Saxicolæ*. It appears to be principally, if not entirely, restricted to the districts between the Oliphant and Orange Rivers, and seldom occurs far from the sea coast. It selects sterile tracts, which are thinly covered with brushwood, for its *habitats*,—and in these situations it is so extremely watchful, that even weeks of incessant toil will scarcely secure the collector a specimen. It seeks its food, which consists of insects, upon the ground; and while on it, as well as when perched, it raises and depresses its half-extended wings, after the other species of this genus." —Dr. A. Smith, loc. cit.

I have received it from Kuruman; and Mr. Atmore found it sparsely at Traka.

199. Saxicola Sperata. (Linn.) Cuv. Vol. 6, p. 437; the *Sibyl Warbler*, Latham; *Le Traquet Familier*, Le Vail., No. 183, f. 1, 2; Sundev. Obser. on Le Vail. Ois. d'Af., p. 44.; *Speck-vreter* of Colonists.

GENERAL colour above, rufous-brown; below lighter, inclining to grey on the throat, and rufous on the flanks and vent; rump bright rufous, as are all the tail-feathers, with the exception of the two centre ones, which are, as are also the wing-feathers, dark-brown, narrowly edged with rufous; the others are broadly tipped with dark-brown, the shafts are of the same colour. The ear-tufts are more rufous than the neck, and being silky, show rather plainly. Length, 6¾″; wing, 3½″; tail, 2″ 8‴.

Received from Colesberg. I have generally found this species, or a variety (for all those procured here are smaller than the Colesberg specimen), during the summer months, about stones in rocky places. It is abundant round the "Lion's Head," at an elevation of about 1500 feet, nesting in crevices, and the young bird is speckled exactly like a young robin. They have a habit of perching on the summit of stones and rocks, opening and shutting their wings and tails. In flight they resemble the wheat-ear, and flit from stone to stone.

From Le Vaillant's description of its habits, which I believe to be much over-coloured, I certainly think this is his *Traquet Familier*. A pair or two frequent every farm-house in the colony, and are accused of picking the grease out of the cart-wheels; hence their colonial name. Dr. A. Smith, in Zool. of S. A., says it differs from Le Vaillant's bird, and calls that species *Sax. Familiaris.* Shaw, Vol. xiii, pt. 1, pp. 241—243, gives a *S. Familiaris,* and describes it thus: Grey-brown wheat-ear, paler beneath, with the breast, flanks, ears, and rump rufous; the two middle tail-feathers entirely brown, the rest brown, edged with rufous. Was this description taken from Le Vaillant without seeing the bird?

200. Saxicola Sinuata, Sund. Observ. on Le Vail.

Ois. d'Af., p. 44, nota.

GENERAL colour above, brown, very slightly tinged with rufous; below light-brown, inclining to grey, particularly on the throat; rump and tail-coverts rufous; vent white; ear-covers slightly rufescent and glistening; tail parti-coloured, white, tinged with rufous and dark-brown. The white portion begins near the point of the outer web of the outer feather, and slopes gradually upwards across all the feathers, till it is hidden on the two centre ones by the coverts. This, and the peculiar appical attenuation of the point of the second quill-feather of the wing, at once distinguishes it from *S. Sperata.* Length, $5\frac{1}{2}''$; wing, $3''$; tail, $2''$.

This plainly-coloured chat is not uncommon at Nel's Poort and to the eastward, having been received from Colesberg and Kuruman. I found it at the first-named place, breeding in November, in holes of buildings, and occupying the place of *S. Sperata.* It makes a nest of hair, roots, and feathers, and lays three to five eggs, of a light-blue, speckled with rufous. These speckles generally assume the shape of a ring at the obtuse end: axis, $10'''$; diam. $7'''$.

201. Saxicola Hotentotta; *Vitiflora Hotentotta,*

Shaw, Vol. 10, p. 573; *Sylvia Hotentotta,* Lath.; *Motacilla Hotentotta,* Gmel.

FULVOUS-BROWN, with the throat and lower part of the belly white; the upper part and rump brown; the upper tail-coverts yellowish; tail black, with a white base and whitish tip. Length, $8''$.

" Found at the Cape of Good Hope," Shaw, loc. cit. This is one of the birds named by this old author, which has eluded all my endeavours to identify with any known species. From the distribution of its colours, I believe it to have been founded upon a specimen of some *Saxicola,* probably a female or young male; likely enough of the common *S. Pileata.*

202. Saxicola Aurantia; *Vitiflora Aurantia,*

Shaw., Vol. 10, p. 527; *Sylvia Aurantia,* Lath.;
Motacilla Aurantia, Gmel.

UPPER parts, dark-brown; beneath, orange-coloured; throat whitish, varied with black beneath; the greater wing-coverts and tail white; the feathers of the latter fuscous; the outer ones tipped with white. Length, 6″.

"Inhabits the Cape of Good Hope."—Shaw, loc. cit. non vidi. Another unknown species, which has baffled my attempts at identification.

203. Saxicola Cursoria, Vieil., Ency. Meth. p,

493; *Le Traquet à calotte et queue blanche,* Le Vail., Pl. 190; *Sax. Leucomelana,* Burch. Trav., 1, p. 335; *Vitiflora Tachydroma,* Boie.

GENERAL colour, black; head, from the bill to the occiput, white; vent and tail white, with the exception of the two middle feathers of the latter, which are black. Length, 8‴ 3‴.

Le Vaillant names the district where he procured this remarkable bird Namcroo. He states that it frequents arid plains, where it lives in the most solitary manner, fleeing from the approach of the hunter, and concealing itself with the greatest dexterity. It lives chiefly on grass-hoppers, and prefers running on the ground to flying. Since Le Vaillant, no one has found this species; and it is doubtful if it exists.

204. Saxicola Mariquensis; *Bradornis Mari-*

quensis, Smith, Zool. S. Af., Pl. 113.

UPPER parts, a dull pale-brown, with, in some lights, a slight shade of green; wing and tail-feathers narrowly edged with rusty-white; under parts white, tinted in some places with wood-brown; bristles at the angle of mouth, liver-brown. Length, 6″ 4‴; wing, 3″ 6‴; tail, 3″ 4‴.

In the young, the colours are much the same as in the adult; but the upper parts are spotted with white, the lower mottled with longitudinal liver-brown stripes.

"Seek their food generally in dense thickets, and when in quest of it, move languidly from branch to branch. Their food consists of insects."—Dr. A. Smith, loc. cit.

I have received a single specimen of this species from Kuruman.

205. Myrmecocichla Formicivora, Hartlaub., Orn. W. Af., p. 65; *Saxicola Formicivora*, Vieill., Nat. Dic. d'Hist. Nat. XXI, p. 421; *Saxicola Leucoptera*, Sw., An. in Menag., p. 292; *Le Fourmillier*, Le Vail., Pl. 186; Cuv., Vol. 1., p. 437.

GENERALLY of a sombre-brown colour; paler on the belly and under the tail; the throat is clouded with reddish, the same colour bordering the feathers of the front of the neck and chest; a prominent white patch adorns the shoulders of the male; the wing-feathers have also white webs. The female wants the white shoulders, and is of a deeper brown. Length, 7″ 6‴; wing, 4″; tail, 3″.

Le Vaillant found this *Saxicola* near the Sunday and Zwartkop Rivers, not far from Algoa Bay; where I also observed it in abundance during a flying visit to that locality. He says it perches on high trees. I only saw it on the ground (but, then, there were no trees for it to perch upon), or on the nests of the white-ants. It appeared to have the same habits as *S. Pileata*. Several specimens have been sent to me from Colesberg and Kuruman.

Le Vaillant found their nests in holes or under rocks. Eggs white.

206. Saxicola Nigra, Vieill.; *Myrmecocichla Nigra*, Ency. Meth., p. 489; *Le Traquet Commandeur*, Le Vail., Pl. 189; Cuv., Vol. 1, p. 437; Lath. Gen. His., Vol. 7, p. 96; Hart. Orn. W. Af., p. 65.

GENERAL colour black, with light rose-coloured epaulettes. Length, 7″.

This species was found, according to Le Vaillant, congregating in small families (consisting of an old pair and their young ones), and feeding upon termites. It was first discovered on the eastern side of the colony, extending from the 28th degree of south latitude as far as the tropic, and only residing there during the breeding season. He never found their nests, but believes they build in holes and caverns. Sundevall states that it is not South African, but from Malimba, on the *East* Coast. Hartlaub, however, has it from the *West* Coast, and it may therefore sometimes approach the extreme northern limits of Le Vaillant's collecting ground.

Genus PRATINCOLA, Koch.

Bill short, with the gape broad, and furnished with short bristles; the culmen slightly curved, and the sides compressed to the tip, which is rather emarginated; the gonys long and ascending; the nostrils basal, lateral, in a small membranous groove, with the opening rounded, and partly hidden by the frontal plumes; wings long and rounded, with the fourth and fifth quills longest and equal; tail short, broad, and slightly rounded on the sides; tarsi slightly longer than the middle toe, slender, and covered by a single scale in front; toes moderate, slender, and the outer toe rather longer than the inner, the hind toe long and rather strong; the claws moderate, curved, and acute.

207. Pratincola Pastor, Strickland; *Motacilla, Sibilla,* Linn., Gmel. Syst., 44; *La Traquet Patre,* Le Vail., Pl. 180; *Bontrochie* of Colonists, literally particoloured.

HEAD, and half way down the throat, black, sprinkled with yellow; lower throat and breast rufous; sides of the neck, shoulders, and rump, white; belly and flanks, pale-rufous; eyes, dark hazel. Female less brilliant in colour. Length, 5″ 6‴; wing, 3″; tail, 1″ 9‴.

It is abundant throughout the colony, inhabiting open plains covered with low brushwood, generally going in pairs, male and female. It is fond of perching on the top of the highest twig or rush, in its neighbourhood, generally selecting one that stands solitary. Le Vaillant found their nests concealed with great care in mole holes, at the foot of thick bushes, in holes in rock, and such like places. Eggs five in number, blotched with yellow-brown. Mr. Atmore says : " Nest on ground, eggs pale-green, spotted."

Genus THAMNOBIA, Swain.

Bill small, slender; base widened; the sides much compressed; culmen slightly arched from the base, but the tip of the upper mandible reflected and entire; gape smooth; wings short, much rounded, the primaries hardly exceeding the secondaries and tertials, which are very broad; tail moderate, broad, much rounded; tarsus much longer than the middle and hinder toes; lateral toes very unequal, the inner shortest; claws slighly curved.

208. Thamnobia Ptymatura, Gr., Ibis Vol. 1863, p. 328; *Ænanthe Ptymatura,* Vieil; *Micropus Leucopterus,* Less., Le Vail., Pl. 188, fig. 2; *Le Traquet a cul roux.*

ALL black, with the exception of the belly, vent, rump, and tail, which are bright rufous; the two centre feathers of the tail black. Length, about 6″.

Mr. Gurney (Ibis loc. cit) states that this bird was sent from Natal by Mr. Ayres; but I cannot help fancying some mistake has occurred, as the bird is an Indian one, as stated by Le Vaillant.

The Sub-Family, PARINÆ, or Titmice,

have the bill short, strong, rather conical, and straight, with the culmen straight, or slightly curved to the tip, which is entire; the nostrils lateral, basal, and generally concealed by the projecting frontal plumes; the wings moderate and pointed, with the three first quills graduated; the tail more or less long, rounded, and even; the tarsi rather long, slender, and covered in front with scales; the toes moderate, with the inner toe the shortest; the claws strong, and much curved.

Genus PARUS, Linn.

Bill short, strong, conic, with the culmen more or less curved, and the sides compressed to the tip, which is entire and acute, the gonys moderate and ascending; the nostrils lateral and basal, with the opening small, rounded, and concealed by the projecting frontal plumes; wings moderate, with the first quill very short; the third rather shorter than the fourth and fifth, which are equal and longest; tail more or less long and rounded, or even; tarsi rather longer than the middle toe, and covered in front with transverse scales; toes moderate, with the lateral ones nearly equal; the hind toe very long and strong, and armed with a long, curved, and acute claw.

209. Parus Cinereus, Vieil.; *P. Atriceps*, Horsf. Pl. Col. 287, f. 2.; *Le Messange Brune a poitrine noir*, Le Vail., Pl. 139, f. 2.

Top of the head, throat, chest, and tail, black; cheeks and lateral tail-feathers edged with white; it has also a small white collar. The under side is a greyish brown, the black of the chest forming a breastplate-shaped mark on it, extending down the belly; the upper parts darker than the lower; irids black. Length, about $5\frac{3}{4}'''$; wing, 3"; tail, 2" 3'".

Although Le Vaillant states that this species is found in the environs of the Cape, I have never obtained it nearer than Beaufort West, where it was first killed in company with the *M. Senegalensis* by Mr. Jackson's son. In this locality it has plenty of rocks and mountain, which Le Vaillant says are its habitual resort, and amid which it breeds, making a voluminous nest, composed of wool and feathers, in the crevices, and laying from 8 to 14 eggs. I never found more than five in a nest. They are white, with red spots and blotches: axis, 9'''; diam., 7'''.

Mr. Andersson brought the species from Damaraland, and I have also received it from Colesberg and Kuruman.

During my visit to Mr. Jackson at Nel's Poort, I obtained several nests in the crevices of an old brick tank or bath, which was constantly used by the members of our household. The entrances to these nests were very small and tortuous, leading to the back of the brick-work, which we had to remove before we could secure the eggs. The nests were large masses of dried bents of grass and feathers. I frequently saw this species on my journey from Nel's Poort to the Swartberg, creeping about the sides of dry water-courses, clinging to the perpendicular banks, and apparently searching for spiders.

210. Parus Cinerascens, Vieil.; *Parus Afer*, Gmel.; *Le Messange Grizette*, Le Vail., Pl. 138.

UPPER parts of back, breast, and belly, dark blue-grey; wings brownish-black, variegated with white, most thickly on the shoulders; tail black; the outer lateral feathers being bordered with white; top of the head, throat, and chest, black; a white patch extends from the corner of the bill, under the eye, and down the neck, dividing the black of the head from that of the chest. In the female, these parts are speckled with white. Length, $5\frac{1}{2}''$.

Le Vaillant found this species in the mimosa forests of Camdeboo. Non vidi.

211. Parus Leucopterus, Swain.; *Le Mésange Noir*, Le Vail., Pl. 137, fig. 2; *Parus Niger*, Vieil.; *Pentherus Niger*, Cab.; *Melaniparus Niger*, Bp.; Cuv. Vol. 2, p. 123.

GENERAL colour, black; below much tinged with brown; a broad white bar commences on the shoulder, and extends down the wing, becoming narrower towards the point; lateral tail-feathers slightly edged with white. Length, $3''$ $9'''$; wing, $4''$; tail, $3''$.

Le Vaillant found this species in the Eastern districts of the colony; he says it retires for the night into holes of trees, where it also breeds, laying six or eight pure white eggs. I have received it from Kuruman, Beaufort, and Damaraland.

Genus PAROIDES, Koch.

Bill moderate and slender, with the culmen straight, or slightly curved to the tip, which is entire and acute, the sides compressed; the gonys long and ascending; the nostrils basal and lateral, with the opening entirely hidden by the projecting frontal plumes; wings rather short, with the first quill very short, and the second nearly as long as the third or fourth, which are equal and longest; tail moderate, and forked, or lengthened and graduated; tarsi as long as the middle toe, and broadly scaled in front; toes moderate,

o

with the lateral ones nearly equal; the outer slightly united at its base; the hind toe long and strong; the claws long and strong, especially that of the hind toe.

212. Paroides Capensis; *Ægithalus Minutus,* Shaw; *Parus Capensis,* Gmel. Sonn. Voy., t. 115; Cuvier, Vol. 2, p. 124; *Le Becque-fleur,* Le Vail., Pl. 134, Figs. 1 and 2.

ABOVE, dull olive-green; greyish on the head and yellowish on the rump; forehead covered with short feathers, partly very dark-brown (approaching to black) and partly white, which give it a mottled appearance; a few white feathers are scattered about the cheeks; under parts of male, with the exception of the chin, all yellow. Female, generally duller underneath; white on the throat and breast; rufous on the belly and vent; forehead as the head. Length, 3" 5"'; wing, 2"; tail, 1" 5"'.

These little birds were found by Le Vaillant in small flocks on the western side of the colony, in the neighbourhood of the "Elephant's River." They are very active, and continually on the move, from flower to flower, feeding on insects. Mr. Andersson brought many from Damaraland, and I have it from various parts of the colony. It is common about Nel's Poort, and is the fabricator of the nest figured by Le Vaillant, plate 131, which is there erroneously attributed to "Le Pinc Pinc," which is clearly the *Drymoica Textrix,* called by the Dutch colonists "Tinc-Tinc."

The nests fabricated by these minute birds are wonderful structures, more like balls of felted cloth than the habitation of a bird. They vary in size from five to eight inches in length, and three to five inches in diameter. Some that I have torn to pieces, appear as if felted in layers, coat upon coat. Each nest is furnished with a tubular entrance, underneath which is a pocket, the use of which is doubtful. Some say the male bird sits therein during the night; others attribute it to the cunning of the birds to deceive snakes, as it draws the tubular neck of the real entrance into the body of the nest, and closes it so tight that the snakes mistake the pocket for the orifice, and vainly try to penetrate the nest thereby. I have conversed with several individuals who had been deceived by this proceeding, and did not find out their error until the parent bird effected her escape through the real opening, of her own accord. They lay from six to ten eggs, pure white: axis, 7"'; diam., 5"'; and the whole brood keeps together for some time after being fully fledged.

Mr. Atmore writes: "I took one in the nest, tied up the mouth, and thought I had my little prisoner safe, nest and eggs, in my cart-box; but on passing a fine *Gladiolus,* which I could not leave behind, I opened the box to get something to dig the bulb out with, when out flew the bird! She had bored a hole through the side, and when the lid opened, 'skedaddled.'"

Genus PARISOMA, Swainson.

Bill very short, straight, considerably compressed; upper mandible arched from the base, and notched; lower thick; gonys ascending; rictus with a few weak bristles; lateral toes equal.

213. Parisoma Rufiventer; *P. Subcœruleum*, Gmel.; *Le Grignet*, Le Vail., Pl. 126.

♂ ABOVE, dark dull greenish-grey, browner on the wings, and black on the tail, the two outer feathers of which are tipped and edged with white; under parts ash-coloured; throat whitish, speckled with long dark ash-coloured spots; vent deep rufous; edges of the wings white. ♀ wants the rufous vent. Length, 5″; wing, 2″ 9‴; tail, 2″ 8‴.

Le Vaillant states that this bird is very common on the banks of the "Gouritz," and "Brak" Rivers, dwelling among the mimosas in small flocks, and incessantly traversing the branches in their search after insects, and the eggs and larvæ of *Lepidoptera*, which form their principal food. I have received specimens from the Cedar Mountains, in Clanwilliam, and from the neighbourhood of "Riebecks Kasteel," in Malmesbury, through the kindness of Mr. Russouw; also from Swellendam, Colesberg, and Kuruman; and Mr. Andersson brought it from Damaraland. Mr. Atmore tells me he finds it plentifully in the Bosjesveldt; and I saw it in great abundance in the Karroo, as far as Nel's Poort. Le Vaillant's description of their habits is correct. While hunting, they utter a clear piping whistle.

214. Parisoma Layardi, Hartlaub. Ibis. Vol. IV, p. 147.

GENERAL colour above, cinereous, with a tinge of green; central tail-feathers approaching to black; the two outer more or less white on their tips and outer edges; under surface, whitish-grey, darkest on the flanks, and mottled on the throat; wing-feathers edged with white; bill and legs dark. Length, 5″ 3‴; wing, 2″ 6‴; tail, 2″ 9‴; tarsus, 10‴; bill, 7‴.

This bird was received from Mr. Russouw, who obtained it in Zwartland, in the Malmesbury division. According to his account, the irides are white. I also procured this species at Nel's Poort, about the mountains. In its habits it resembled *P. Rufiventer*, for which at first we mistook it. It is difficult to shoot, as it creeps about dense bushes, and on being hunted, conceals itself in the thickest parts and remains perfectly still. My friend, Mr. Henry Jackson, calls it the "Mocking Bird," from its habits of imitation, and informs me that it makes a cup-shaped nest in a bush, and lays three eggs, which are pure white, blotched chiefly at the obtuse end with greenish-brown and faided purple spots : axis, 9‴; diam., 7‴.

The Sub-Family, MNIOTILTINÆ, or Bush-creepers,

have the bill more or less acutely conical, with the culmen sloping, and the sides compressed to the tip, which is slightly emarginated ; the gonys long, and advancing upwards ; the nostrils basal, with the openings more or less large and exposed ; the wings rather long, and generally pointed ; the tail moderate ; the tarsi as long as or longer than the middle toe, covered in front with broad scales ; the toes long and slender, with the outer toe usually longer than the inner, and united at its base.

Genus ZOSTEROPS, Vigors and Horsfield.

Bill moderate and slightly curved, with the culmen curved, and the sides compressed to the tip, which is acute and emarginated ; the gonys long, and slightly ascending ; the gape furnished with a few very short weak bristles ; the nostrils basal, and placed in a broad groove, with the opening closed by a lunate scale ; wings moderate, with the first quill very small, and the fourth and fifth equal and longest ; tail moderate, broad, and slightly emarginated in the middle ; tarsi rather longer than the middle toe, and covered in front with broad scales ; toes rather long ; with the outer toe rather longer than the inner, and united at its base ; the hind toe long, strong, and armed with a long curved claw.

215. Zosterops Capensis; Z. *Madagascariensis,* Linn.; *Le Tcheric,* Le Vail., Pl. 132 ; *Sylvia Annulosa,* Swain. Zool. Ill., Pl. 164 ; *Z. Flavigula,* Swain. *(Witteoogje,* lit. white eye; and *Glasoogje,* lit. glass eye.)

ABOVE, olive-green ; yellowish towards the rump ; below grey, with throat, vent, and flanks greenish-yellow ; between the bill and the eye a black line ; eye surrounded with a fringe of white feathers. Length, 4″ 9‴ ; wing, 2″ 3‴.

The "white-eye" is common throughout the whole of the colony, roaming about in small families of from 5 to 20 in number. During the fruit season they do great damage to the apricots, peaches, plums, &c. ; they also destroy the buds, to get at the insects that lurk therein. While on the wing, or feeding, they utter incessantly a stridulous chirp, which generally is the first thing that reveals their presence. I never saw them on the ground ; but they sometimes creep about low bushes. They place their nests, which they conceal with great care, in a fork caused by the union of several small twigs. It is composed of moss and fibres, covered with cobweb and lichens, and lined with hair, and is shaped like a cup, about 3 inches across by 2½ deep. The eggs, five in number, are of a beautiful, spotless blue, rather sharp at the ends: axis, 8‴; diam., 6‴.

216. Zosterops Pallida, Swainson's Animals in Menag., p. 294.

PALE, greyish olive; beneath yellowish-white, tinged with isabella on the body and flanks; wings and tail very light-brown; inner wing-covers white. Length, 4″ 3‴; wing, 2″ 2‴; tail, 1″ 11‴.

This very rare species was first described from a specimen collected by Dr. Burchell. The example in the S. A. Museum was probably sent from Swellendam by Mr. Cairncross, as it bears traces of his stuffing. A second example has been received from Mr. Atmore.

217. Zosterops Virens, Sundev., Ibis Vol. 1861, p. 360.

YELLOWISH-GREEN; belly yellow; flanks green; lores black; above yellow. Length, 4½″; wing, 2″ 2″.

Inhabits Upper Kaffraria (Wahlb.), loc. cit. Found also by Mr. Ayres in Natal, where they are plentiful in the spring of the year.

218. Zosterops Lateralis, Sund., Ibis Vol. 1861, p. 358; *Z. Abyssinica*, Guérr., Rev. Zool. 1843, p. 162.

ABOVE, pale olive-green; beneath whitish; flanks fulvescent-grey; throat and vent yellow; neck tinted with fulvus; ring round the eye white. Length, 4″; wing, 2″ 2″.

Upper Kaffraria, Wahlberg.

The Sub-Family, MOTACILLINÆ, or Wagtails,

have the bill moderate, generally slender and straight, with the culmen curved at the end, and the sides much compressed to the tip, which is emarginated; gonys long and ascending; the nostrils lateral, and placed in a small groove; the wings long and pointed; the tail more or less long, and usually even; the tarsi long, slender, and covered in front with transverse scales; the toes more or less long, the lateral toes unequal, and the outer slightly united; the claws long, slightly curved, and acute, that of the hind toe sometimes very long.

Genus MOTACILLA, Linn.

Bill moderate, straight, and slender, with the culmen slightly curved, and the sides much compressed to the tip,

which is emarginated, the lateral margins nearly straight
and inflexed, and the gonys long and ascending ; the nostrils
lateral, with the opening oval, and partly concealed by a
membrane ; wings lengthened and pointed, with the first
quill almost as long as the second and third, which are equal
and longest ; tail very long, generally even, but·sometimes
forked ; tarsi longer than the middle toe, slender, and covered
in front with slightly-divided scales ; toes moderate ; with
the lateral toes nearly equal, the outer one slightly united
at its base ; the hind toe long, and armed with a long, more
or less curved claw.

219. Motacilla Capensis, Linn. ; Shaw., Vol. 10, p. 349 ; *La Levandiére Brune*, Le Vail., No. 177 ; *La Bergeronnette du Cap de Bonne Esperance*, Brisson ; *Motacilla Afra*, Gmel., Pl. Enl. 28, f. 2.

GENERAL colour, grey-brown ; the wing-feathers edged with
light-brown ; under parts white, with a yellowish tinge ;
throat pure white ; chest with an ash-coloured collar, extending downwards in the shape of a peak ; eyebrows dirty-
white ; two outer feathers of the tail on each side, brilliant
white ; the rest dark, indian-ink brown. Length, 6" 6''' ;
wing, 3" 1''' ; tail, 3" 7'''.

The common wagtail is abundant throughout the colony, frequenting the crowded cities equally with the outlying farmsteads ; everywhere it is petted and protected on account of its trusting, confiding
manners, and the gentle way in which it makes itself one of the family.

Of course, in the towns, where boys, those natural enemies of birds,
abound, the wagtail is more shy than in the country ; but even there
they know the dwellings of those who protect them, and take up their
abode accordingly.

In the country, each farm-house and "pondok" (mud-dwelling of
Hottentot labourer) has its well-known pairs of this engaging bird ; and
woe to the unlucky urchin who dares to meddle with them or their
nests ! ! To say that "the angels won't love them," would be a bless-
ing compared to the fate that would be prophesied for the wicked
child.

I have often seen the master of the house sitting in his chair in the
cool of the evening, and, perhaps, while one bird perched on the rail
of his chair, another would jump at the flies on his soil-stained shoes,
while two or three more stand pecking at those that plagued the old
dog lying at his master's feet. Perhaps one or.two would have found
their way into the *voorhuis*, or entrance-hall, where a rich harvest
awaited them in the bodies of those flies slain by the attendant dark
urchins, who, often ignorant of breeches or petticoats, guard their
master's viands with a plume of dirty ostrich-feathers or leafy bough
torn from the nearest tree.

These birds consort much with cattle, and jump up against their sides as they stand lazily chewing the cud, to catch the small flies that keep about them; they also congregate in considerable numbers on the sea-beach, to feed on the flies bred in the putrifying sea-weed: they run along the sand with great agility, or walk with a stately, swaggering gait, which is very amusing. They also congregate in flocks upon favourite trees for the purpose of roosting; and this may chiefly be observed in towns.

The nest is generally constructed in a bank if in the fields; but when in the town they select a hole in the wall, or a dense mass of leaves in some plant creeping up a wall or tree. The nest is composed of leaves, small roots, and horse-hair, with which the structure is lined. The eggs are four or five in number, of greyish white, minutely freckled with brown, chiefly at the obtuse end: axis, 9'''; diam., 6½'''.

A yellow wagtail was sent to me by Mr. J. Reitz, of Swellendam, which was unfortunately lost through the carelessness of a person then attached to the Museum, before I had time to describe it. I have never received another specimen; but if my memory serves me rightly, it closely resembled the species found in Damaraland.

220. Motacilla Longicauda, Rupp., Neue Wirb., Pl. 29, Fig. 2.

GENERAL colour above, cinereous; below white; gorget narrow, dark bluish-grey; eyebrows white; wings black, with some of the lesser quills and the secondaries edged with white; tail white, with the exception of the four centre feathers, which are coloured like the wings. Length, 7" 6'''; wing, 3" 3'''; tail, 3" 6'''.

This very elegant wagtail has not yet occurred to me within the bounds of the colony. Mr. Ayres has procured it at Natal, and Ruppell in Abyssinia.

221. Motacilla Aguimp, Tem.; *M. Vidua*, Sund.; *Le Bergeronnette "à guimpe,"* Le Vail., Pl. 178; *M. Levaillantii,* Cab., Cuv., Vol. 1, Pl. 475.

SHINING black; eyebrows, throat, and belly white; chest-band black; two outer tail-feathers and wing-band white. Length, 8"; wing, 3" 9'''; tail, 3" 11'''.

Le Vaillant found this species first on the borders of the Orange River, about the 28th degree of south latitude, and thence to the tropic. I have received one or two specimens from Beaufort West, more from Colesberg, and several from Kuruman and the Free State. Le Vaillant describes its habits as similar to those of the common wagtail. Its nest appears to be usually built near water, in a hole in a rock, or drift tree, and composed of moss, &c., and lined with hair and feathers.

222. Motacilla Indica, Gmel.; *Motacilla Variegata,* Vieil., Cuv., Vol. 1, Pl. 475; *Nemoricola Indica,* Bp.; *Le Levandiére varieé,* Le Vail., No. 179.

HEAD and back, olive-brown; beneath the same, varied with yellow, and a black stripe across the throat; quills black, varied with yellow and white; four middle tail-feathers black, the rest becoming more white as they become lateral. Length, 6" 9‴.

"Rare, and only found in Kaffraria," according to Le Vaillant; but in reality an Indian species, found also in Ceylon, where I was very familiar with it. It could not, therefore, have escaped my notice if it had occurred in any of the numerous collections of Cape birds which I have examined.

Genus ANTHUS, Bechstein.

Bill more or less straight and slender, with the culmen almost straight, or slightly curved, and the sides compressed to the tip, which is emarginated, the lateral margins straight and inflexed; the gonys long and ascending; the nostrils lateral, placed in a short, broad groove, with the opening rounded, and partly closed by a membrane; wings moderate, with the first three quills equal and longest; tail moderate, and emarginated; tarsi longer than the middle toe, rather slender, and covered in front with broad, transverse scales; toes long, and rather slender; with the lateral toes equal, and the outer one slightly united at its base; the hind toe long; the claws of the anterior toes rather short and curved, and that of the hind toe very long and acute.

223. Anthus Capensis, Linn., Pl. Enl., 504, Fig. 2; Cuv., Vol. 6, p. 479; *Macronyx Capensis*, Swain., Nat. Lib., Vol. 9, p. 216; *L'Alouette Sentinelle*, Le Vail., Pl. 195--6; *Kalkoentje* of Colonists, lit. Little Turkey.

UPPER parts, variegated dark-brown, the feathers having light edges; wing and tail-feathers the same; the edges of some of the former being yellow, others white; three outer tail-feathers more or less tipped with white; over the eye a bright orange line; from the corners of the bill rise two black stripes, which extend down the sides of the chin, and unite in front on the lower part of the throat; all within this is a brilliant, shining, crimson-orange; below it the centre of the breast and belly is orange; sides and flanks cinereous; edges of shoulders bright orange; toes very long and strong, the hinder one armed with a claw 9‴ long. Length, 7" 9‴; wing, 3" 9‴.

This handsome lark is common throughout all the open country of the colony: it would be better to say the forest itself is the only place where it is not to be found. In its habits it resembles the sky-lark, except that it does not soar, sing, or congregate in flocks. Instead of perching on clods of earth,—such luxuries not being common in this land,—it mounts a white-ant's heap, and keeps a look-out for its enemies from that eminence. It is usually found in pairs, and when one flies off, uttering its peculiar *"mewing"* cry, the other is sure to follow. It often perches on low bushes, amid which it runs with great rapidity, leaving a scent so strong that even the best pointers will "draw" after it for a considerable distance.

The flesh of this bird is delicious eating,—and on the high plateau about the Knysna, where they abound, a good shot might secure a sumptuous dish in a very few hours. But powder and shot is too expensive in these regions to be wasted on such "small fry," and perhaps the report of your gun will spring two or three bustards within a few hundred yards.

224. Anthus Flavigaster; *Alauda Crocea,*

Vieil. ; Hartlaub, Orn. W. Af., p. 73; *Macronyx Flavigaster,* Swain., Nat. Lib., Vol. 11, p. 215 ; Jard. and Selby, Ill. Orn., n. s., Pl. 22.

ABOVE, varied with black and brown ; beneath, fine yellow, with a broad black gorget on the breast; bill slightly curved. Length, 7″ 6‴ ; wing, 3″ 4‴ ; tail, 2″ 5‴.

Natal.

225. Anthus Cathropæ, Layard.

GENERAL appearance, lark-like ; colour, rufous-brown ; the centres of the feathers being dark and the edges lighter, give a mottled appearance ; wing and tail primaries very dark-brown, with light rufous edges ; head finely mottled ; stripe over the eye, light rufous-brown ; throat grey ; under parts immaculate, dusky grey, ruddy on the flanks ; bill dark horn-colour, the lower mandible lightish ; legs flesh-colour ; eyes black. Length, about 4″.

Such is a description taken from a little pet—the only specimen I have seen—of a small *Anthus* that is merrily hopping about in my aviary, and known to the household by the familiar name of "*Brownie.*" "*Brownie,*" from his engaging ways and sprightly song, is a general favourite. He came into my possession more than six years ago, and was brought from Swartland. He usually commences his song the earliest, but one, of all my birds : the "early bird" is a Java sparrow, who with the faintest dawn begins a low guttural gobbling, ending in a mellow but short pipe. As soon as I remove the cover from the cage, "*Brownie,*" who roosts on the ground, sometimes in a corner, at other times behind the seed-box, mounts a large stone, placed in the cage for his special benefit, and pours out his voluble song, short, certainly, but oft-repeated. Occasionally he will hop on the edge of the cage, or mayhap on a perch, and then treat us to a stave ; but his favourite singing-place is the stone. Sometimes, when I am going to bed, without any warning, "*Brownie*" will start off in full tide of song : he is then usually on the ground. "*Brownie,*" however, has more than once been in disgrace. Among the many birds confined

P

with him, are a pair of doves from Java : these he almost stripped of their feathers, for the sake of nibbling the quill-ends, which are rapidly passed through his little sharp bill, like canes through a sugar-crusher, and with the same results. From this propensity, I fancy "*Brownie*" must like a meat diet—worms and insects, perhaps ; he, however, feeds upon canary-seed, and will eat groundsel and chick-weed, and all that the canaries that are inhabitants of the same large cage feed upon. I dedicate this, to me, new species to perpetuate the name of the faithful companion of my labours for upwards of twenty years, who has aided me with pen and pencil, and shared the pleasures I have experienced in the study of the works of Nature.

226. Anthus Sordidus, Rupp. ; Faun. Abyss., t., 39 ; *Agrodroma Sordida ; Corydalla Sordida*, Blyth ; *Anthus Gouldii*, Fr.

ABOVE, variegated lightish hair-brown ; below, light buff, tinted with rufous ; chest with some obscure dark-brown markings ; vent whitish ; outer edge of exterior tail-feather white.   Length, 8″ ; wing, 4″ 6‴ ; tail, 3″.

Several specimens of this pippit, received from Capt. Bulger, at Windvogelberg, are the only examples that have fallen under my notice.   It has also been procured at Natal by Mr. Ayres.

227. Anthus Brachyurus, Sund. ; Ofvers. Kongl. Vet. Ak., Förhandl.

DARK greyish-brown, variegated with blackish markings ; belly white ; tail short, the outer feather half-whitish.

A small species, found at Natal.   Ibis, Vol. 1863, p. 327.

228. Anthus Leucophrys, Vieill. ; Gal. Ois., t., 262 ; N. Dict. d' Hist. Nat. XXVI., p. 502 ; *Enkelde Lewerk* of Dutch Colonists.

ABOVE, immaculate greenish-brown ; in some specimens ashy ; below, pale yellowish-ash ; chest obscurely freckled with brown ; chin white ; eyebrow yellowish-ash ; wing-feathers, dark-ash, with pale margins ; tail, the same, the two outer pair being the same colour as the under parts ; eye brown. Length, 7″ ; wing, 3″ 8‴ ; tail, 2″ 7‴.

This pippit is abundant throughout the colony.   It frequents open country, either quite bare or covered with bush, on which it perches readily.   Feeds on insects and seeds.   Mr. Atmore writes : " Plentiful on karroo flats, particularly near Oliphant's River,—perches on bushes, &c."

229. Anthus Raaltenii, Temm.; Licht. Cat., 1842, p. 54.

LARK-LIKE; above variegated, very dark hair-brown and yellow-brown, the latter colour being on the edges of the feathers; tail dark hair-brown, outer feather on each side white; the next white, with half the inner web brown; under parts light-yellow buff; breast streaked with dark umber-brown. Length, 5″; wing, 2″ 6‴; tail, 1″ 11‴.

The only specimens of this bird that I have seen came from Messrs. Cairncross and Atmore, of Swellendam. The latter has forwarded the eggs, which are three in number, white, and profusely freckled with darkish, grey-brown markings: axis, 10‴; diam., 7‴. He observes that in habits *A. Raaltenii* resembles the "wagtails, running swiftly along the ground, snapping at insects."

230. Anthus Caffer, Sundevall; Mus. Holm.

ABOVE, brown, the edges of the feathers being a dirty nankin, imparting to the bird a mottled appearance; rump tinged with rufous; eyebrows, clear bright nankin; a series of dark-brown spots form a moustache on each side of the chin, which is dirty white; under side, rufous-nankin, and, with the exception of the breast, which is mottled with brown spots, immaculate; outer tail-feathers, with the outer web, and a considerable portion of the inner web, clear nankin; central pair edged with the same, as are also the feathers of the wings; eyes brown. Length, 6½″; wing, 3½″; tail, 3″.

Rare; but widely distributed. I have received it from Kuruman and Swellendam, and shot a single specimen on the Cape Flats. The stomach of this last contained seeds and the shells of *Succinea Delalandi*. I killed it near a vley.

231. Anthus Lineiventris, Sundevall.

ABOVE, dull-brown, with a greenish tinge, darkest on the centre of the feathers; eyebrow nankin; chin white, slightly speckled at the angles of the bill; under parts ashy-brown, minutely lineated down the shaft of each feather; wing and tail feathers dark-brown, faintly bordered with the ashy-brown of the breast, of which colour also is a triangular spot at the apex of the inner web of the exterior tail-feathers. Length, 7½″; wing, 3½″; tail, 3″.

A solitary specimen of this bird was received from Mr. Arnot, at Colesberg.

232: **Anthus Campestris,** .Bechst.; *A. Rufescens,* Temm. ; *A Rufus,* Vieil, Pl. Enl. 661, f. 2 ; Gould's B. of Eur., Pl. 137 ; *Alauda Montana,* Gmel.

GENERAL colour above, brown, much mottled, the edges of the feathers being very light ; under parts whitish, tinged with reddish ; breast mottled with dark-brown spots ; outer pair of tail-feathers on each side more or less white ; the centre pair of the colour of the back, the rest dark-brown. Length, 7" 6'''; wing, 3" 6'''; tail, 3".

Sent with No. 226 by Capt. Bulger. Mr. Victorin also seems to have found it at the Knysna.

233. **Anthus Ameliæ** ; *Macronyx Ameliæ,* De Tarragon Mag. de Zool., 1845 ; Ibis. Vol., 1862, p. 28.

ABOVE, lark-like, variegated with brown ; throat, chest, and belly, salmon-coloured ; a narrow dark-brown collar extends from the bill across the throat, more or less broken by the spots on the chest ; vent, cinereous brown, spotted ; two outer tail-feathers white.

The above description is taken from the Plate in Mr. G. R. Gray's "Genera of Birds," as the species has never occurred to me ; nor have I access to any work wherein it is described. Mr. Ayres (Ibis. loc. cit.) says it is found along the coast of Natal, on marshy flats.

The Second Family, TURDIDÆ, or Thrushes,

have the bill of various lengths, and more or less strong, with the culmen generally keeled, curved, and the sides compressed to the tip, which is emarginated ; the nostrils lateral, basal, and generally protected by a membranous scale ; the wings more or less long, and rounded or pointed ; the tail mostly of moderate length ; the tarsi more or less short, and usually covered with transverse scales ; the toes of various lengths, with the outer toe generally longer than the inner one.

The Sub-Family, TURDINÆ, or Thrushes,

have the bill as long as or longer than the head, and strong, with the culmen curved, and the sides compressed to the tip, which is generally emarginated ; the nostrils lateral and placed in a small groove, with the opening exposed ; the wings moderate, with the first quill very short, the third and fourth, and sometimes the fifth, the longest; the tail long, broad, even, or graduated; the tarsi long, and covered with an entire scale, or several broad scales, in front; the toes long, the outer one longer than the inner, and united at the base; the hind toe long and strong.; the claws moderate, curved, and acute.

Genus CHÆTOPS, Swainson.

Bill moderate, thrush-like, notched; nostrils basal, large, naked, membranaceous, the aperture lateral and linear; frontal feathers rigid, the shafts composed of bristles; chin-feathers the same, but weaker; rictus bristled; wings very short and rounded; tail rather lengthened, broad, convex, soft, and slightly rounded; tarsi very long and strong, anterior scales divided; lateral toes unequal; claws small, obtuse, and slightly curved, the three anterior of equal size.

234. Chætops Frenatus, Temm.; *Chætops Burchelli*, Sw.

UPPER parts of head and back, ash-coloured, the centre of each feather being very dark; rump, chest, and belly, bright red-brown; vent and thighs ash-coloured; head with a white eyebrow and moustache; throat and cheeks black; wings black-brown, some of the feathers edged with rufous ash; secondaries tipped with white, causing the wing to appear crossed with a white bar; shoulders spotted with white, a few spots of the same colour between the white bar and the shoulder; tail-feathers, with the exception of the centre pair, tipt with white; legs, claws, and bill, black; eye bright-red; plumage very lax and fluffy, particularly about the rump: on being raised on this spot, it will be found to be ash-coloured, like the back, the rufous tint being quite superficial. Length, 9″; wing, 3″ 7‴; tail, 3″ 7‴; tarsus, 1″ 7‴.

This bird is, as far as I yet know, peculiar to the mountain ranges between Caledon and Swellendam. It frequents the tops of the hills and high elevations on their stony sides, and seeks its food, consisting of insects, about stones and rocks. In habits it much resembles the *Rock Thrush*, and, like it, is fond of perching on the summit of some conspicuous stone or ant-hill, from which it surveys the surrounding prospect, and seeks for safety by immediate flight on perceiving the approach of a foe. It progresses by a series of enormous hops, its powerful legs being well suited to this end; while its strong pointed claws enables it to traverse with ease the inclined surfaces of slippery rocks. It conceals itself readily in holes, and, if wounded, seeks such a retreat in which to die. Generally found in small families of three or four individuals.

235. Chætops Aurantius, Layard.

HEAD and shoulders above, ashy-grey, streaked with black ; back and rump brick-red ; tail graduated, black, each feather, with the exception of the two central, tipped with white ; this colour extending farther up each feather, until it attains its maximum in the short outside pair ; throat jet-black, with a white band extending from the lower mandible to the breast ; another smaller white line passes from the nostril over the eye, and terminates a little beyond it ; breast and belly, a fine clear orange-rufous, or burnt sienna ; wings black, each feather more or less edged with ashy-grey ; a white bar extends across wing ; the plumage very lax and decomposed, particularly over the rump. Length, 8″ 9‴ ; wing, 3″ 7‴ ; tail, 4″ ; bill, 9‴ ; tarsus, 1″ 8‴.

This handsome species was obtained by Mr. J. O'Reilly in some abundance in the mountains- near Graaff-Reinet. He describes it as extremely wary and difficult of approach, and feeding on insects, for which it seeks among the low brushwood. It has also been received from Capt. Bulger, at Windvogelberg, and from Mrs. Barber.

Mr. J. O'Reilly writes as follows :—" Graaff-Reinet, January 2nd, 1863. Inhabits rocks in high mountain ranges. Scarce ; very shy and cunning, usually frequenting places assimilating to its plumage. Continually on the hop, and seldom takes wing ; when it does so, flies but a short distance. Always on the watch, seldom showing much more than its head above the stones. Found about Graaff-Reinet all the year round, and in pairs. Food consists of small insects of any sort ; drinks in the evening. Note, a sharp chirp, particularly when surprised. Breeds in December. Nest built of grass and rock-mosses, in crevices among rocks. Eggs, three to four, green, with brown speckles."

When this bird was sent home, Dr. Hartlaub and Mr. Sclater identified it as *C. Frenatus*, Temm. Since then, specimens have been obtained corresponding entirely with Temminck's figure of that bird, and I am convinced that this species is distinct. I have male, female, and young birds of each ; and Mr. O'Reilly describes the nests and eggs of the orange-bodied species, which Dr. Hartlaub supposes to be the young of *Frenatus*. *C. Frenatus* breeds about Caledon ; and our *C. Aurantius* never appears there by any chance.

Genus TURDUS, Linn.

Bill as long or nearly as long as the head, with the culmen curved, and the sides gradually compressed to the tip, which is emarginated ; the gonys long, and rather ascending ; the gape furnished with short weak bristles ; the nostrils basal and lateral, with the opening oval and exposed ; wings moderate, with the first quill very short, the second rather shorter than the third and fourth, which are equal and longest ; tail moderate, broad, and slightly emarginated ;

tarsi as long as the middle toe, and covered in front with an entire scale; toes moderate, the outer rather longer than the inner, and slightly united at the base; the claws moderate, curved, and acute.

336. Turdus Libonyana, Smith; Zool. S. Af., Pl. 38.

ABOVE, deep yellowish-grey; top of the head strongly shaded with brownish-red, margins of some of the feathers of the wing showing a pale buff-orange; middle tail-feathers and outer webs of the rest, deep broccoli-brown; the inner webs, brownish-red, distinctly tipped with white; throat white, tinged with ochre-yellow, and margined on each side by a series of small spots, of a dark brown, forming two longitudinal bands; breast, light yellowish-grey, with an orange tinge, which becomes brighter on the belly and inner surface of the shoulders; middle of belly and vent, white; bill, feet, and claws, yellow. Length, 9″; wing, 4″ 6″; tail, 4″.

Dr. Smith first procured specimens of this thrush in the neighbourhood of Kurichane, actively engaged, in the thickets which occur in the banks of the river, scraping among the decaying leaves for its food. I have never seen a specimen from any part of the colony, or as far north as my friends have penetrated: it must be very rare.

237. Turdus Strepitans, Smith; Zool. S. Af., Pl. 37; *Tardus Simensis*, Rupp.; *Turdus Crassirostris*, Lichten.

FRONT and top of head, brownish-grey; back of head, neck, back, and rump, ash-coloured; under parts ochraceous-white, dotted all over with dark-brown spots; under the eyes are two black bands, one semi-lunate, extending from the base of the bill to the back of the eye, the other downwards from the eye to this lunate band. Length, 8″ 6‴; wing, 5″ 2‴; tail, 3″.

Dr. Smith (loc. cit.) says: "Immediately upon reaching Kurichane this thrush began to appear in the thickets, and we continued to acquire occasionally a specimen even in the vicinity of the tropic. It seeks its food upon the ground, and when so occupied, its resort is readily discovered by the natives from the noise it makes in scratching the ground, or in displacing rubbish and decayed leaves which conceal the insects it is seeking." I have received specimens of this bird from Kuruman, Colesberg, Otjimbinque; and Mr. Andersson procured it all over Damaraland.

238. Turdus Guttatus, Vigors ; Proc. Z. S., 1832, Pl. 62 ; Smith, Z. S. A., Pl. 39.

UPPER parts, rich yellowish-brown; eyebrows and sides of the head dirty white; the latter with three umber-brown stripes ; under parts white, spotted with good-sized umber-brown spots; two white bars across the wings when at rest; tail yellowish-brown; the three lateral feathers on each side broadly tipped with white. Length, 9″ 6‴ ; wing, 4″ 10‴ ; tail, 4″.

Dr. A. Smith procured a few specimens of this thrush in the forests which border upon Port Natal.

239. Turdus Obscurus, Smith ; Zool. S. A., Pl 36 ; *T. Smithii*, Bp. Cons. Av., p. 274.

UPPER parts, greyish-brown, of different shades ; belly reddish-orange ; breast rather lighter than the back ; throat still lighter ; bill, legs, and claws, yellow. Total length, 9½″ wing, 5″ 2‴.

Dr. And. Smith procured this species towards Delagoa Bay, on the East Coast. He says " it is a solitary bird, inhabiting damp, secluded thickets and brushwood skirting the banks of rivers, and is generally seen in such localities actively employed in displacing the decayed vegetation, which occurs upon the surface of the soil, covering the insects and worms upon which it feeds."

I have received specimens from Mr. R. Moffat, killed in the neighbourhood of Kuruman.

240. Turdus Olivaceus, Cuv., Vol. 1, p. 387 ; *T. Ochrogaster*, Sparm. ; *Le Griverou*, Le Vail., Pl. 98, 99.

ALL the upper parts, very dark olive-brown ; throat underneath, dirty white, speckled with dark-brown ; breast cinereous-brown ; belly and flanks, deep rufous ; vent, dirty white ; a few dark-brown spots are distributed on the chest, flanks, and vent ; upper mandible horn-colour ; lower mandible and legs orange. Length, 10″ ; wing, 4″ 9‴ ; tail, 3″ 9‴.

At some seasons of the year, when fruit is ripe and apples and pears fall off the trees, and rot on the ground, this thrush may be found in great abundance in the orchards about Rondebosch. It utters no song, but only a short hurried cry when alarmed, and flying upward to the topmost branches of the thickest trees, amid which it endeavours to conceal itself. If unsuccessful in this, it launches itself off and makes for the nearest thicket,—its voice and manner reminding the observer of the European blackbird. It breeds in thick bushes, making a nest like that of the blackbird, and lined with fibres and roots. The eggs, large for the size of the bird, are generally four in number, of a light verditer blue, mottled with irregular patches of brown, thickest on the obtuse end: axis, 15‴; diam., 11‴.

241. Turdus Gurneyi, Hartlb.; Ibis Vol. 1864, p. 349.

ABOVE, olive-brown; spot between the eye and the bill, throat, chest, and flanks, cinnamon; belly and under tail-covers, pure white; tail coloured as the back; two rows of white spots across the wing. Length, 6″ 5‴; wing, 4″ 5‴; tail, 3″ 3‴.

Natal.—Mr. Ayres. Non vidi.

242. Turdus Senegalensis, Cuv. nec Gmel.*
Vol. 6, p. 379; *T. Nigerrimus*, Gmel., Pl. Enl., t. 539, f. 2.

"SHINING black; feathers yellow-edged; throat, quills, and tail black."

South Africa.—Cuv., loc. cit.; but in reality from Madagascar.

Genus PETROCINCLA,† Vigors; *Rock-Thrushes.*

Bill notched; culmen curved to the tip, which is bent, but not hooked over the lower mandible, nearly entire; wings moderate; first quill spurious, second shorter than the three next, which are equal and longest; tail even; anterior scales divided; lateral toes equal; claws small, slightly curved.

243. Petrocincla Rupestris, Cuv., Vol. 6, p. 376; *Turdus Rupestris*, Vieill.; *Pet. Montana*, Sw.; *T. Rupicola*, Licht.; *Pet. Rocar*, Steph.; *Le Rocar*, Le Vail., Pls. 101, 102.

THE whole of the head and throat, dark ashy-blue; back and wings dark-brown; the former with a deep rufous tinge, increasing towards the rump, which is entirely deep rufous; the tail is also rufous, with the exception of the two centre feathers; all the under parts deep rufous, lightest towards the vent; ♀ wants the blue head, otherwise she is like the male; irides dark-brown. Length, 9″; wing, 4″ 6‴; tail, 3″ 8‴.

This rock-thrush is seldom found except among stones covering the sides of mountains. In habits it closely resembles *P. Explorator*.

I have seen it at Kalk and Simon's Bays, and on Table Mountain. I have also received it from Swellendam, Beaufort, and Colesberg.

Mrs. Barber informs me that it breeds in holes under rocks, and

---

* *T. Senegalensis* of Gmelin = *Phyllastrephus Lugubris*, Bodd.
† Mr. Gray (Genera of Birds, p. 218) makes this genus synonymous with *Turdus*.

Q

sends a nest, composed of fine roots and hair. Eggs, 3—5, of a light buff or dirty white, minutely freckled throughout with pale rufous : axis, 13‴; diam., 9‴. Of its habits, she writes: "The ♂ sings a rather lively song ; he does not commence early in the morning, at or before daylight, as the robins do; but after the sun has warmed him he begins to pour forth his cheerful melody : his notes are clear, and rather pleasing, and he is fond of choosing a conspicuous situation whence he can be heard and seen to the best advantage."

244. Petrocincla Explorator; *Turdus Perspicax*, Shaw ; *Turdus Explorator*, Vieil., Cuv., Vol. 6, p. 377 ; *L'Espionneur*, Le Vail., Pl. 103.

HEAD, neck, breast, throat, back, and shoulders, blue-grey ; rump, belly, and vent, orange-red, inclining to buff on the vent and thigh ; wing and outer tail-feathers deep brown, the former edged with light rufous ; outer tail-feathers orange-red, more or less marked with brown. Length, 8″; wing, 4″ ; tail, 2″ 0‴.

Not very common, but widely distributed. I have seen it principally in rocky places about Table Mountain, and the face of the hills overhanging Simon's Bay and Kalk Bay. It appears to be a frequent species at Swellendam ; and I have received it from Beaufort and Colesberg. It is a shy and wary bird, keeping well out of gun-shot when pursued, flitting from rock to rock, running up their sloping surfaces, and keeping a good look-out from the extreme summit. It feeds on insects, but often takes a little vegetable food.

Genus BESSONORNIS, Smith.

Bill more or less short, with the culmen gradually curved, and the sides compressed to the tip, which is emarginated ; the gonys moderate and ascending ; the lateral margins nearly straight ; the nostrils lateral and basal, with the opening oval, placed in a small membranous groove, and partly covered by the projecting feathers ; wings moderate, and rounded, with the first or fifth quill the longest ; tail more or less long, broad, and rounded ; tarsi rather longer than the middle toe, slender, and covered in front with an entire scale, or with several slightly divided scales ; toes moderate and weak, with the outer longer than the inner one, and united at the base ; the hind toe long ; the claws moderate, curved, and acute.

245. Bessonornis Vociferans, Swain.; Zool. Ill., Pl. 179 ; *Turdus Reclamator*, Vieil., Cuv., Vol. 6, p. 379 ; *Muscicapa Bicolor*, Sparm. ; *M. Dichroa*, Gmel. ; *Le Reclameur*, Le Vail., Pl. 104 ; *Piet-myn-Vrouw* of Colonists.

ABOVE, dark blue-grey ; rump reddish ; wing-feathers edged with light blue-grey ; tail red, with the exception of the two centre feathers, which are coloured like the back ; the outer webs of the two outer feathers, and some portions of the rest, are also similarly coloured ; under parts red, as in the tail ; iris light-brown. Length, 7″ 3‴ ; wing, 4″ ; tail, 3″ 8‴.

This bird—which, from its singular cry, has acquired the name by which it is known to the colonists—is common in the Knysna district, but is very difficult to procure, owing to its shy, retiring habits, and the thickets in which it lives. Its loud call may be heard on every side of the traveller, but not a single bird will show itself, and I only procured the few specimens which I did obtain, by remaining quiet in one spot and imitating their cry until one came within range ; then the difficulty often was to avoid blowing the bird to pieces, as the thickets are usually so dense, that the bird was almost within reach of my gun-barrel ere I saw it.

It feeds upon worms and insects, and has the general habits of a thrush. Le Vaillant says he found it also in Kaffraria.

246. Bessonornis Signatus ; *Cossypha Signata,* Sundev., Ofvers. Kongl. Vet. Ak. Förhandl.

YELLOWISH-BROWN ; beneath white ; a white stripe, bordered with black, extends from the nostrils over the eye ; tips of the tail-feathers and spot on the wing white. Length about 7″.

Kaffraria (Wahlberg). Non vidi.

247. Bessonornis Natalensis ; *Cossypha Nata-lensis,* Smith, Zool. S. A., Pl. 60.

HEAD and neck, orange-brown ; back and wings lavender-purple ; posterior part of the back, rump, upper tail-coverts, inner vane of the outermost tail-feather of each side, and the four next to it, orange-yellow ; under parts, bright gallstone-yellow. Length, 7″ 6‴ ; wing, 3″ 6‴ ; tail, 3″ 3‴.

Dr. A. Smith procured a single specimen of this bird in the neighbourhood of Port Natal. It was shot close to the edge of an extensive forest, and was said by the natives to be a shy bird, generally observed in trees, though often seeking its food on the ground. It does not, that I know of, extend into the Cape Colony ; and no specimen has ever fallen under my notice.

248. Bessonornis Phœnicurus, Gmel. ; *Cossypha Pectoralis,* Smith, Zool. S. A., Pl. 48, letter-press ; *Turdus Phœnicurus,* Vieil., Cuv., Vol. 6, p. 377 ; *Saxicola Superciliaris,* Licht., Cuv., Vol. 6, p. 438 ; *Le Janfredric,* Le Vail., Pl. 111 ; *Petrocincla Superciliosa,* Swain.

ABOVE, cinereous brown, inclining to rufous on the back and shoulders ; rump deep rufous ; wings and two centre feathers of tail, dark-brown ; the rest deep rufous, more or less edged and tipped with dark-brown ; a white line extends from the forehead over the eye ; space between the eye and the bill and ears black ; throat, breast, and chin, orange ; the rest of the under parts bluish-grey, inclining to white in the centre of the belly, and rufous on the vent. Length, 7″ ; wing, 3″ 9‴ ; tail, 3″ 3‴.

This is the Cape "robin," and decidedly deserves the name. It is common in all the gardens, even in the midst of Cape Town, flitting along the gravel paths, scraping in the flower-beds, perching on the leafless summit of some deciduous tree, or the ridge of the housetop, and pouring out a short, robin-like song, which, when heard in the darkening twilight, reminds the listener of the familiar note of the "household bird with the red stomacher." Its nest is placed in much the same situations, and built of the same materials, and the eggs equal in number and resemble in form, size, and shape, those of the European bird, only rather less coloured, being of a dirty white or buff ground, more or less freckled with dull pale rufous : axis, 11‴ ; diam., 7‴. Its food consists of insects, worms, &c., and when it hops from place to place in search of it, it droops its wings and jerks its tail, just like the robin.

249. Bessonornis Humeralis; *Cossypha Humeralis*, Smith, Zool. S. A.; Pl. 48.

ABOVE, including two centre tail-feathers, very dark grey, approaching to black ; rump and lateral tail-feathers, bright rufous, the latter with dark-grey tips ; a white stripe extends over each eye from the nostrils ; another stripe of the same colour extends from the shoulder almost to the top of the closed wing ; throat and breast white ; belly, flanks, and vent rufous. Length, 7″ 1‴ ; wing, 3″ 3‴ ; tail, 3″ 2‴.

Dr. A. Smith first discovered this species in latitude 26° south. I have received it from Kuruman, but in very limited numbers. In its habits it is said closely to resemble *B. Phœnicurus*, except that it is but rarely seen on the ground.

250. Bessonornis Caffra; *Motacilla Caffra*, Linn. ; *Sylvia Caffra*, Shaw, Vol. 10, p. 669.

UPPER parts, ashy green ; rump and tail rufous, with the exception of the two centre feathers, which are brown ; chin and throat rufous ; chest and belly ashy-grey, inclining to rufous on the vent ; a white stripe extends from the nostrils over the eyes. Length, 8″ 3‴ ; wing, 3″ 6‴ ; tail, 3″ 9‴.

This bird, the only one which has fallen under my notice, was received from Mr. R. Moffat, at Kuruman.

The Sub-Family, TIMALINÆ, or Babblers,

have the bill moderate, with the culmen much curved, and the sides compressed to the tip, which is generally entire, or only slightly emarginated ; the gonys long and ascending ; the nostrils basal, and more or less exposed ; the wings short, and much rounded ; the tail of various lengths, and gra- duated ; the tarsi lengthened, robust, and usually covered with an entire scale ; the toes long, strong, and strongly scutellated above ; the claws more or less long, compressed, and acute.

Genus CRATEROPUS, Swain.

Bill nearly as long as the head, more or less straight from the base, much compressed, obsoletely notched ; rictus bristled ; frontal feathers rigid ; wings short, and rounded ; tail large, broad, soft, and rounded ; feet very large and strong ; tarsus lengthened ; anterior scales divided ; lateral toes nearly equal ; hind toe large, nearly as long as the middle toe ; plumage lax, soft.

251. Crateropus Jardinii, Sm.; Z. S. A., Pl. 6.

ABOVE, olive-brown ; feathers of the head dark-brown, edged with greyish-white, imparting to the bird a scaled appear- ance ; chin, throat, breast, and anterior portion of belly, ashy-brown, marked with elongated white spots ; posterior portion of belly and vent, yellowish-grey ; tail dark-brown ; centre pair of feathers strongly tinged with grey ; all crossed by dark-brown bars, only visible if held in certain positions. Length, 10″ ; wing, 4″ 6‴ ; tail, 4″ 3‴.

The expedition under Dr. A. Smith first obtained this bird in lati- tude 25° 24′ south, in spots covered with reeds, such as are seen along the margins of the rivers of that country. It is described as being of a restless, noisy disposition, and congregating together in considerable numbers, preying upon insects, &c.

I have received a few specimens from Kuruman ; and Mr. Andersson obtained it in Damaraland.

252. Crateropus Bicolor, Jardine ; Edin. Jour. Nat. and Geo., Pl. 3 ; Smith, S. Af. Q. J.., Vol. 2, p. 85.

GENERAL colour, white ; with black wings and tail, bill, and legs ; feathers of the back very thick and fluffy ; tail rounded. Length, 10″ ; wing, 5″ ; tail, 4″ 7‴.

Procured by Dr. A. Smith between the Ky-Gariep and Kurrichane. I have also received many specimens from Damaraland, through Messrs. Andersson, Kisch, Green, and Chapman ; they appear to be

very common, scraping among decaying vegetation for their food.
Eggs said to be of this species were brought me from the Interior by
Mr. Chapman, in shape and size like those of the thrush, but of a pale
spotless blue, covered throughout with minute raised tubercles.

253. Crateropus Pectoralis, Gmel., Zool., 13,
2, p. 200 ; *Le Hausse-col Noir*, Le Vail., Pl. 110.

GENERAL colour, black ; under parts pure white, with a black
gorget ; tail tipped with white ; bill and legs black. Length,
about 10″.

Great Namaqualand, according to Le Vaillant; but a doubtful species,
not seen by any one but himself.

The Sub-Family, ORIOLINÆ, or Orioles,
have the bill as long as the head, broad at the base, and
compressed on the sides, with the culmen more or less
elevated at the base, and curved to the tip, which is emarginated ; the nostrils lateral, basal, exposed, and partly closed
by a membrane ; the wings long, with the three first quills
equally graduated, and the third and fourth longest ; the
tarsi short and strongly scaled ; the toes moderate, and the
lateral toes usually unequal.

Genus ORIOLUS, Linn.
Bill the length of the head, broad at the base, with the
culmen slightly depressed, and curved to the tip, which is
emarginated ; the sides compressed, and the lateral margins
straight ; the nostrils basal, lateral, ovoid, exposed, and the
opening partly closed by a membrane ; wings long, with the
first quill half the length of the third and fourth, which are
equal and longest ; tail moderate and rounded ; tarsi short,
nearly the length of the middle toe, and covered in front
with broad scales ; toes moderate, free at the base, and the
outer one longer than the inner; the hind toe long and
strong ; the claws long, strong, and curved.

254. Oriolus Capensis, Swain. ; *Le Loriot Cou-*
dougan, Le Vail., Nos. 261, 262 ; *O. Radiatus*, Gmel. ;
O. Larvatus, Licht., Cat. Dup. Ber. Mus., p. 20. ;
G. Monachus, Wagl., Cuv., Vol. 6, p. 397 ; *O. Cou-*
dougnan, Temm. ; *O. Melanocephalus*, var. Vieill. ;
O. Chloris, Cuv.

GENERAL colour, a dirty, greenish yellow ; head, throat,
chin, and breast, black ; back of neck yellow ; wing-feathers
black, with whitish edges, and white marks on the shoulders ;

the black and yellow of the tail distributed as in *O. Auratus*, the two centre feathers being dark green-yellow ; irids red. Length, 9″ 6‴ ; wing, 5″ 8‴ ; tail, 4″ 4‴.

This is the only species of oriole that builds in South Africa. It is not uncommon in the forests of the Knysna, and along the south-east coast of the colony in general. Le Vaillant met with it in the same locality, and states that its nest is placed in very high trees, and composed of twigs and fibres, covered with moss, and lined with feathers. Eggs four, of a dirty white colour, with brown blotches.

On the left bank of the Keurboom's River, which falls into Plettenberg's Bay, about half a mile from where the mountains narrow down to the river, there is a lovely kloof, which opens to the water's edge, and stretches back inland for about a couple of miles. A clear running stream flows through the centre of it, and on each side rocky, inaccessible precipices hem in a splendid forest. In this lovely spot, the silence was only broken by the babbling brook and the loud pipe of the oriole, which frequented the summits of the gigantic yellowwood trees, whose mighty heads, hung with dense masses of a grey moss, seemed, like vegetable Titans, to watch over the solitude around them. Thinking this a very paradise for birds and game, I twice visited it; but, with the exception of the oriole, the little black swallow (*H. Holomelas*), and the loury (*Corythaix Persa*), not a creature was to be seen.

255. Oriolus Galbula, Linn.; Pl. Enl. 26 ; *The Golden Oriole*, Bewick's Brit. Birds, Vol. 1, p. 96.

ENTIRELY yellow, with a shade of green on the lower portion of the back ; a black patch extends from the bill to the eye ; wings black, with a yellow patch near the centre of the outside edge when closed, and the feathers more or less edged and tipped with white ; two centre tail-feathers black ; the bases greenish ; the rest black with bright yellow tips, the dimensions of the yellow increasing towards the outer feathers ; bill red ; feet black. Length, 9″ ; wing, 5″ 9‴ ; tail, 3″.

The ♀ of all our orioles are greenish, in the place of being yellow, and all the colours are duller.

Inhabits South Africa, teste Hartlaub, Orn. W. Af., p. 80. Mr. Andersson obtained it in Damaraland ; Mr. Ayres in Natal.

256. Oriolus Auratus, Vieil.; Nat. Lib., Vol. 8, p. 33; *Le Loriodor*, Le Vail., Pl. 260 ; *O. Bicolor*, Temm. ; *T. Flavus*, Gmel. ; Cuv., Vol. 6., p. 397.

GENERAL colour of male, rich golden-yellow ; eye stripe black, the eye being in the centre ; wing-feathers black, bordered with yellow ; tail black ; the two centre feathers just tipped with yellow ; the next pair with more yellow ;

the succeeding still more, till the outermost are reached.
Length, about 9" 6"' ; wing, 5" 4"' ; tail, 3" 9"'.

Le Vaillant says he only found this oriole in its migration; this being
the most southerly point reached by it. It frequented the summits of
the highest trees in the forests of Kafirland.

Mr. Andersson brought a species from Damaraland, agreeing with
this in every particular, except in having the tail yellow, with the two
centre-feathers black. *O. Auratus* is known to be an inhabitant of
Tropical Africa; but it is doubtful if it extends to South Africa.

257. Oriolus Melanocephalus, Linn.; Pl. Enl., 79 ; *O. Galbula*, Lath.; Cuv., Vol. 6, p. 397 ; *L'Oriot Rieur*, Le Vail., Pl. 263 ; *Oriolus Maderaspatanus*, Franke, Ed. Birds, Pl. 186.

GENERAL colour, fine orange-yellow ; head, throat, and breast
black ; wings black, with broad yellow edges to some of the
secondaries ; tail yellow ; with a black patch across the
middle. Length, about 9" ; wing, 5½" ; tail, 3" 9"'.

Procured, according to Le Vaillant, among the forests that clothe the
sides of the mountains near the "Great Fish" and "Gamtoos"
Rivers ; but it is an Indian, not an African species, and must be
classed among Le Vaillant's fables.

The Sub-Family, PYCNONOTINÆ, or Bulbuls,

have the bill usually short, with the culmen curved, and the
sides compressed to the tip, which is marginated ; the gape
furnished with more or less lengthened bristles ; the nostrils
basal, and placed in a short membranous groove ; the wings
moderate and rounded ; the tail long, broad, and generally
rounded at the end ; tarsi as long as, or shorter than, the
middle toe, and generally covered by an entire scale ; the
toes moderate ; the outer toe sometimes longer than the
inner, and united at the base ; the hind toe long, and strong.

Genus CRINIGER, Temminck.

Bill short, strong, broad, and elevated at the base, with
the culmen curved, and the sides compressed to the tip,
which is emarginated ; the gonys short, and slightly ascending ; the gape furnished with very long bristles ; the nostrils
basal, and placed in a short, broad groove, with the opening
oval and exposed ; the nape is furnished with long slender
hairs ; wings moderate and rounded, with the fourth quill
nearly as long as the fifth and sixth, which are equal and
longest ; tail rather long and broad ; tarsi shorter than the
middle toe, and covered in front with an entire scale ; toes

moderate ; the lateral toes unequal, the outer one slightly united at the base ; the hind toe long, strong, and armed with a prominent claw.

258. Criniger Importunus; *Trichophorus Importunus; Turdus Importunus,* Vieil., Ency. Meth., p. 662 ; Cuv., Vol. 6, p. 378 ; *L'Importun,* Le Vail., Pl. 106 ; *Boschvogel* of Colonists ; *Turdus Clamosus,* Steph. ; *Andropadus Vociferus* et *A. Familiaris,* Swain. ; *T. Brachypolioides,* Jard.

GENERAL colour, uniform dull olive-green ; lightest on the under side ; feathers of the wings and tail edged with yellow ; some thin, long, hair-like bristles protrude from the back of the head ; eye very pale yellow. Length, 7″ 9‴.; wing, 3″ 9‴ ; tail, 3″ 9‴.

" L'Importun " of Le Vaillant is not very rare in the neighbourhood of Cape Town, but does not manifest its presence by its importunity, as indicated by that author. I have generally found it in thickets, moving quietly about in search of insects, and darting upon them while at rest. I have sometimes, but rarely, seen it on the ground, scraping among dead leaves.

Le Vaillant found it very plentiful from Duivenhok's River to Algoa Bay ; and Messrs. Atmore and Cairncross have sent it from Swellendam. Le Vaillant states that it builds on branches of large trees, and lays four or five eggs, marked with olive-colour. It has a clear, shrill, single piping note, and will readily come to any one who knows its call and can imitate it.

259. Criniger Flaviventris ; *Trichophorus Flaviventris,* Smith, Zool. S. A., Pl. 59.

UPPER parts of head, back. rump, and tail. olive-brown ; outer vanes of some of the quill-feathers slightly edged with green ; chin and throat, pale straw-yellow ; breast and flanks, greenish-yellow ; centre of belly and vent, yellow ; bill, legs, and toes, a dark horn-colour. Length from point of bill to tip of tail, 9″ ; wing, 4″ ; tail, 4″ 2‴.

Inhabits the country near Port Natal ; is entirely confined to thickets, among which it searches for insects, never descending to the ground.—Dr. A. Smith loc. cit.

Genus PYCNONOTUS, Kuhl.

Bill more or less short, and broad at the base, with the culmen curved, and the sides compressed to the tip, which is emarginated ; the gonys short and ascending ; the gape furnished with a few short, weak bristles ; the nostrils basal,

R

and placed in a groove, with the opening oval and exposed; wings moderate and rounded, with the fourth, fifth, and sixth quills nearly equal and longest; tail moderate and rather rounded; tarsi as long as the middle toe, and rather strong, with the front covered with an undivided scale; toes moderate, the lateral toes nearly equal, the hind toe rather long, strong, and armed with a long curved claw.

260. Pycnonotus Capensis, Gmel.; *Turdus Capensis*, Gmel.; *Le Brunet*, Le Vail., Pl. 105; Briss. Orn., III., t. 27, f. 3.

ABOVE, entirely of a rich hair-brown; slightly darker on the forehead and round the eye and ears; eye-lids surrounded by a white fringe; under parts light; the edges of the feathers being still lighter, give a scaled appearance. Vent, light citron-yellow; edge of shoulder white. Length, 7" 6'''; wing, 3" 9'''; tail, 3" 6'''.

These birds are found in great abundance in the neighbourhood of Cape Town, and indeed throughout the whole colony. They migrate according to the fruit season, and are especially partial to figs and grapes. They also feed largely on the berries of the " Persian Lilac," and when that tree is in fruit, any number might be shot by a person lying in ambush near. When feeding, they keep up a continued chattering, and as they usually go in flocks of ten or fifteen in number, their presence is soon detected.

These birds conceal their nests so skillfully, that they are rarely detected, notwithstanding their numbers. It is composed of rootlets, lined sometimes with hair and feathers, and is generally placed in the fork of a tree or large bush. The eggs, three or four in number, are a lovely pale pink, densely spotted and blotched with dark pink and pale purple, presenting a most beautiful appearance: axis, 11'''; diam., 7½'''.

261. Pycnonotus Nigricans, Vieil.; Cuv., Vol. 6, Pl. 389; *Turdus Levaillantii*, Tem.; *Turdus Capensis*, Lath., Pl., Enl., 317; *Ixos Xanthopygius*, Ehrenb.; *Le Brunoir*, Le Vail., No. 106, f. 1.

GENERAL colour, brown; slightly darker on the wing and tail feathers; the latter pale at the tips; head slightly crested, and black; as is also the chin and upper part of throat; eye-lids surrounded by a red fringe; lower parts brown, like the back, fading into white on the belly; vent, brilliant citron-yellow. Length, 7" 9'''; wing, 3" 9'''; tail, 3" 5'''.

Le Vaillant procured this species in Namaqualand, particularly after passing the Orange River.* He states that they lay four eggs, of an olive-green colour. I have received one single specimen from Mr. Moffat, of Kuruman ; several from Mr. Arnot (Colesberg), and Capt. Bulger, at Windvogelberg : and found it abundant at Nel's Poort, replacing *P. Capensis*, which it exactly resembles in habits and note. Mr. Henry Jackson sends eggs of this species, precisely similar to those of *P. Capensis*, and writes that " the nest is a cup, slightly formed, in a tree."

262. Pycnonotus Aurigaster, Vieill. ; *Turdus Aurigaster*, Vieill.; *F. Chrysorhoeus*, Temm. ; Le Cudor, Le Vail, No. 107, f. 2 ; Cuv. Vol. 6, p. 389.

CLOSELY resembles *Pycnonotus Cafer*, but wants the crest ; the head and throat are black ; back of the neck, back, and scapulars are of a uniform grey-brown ; the wings are also brown, but of a duller hue ; the tail is dark-brown, tipped with white, as are also the front of the throat, the breast, flanks, and belly ; vent, a beautiful golden-yellow.

Le Vaillant says his faithful attendant, the Hottentot Klaas, shot the only specimen of this bird procured, on the banks of the Great Fish River, in Kaffraria. Another, probably the ♀, was in company with it, but escaped. Sundevall, however, denies this statement, and shows that it is a Javanese species.

263. Pycnonotus Cafer, Linn.; *Turdus Cafer*, Lath. ; Cuvier, Vol. 6, p. 388 ; *Muscicapa Hæmorrhousa*, Lath., Pl. Enl. 563, f. 1 ; *Merle Curouge*, Le Vail., Pl. 107, f. 1.

THROAT and top-knot black ; neck, back, wings, and tail, clear-brown, but each feather bordered with white, which gives the bird a scaled appearance ; the chest is paler ; the

---

* I have been at some little pains to trace Le Vaillant's footsteps in Southern Africa, in order, if possible, to identify such of the birds as have been introduced into his great work as South African, but which are supposed by some to have been obtained from other countries. A statement which appeared some time ago in the serial " Household Words," to the effect that Le Vaillant never was in South Africa, also stimulate my desire to obtain full information regarding him.

I need not follow him through all his wanderings at this moment : this I may perhaps do at some future time. Suffice it for my present purpose to say, that I do not believe that he ever crossed the Orange River and procured this bird there.

He describes in his travels how he was floated across the swollen river, and his chase after the giraffe. I question much if this account is true. There was living at Camiesberg, within the last few years, an aged woman named Van Zyl, who related to my informant that she well remembered the " Kleine Franschman " (*little Frenchman*), as she called him ; that during his stay in that part of the country he lodged entirely at her house ; and that he never crossed the Orange River, being too much of a coward so to do. When told that he stated he had shot the giraffe, she scouted the idea, and declared that the skin which he took away was brought piece-meal from the opposite side of the river by his Hottentots. Mrs. Van Zyl was a huge, rawboned woman, who stood upwards of six feet, and usually wound up her narrations concerning Le Vaillant by laughingly relating how she had horsewhipped the "little Frenchman " for attempting some liberties with her.

rump, belly, and thighs, white; vent brilliant-red; tail-feathers brown, darkening towards their points, which are all white-tipped.

Le Vaillant says he found this bird only in Great Namaqualand, and then as a bird of passage. This is more than doubtful: probably it is the well-known Indian species; though Bonaparte, in his " Conspectus Generum Avium," admits it as distinct.

264. Pycnonotus Nigricapillus; *Turdus Nigricapillus*, Vieil., Cuv., Vol. 6, p. 377 ; *Lioptilus Nigricapillus*, Caban. Mus. Hein. ; *Le Merle à calotte noire*, Le Vail., Pl. 108.

Top of the head and back of the neck, black; the rest of the upper plumage olive-brown; under parts, ashy blue-grey, paler on the belly and vent; bill orange; legs flesh-colour. Length, 7"; wing, 3" 1"'; tail, 3" 2"'.

Le Vaillant found this bird only in the forests of Bruintjes Hoogte, and even there sparingly. They feed on insects and wild berries.

Capt. Bulger, of Her Majesty's 2nd-10th Regiment, to whom I am indebted for numerous very interesting birds, procured it in the neighbourhood of Windvogelberg. I have not seen it from any other locality.

Cabanis (Mus. Heineanum, p. 88,) forms this into a new genus, *Lioptilus*, but does not give his characters. From my observations on the Ceylon species, *P. Atricapillus*, I believe it to be closely allied to the genus *Pycnonotus*, in which I therefore place it.

265. Pycnonotus Aurigularis; *Sylvia Auraticollis*, Vieil. ; *Ixos Aurigularis*, Voigt. ; *Turdus Auraticollis*, Bp. ; *Le Col d'Or.*, Le Vail., No. 119 ; Cuv., Vol. 1, p. 391.

Brown; breast and under yarts white; throat and eyebrows yellow; mustachios black; lateral tail-feathers, and some of those on the wing, bordered with yellow.

Rare: according to Le Vaillant, inhabiting the forests of Outeniqua, where he procured it during the winter. Sundevall, however, says it is a doubtful species, not known to any one.

Genus PHYLLASTREPHUS, Swainson.

Bill as long as the head, strong, the tip rather hooked ; rictus strongly bristled ; frontal feathers small, compact, directed forwards, and compressed on the base of the bill ; wings and tail moderate, rounded ; feet short, strong, and robust ; tarsus and middle toe equal ; lateral toes unequal ; the inner shortest, hind toe shorter than the inner ; anterior tarsal scales divided.

266. Phyllastrephus Capensis, Bp. Consp.,
p. 28 : *Phy. Terrestris*, Swain., Nat. Lib., Vol. 11, p. 271 ; *Le Jaboteur*, Le Vail., Pl. 112., f. 1.

GENERAL colour, dull olive-green ; chin, throat, and centre of belly, pure white ; flanks olive-green ; shafts of wing and tail-feathers, white.   Length, 8″ ; wing, 3″ 8‴ ; tail, 4″.

" Dwellers in woods, they love the deepest and most shady parts of the forest, where they are seldom noticed ; they build their nests in low situations, amid the thick underwood that grows beneath the tall forest-trees, and seldom lay more than two eggs ; if the female is disturbed, she utters one shrill note, as she flies away, letting her mate know danger is near.   The notes of this bird are low and croaking, somewhat resembling those of a frog ; there is no perceptible difference between the ♂ and ♀ ."—(Mrs. Barber *in epistolâ*.)

A nest forwarded by this lady is composed of coarse grass, moss, roots, and hair.   The eggs are creamy white, densely blotched with purple-brown at the thick end, the blotches being confluent : axis, 11‴ ; diam., 8‴.

This species has never been seen by me from any place to the westward of Graham's Town ; to the east of that locality it appears not uncommon, though from its retiring habits it is seldom procured.

The Third-Family, MUSICAPIDÆ, or Fly-Catchers,

have the bill of various lengths, generally broad, and depressed at the base, with the culmen more or less curved, and the sides compressed to the tip, which is emarginated ; the gape usually furnished with long and strong bristles ; the wings generally long ; the tail more or less long ; the tarsi mostly short and weak ; the toes more or less long, and the outer one generally united at the base.

The Sub-Family, MUSICAPINÆ, or Fly-Catchers,

have the bill moderate, broad at the base, and narrowing to the tip, which is emarginated ; the culmen more or less depressed, and curved at the end ; the gonys usually long and ascending, and the gape furnished with bristles ; the wings generally long, and more or less pointed ; the tarsi usually short and slender, and the toes short, with the outer toe longer than the inner one ; the hind toe prominent, and the claws moderate, compressed, and acute.

Genus PLATYRHYNCHUS, Desmarest.
Bill moderate, straight, and broad at the base, with the culmen more or less depressed, and slightly curved to the tip, which is emarginated, the sides compressed, and the

lateral margins straight to the tip ; the gonys rather. long, and ascending ; the gape furnished with long slender bristles ; the nostrils basal, lateral, rounded, and exposed ; wings moderate, with the third quill nearly as long as the fourth and fifth, which are nearly equal and longest ; tail more or less long, and rounded on the sides ; tarsi slender, rather short, longer than the middle toe, and covered in front with scales that are slightly divided ; toes rather short, with the outer toe longer than the inner ; the hind toe rather long, and armed with a long and slightly-curved claw.

267. Platyrhynchus Capensis, Smith ; Z. S. A., Pl. 27.

MALE.—Top of head black, with a narrow yellow frontlet ; upper parts yellowish-brown, inclining to green, each feather with an oblong umber-brown stripe along its centre ; underparts, dirty-white, with narrow, longitudinal, greyish-brown streaks ; head large and broad ; bill depressed, very wide at the base, narrow at the tip, and hooked. Total length, 5″ 10‴ ; wing, 2″ 11‴ ; tail, 2″ 11‴.

Dr. A. Smith, who found this species, did not notice it beyond the forests which exist upon the South-East Coast, towards Delagoa Bay. It chiefly affects brushwood, and feeds on insects, seizing small *Lepidoptera*, &c.. even when on the wing.

Genus PLATYSTEIRA, Jardine and Selby.

Bill more or less long, broad at the base, and depressed, with the culmen straight at the base, and then curved to the tip, which is emarginated and hooked ; the lateral margins straight, and the sides compressed to the tip ; the gonys long and ascending ; the gape furnished with moderate bristles ; the nostrils basal, lateral, covered by a membranous scale, and partly by the frontal plumes and bristles ; wing moderate. with the third quill nearly as long as the fourth, fifth, and sixth, which are equal and longest ; tail more or less long, and rounded at the end ; tarsi longer than the middle toe, and covered in front with broad scales ; toes moderate, with the outer toe longer than the inner, and the hind toe long ; the claws moderate, curved, and acute.

268. Platysteira Peltata, Sunde. ; Ofvers. Kongl. Vet. Ak. Forhandl., p. 195.

BACK, cinereous ; below white ; breast with a narrow black band ; rump whitish ; back of head blackish ; tail-feathers black, the tips and the external pair white. Length, 5″.

Kaffraria (Wahlberg). Non vidi.

269. Platysteira Longipes, Swain.; Nat. Lib., Fly-Catchers, p. 185 ; *Mus. Scita*, Vieil. ; *Le Mignard*, Le Vail., Pl. 154 ; *Muscicapa Tenella*, Licht.

ABOVE, very dark-grey, cinereous beneath ; the chin and middle of the breast tinged with rose-colour ; wings, tail, and sides of the head, black, varied with white ; legs long ; bill slender, triangular. Total length, 4¼″ ; wing, 1″ 11‴ ; tail, 2″.

Le Vaillant states it has the manners of the true fly-catchers, sitting in ambuscade and darting at passing insects, as well as searching for larvæ and apterous insects among foliage. In this he is correct. It is very abundant in the neighbourhood of Nel's Poort (Le Vaillant says Kaffraria and Namaqualand, about the Orange and Groote Rivers), frequenting mimosa thickets along the borders of the rivers.

270. Platysteira Pristrinaria ;* *Muscicapa Pristrinaria*, Vieil., Cuv. Vol. 1, p. 329 ; *Le Molenar*, Le Vail., Pl. 160 ; *Mus. Molitor*, Licht. ; *Muscipeta Pulsator*, Steph.

ABOVE, brownish ; head dark-grey ; tail black, a black stripe through the eyes ; wings ochraceous ; under parts (of male), throat, and chin, pure white ; a broad black band across the chest ; centre of belly white ; flanks rufous. In the female, the chin and throat are light rufous ; the chest dark rufous ; tail, in both, tipt with white, and the outside feather margined with white. Length, 4½″

This little fly-catcher is rather abundant in certain favourite localities in the neighbourhood of Cape Town, and is generally distributed throughout the colony. I have shot it at the Knysna ; and received it from Swellendam, Colesberg, and Beaufort ; and seen it in collections from Natal. Mr. Andersson also brought it from Damaraland. It frequents wooded places, generally hunting about thickets, high or low, for its insect prey, upon which it darts, when at rest, on the under side of leaves and on the branches. I have occasionally seen it fly out and capture an insect on the wing ; but this is rare. It generally makes known its presence by its curious note, which sounds as if two stones were ground together. I have seen the bird about at all seasons, but never could find a nest. Le Vaillant also was equally unsuccessful.

* In the Museum copy of Le Vaillant, this bird is marked, in Swainson's own hand-writing, *G. Todus*, Sw.; sub-genus *Platysteria*, Jardine. This copy formerly belonged to Mr. Swainson, but was wrecked in Table Bay, in the ship conveying his effects to New Zealand. Several of his books were recovered and bought up by a number of gentlemen who admired his talents, and he was informed that they would be forwarded to him if he would indicate his address. This he never cared to do, and the books remained here. Of them, the Museum has Temminck's Planches Coloriées, Le Vaillant's Oiseaux D'Afrique, and the 1st Vol. of his Histoire Naturelle D'Oiseaux Nouveaux de L'Amerique et des Indes, and Wilson's American Ornithology. They contain many curious manuscript notes on the plates and margins, all bearing on his ideas of the affinity of species.

271. Platysteira Strepitans, Licht. ; *Le Pririt*, Le Vail., Pl. 161 ; *Mus. Pririt*, Vieill. ; *Mus. Strepitans*, Licht.

THIS bird is closely allied to the preceding *(Le Molenar)* in shape and general appearance ; it is, however, differently coloured. In the male, the upper parts are all grisly, variegated with white ; the sides of the head, tail, shoulders, and breast are black ; throat and chin pure white ; belly variegated with black and white. In the female, the upper parts are lighter, and the under parts rufous. Length, $4\frac{1}{2}''$; wing, $2''$ $2'''$; tail, $1''$ $5'''$.

I have received this bird from Beaufort, where it was discovered by the son of my valued correspondent, Mr. A. V. Jackson, from Colesberg and from Damaraland. Le Vaillant says it is common on both sides of the colony. He never saw any nests ; and states their food consists of insects, spiders, and caterpillars ; that they are always seen in pairs ; and that, though abundant, they are difficult to procure, on account of their wildness and restless nature.

272. Platysteira Perspicillata, Swain. ; *Platyrynchus Perspicillatus*, Vieil. ; *Le Gobe mouche à lunettes*, Le Vail., Pl. 152 ; Cuv., Vol. 1, p. 329.

DEEP brown above ; white beneath ; male with a dark collar ; eyebrows and orbits white ; tail long and graduated ; three outside feathers tipped with white.

Inhabits the forests in the neighbourhood of the Gamtoos River, lurking amid the high tree-tops, and darting on the passing insects. Le Vaillant's statement that it has the habit of spreading out its fan-tail and turning it over its back, reminds one of the manners of *Rhipidura*.

Mr. Ayres has found it plentifully at Natal. Mr. Swainson's type specimen is in the Cambridge Museum.

Genus TCHITREA, Lesson.

Bill more or less long, broad, and depressed at the base, with the culmen slightly curved, and the sides compressed towards the tip, which is emarginated ; the lateral margins nearly straight, and the gonys long and ascending ; the gape furnished with long slender bristles ; the nostrils basal, lateral, exposed, and rounded ; wings long, with the first four quills graduated, and the fourth and fifth equal and longest ; tail long and graduated, with the middle feathers sometimes prolonged ; tarsi as long as the middle toe, and covered in front with scales that are slightly divided ; toes moderate, with the outer longer than the inner toe, and the hind one long ; the claws long, compressed, curved, and acute.

273. Tchitrea Cristata, Linn. ; Pl., En., 373, 2 ;
Le Tchitrec ; Le Vail., Pl. 142, 3 ; Cuvier, Vol. 1, p. 328 and 330.

HEAD and neck, brilliant blue-black ; the former crested in the male ; less so in the female ; back and tail red ; the two centre feathers of the latter being greatly prolonged in the male ; under parts, slate colour, darkest on the breast. Length, $14\frac{1}{2}''$; wing, $3\frac{1}{4}''$; tail, $10\frac{1}{2}''$.

This bird has a great range in South Africa. We have received specimens from most of our correspondents. It is found not far from Cape Town, at Somerset West, and in Drakenstein. It forms a deep nest, in the fork of some small branch of a tree, and generally in the neighbourhood of water. It is composed of fibres and dead leaves, stuck over with bits of bark, cobwebs, and lichens, to resemble a knot in the tree. I have not seen the eggs.

It is a thicket-loving bird, frequenting trees, and feeding on insects, darting at them from some favourite bough, to which it returns after making a capture.

274. Tchitrea Paradisii, L.; Pl. Enl., 234,
f. 2 ; *M. Paradisii*, Lath. ; *Todus Paradisiacus*, Gmel. ; *Le Tchitrebec*, Le Vail., Pl. 144, 145, 146 ; Cuv., Vol. 1, p. 330 ; *M. Castanea*, Temm.

THIS bird resembles *M. Cristata*, but is whiter beneath. In the second or third year it assumes a beautiful white garb, each feather having a black quill. I have shot them in Ceylon, putting on this livery, which is effected not by a renewal of the feathers, but by the loss of colour. I have observed this same loss of colour in many cage birds; and in none more plainly than in a male *Loxia Capensis*, which is at this moment changing his fine black velvet jacket for the sombre grey of winter.

I introduce this species on the authority of Cuvier and others ; but Le Vaillant denies its being found in South Africa. I agree with him.

275. Tchitrea Rufa; *Muscipeta Rufa*, Swain.;
Le Schet Roux, Le Vail., Pl. 147.

MUCH resembles *Cristata*. The head is dark steel-blue, or green, according to the light in which it is held ; throat, body, above and below, and tail, red ; the two middle-feathers of the latter much elongated, and slightly spatulate at the end ; wings black and white.

Le Vaillant says he found this species " Sur la côte Natal, vers le pays de Caffres ;" but it is a Madagascar bird, and must be enume-

S

rated among those which only exist in South Africa, in Le Vaillant's book. I took its nest, or that of a closely allied species, in Mauritius, situated in the fork of a tree—a deep, warm, thickly-lined, cap-shaped structure, containing three white eggs, spotted profusely with purpleish-brown.

276. Tchitrea Melanogaster, Swain.; Nat. Lib., Vol. XII, p. 55.

CREST, head, neck, and breast, black, glossed with chaly-beate-blue; body beneath, in the young bird, grey-black; back, tail, and under-covers, rufous; wings, black and rufous, the colours divided by a white stripe; tail long; lateral feathers graduated; the two centre being, when full-grown, six inches longer than the others. Length, exclusive of two centre tail-feathers, 8"; wing, 3" 4'''; tail, 9".

Inhabits Kaffraria.—Hartlaub, Orn. W. of Af., p. 90.

277. Tchitrea Cyanomelas; *Muscicapa Cyano-melas*, Vieil.; Cuv., Vol. 1, p. 329; *Philentoma Cyanomelas*, Bp. Consp., 324; *Muscipeta Scapularis*, Steph.; *Le Gobe Mouches Mantelé*, Le Vail., Pl. 151.

♂ —HEAD crested, shining blue-back; neck above and below the same; upper parts bluish-grey, with a white wing stripe; under parts white. ♀: head, crest, neck, and under parts, bluish-grey; wings, tail, and back, clear brown. Length, 6½"; wing, 2" 9'''; tail, 3½".

Le Vaillant found this bird in the forests of Onteniqua, and in the mimosa groves of Kaffraria: a pair frequently entered his tent, and sat on his chair or a box, catching flies. He often handled them without opposition. The male had a sharp stridulous cry, with which it would summon its female if she lingered behind. Neither Le Vaillant nor his hunters ever found the nest.

Mr. Gurney has received this species from Natal (Ibis Vol. 1862, p. 30); and Mr. G. Rex procured a single specimen of the ♂ at the Knysna.

278. Tchitrea Torquata; *Muscipeta Pectoralis*, Swain.; *Sylvia Melanoleuca*, Vieil.; *Le Cordon Noir*, Le Vail., Pl. 150; *M. Torquata*, Gm.; *M. Capensis*, Kuhl.; Cuvier, Vol. 1, p. 341.

ABOVE black; beneath white; chest, black in the male; red in the female; back and wings blackish-brown, with a white patch; tail in the male, black, with the lateral feathers bordered with white, the centre feathers white, and prolonged to nearly three times the length of the body, very narrow and pliant.

Le Vaillant states that he found this species in the Coraquoi and Kaboquoi countries, near the tropic; but no one has ever seen anything like it since, and it is thought to be one of his fictitious species.

279. Tchitrea Mutata; *Muscipeta Mutata*, Lin. ; *Le Schet Noir*, Le Vail., Pl. 148 ; Cuv., Vol. 1, p. 330 ; *M. Bicolor*, Swain.

HEAD, neck, upper parts of back, and chest, bluish-black; under parts, except the vent, white ; wing-feathers black and white ; tail black, with the exception of the two centre plumes, which are white, with black shafts. The female is less brilliant in colour, not crested, and has not the long tail-feathers.

Found, according to Le Vaillant, with the preceding. The males of this, like those of the other two species, lose their long tail-feathers when not in breeding plumage. Sundevall states that it is not a South African bird, and I agree with him.

280. Tchitrea Nebulosa, Veiel. ; *Muscipeta Fluvicola*, Swain. ; *Sylvia Nebulosa*, Veiel, ; *Le Ne-buleux*, Le Vail., Vol. 3, Pl. 149.

GENERAL colour of the body, pure white ; tail and wings' jet-black, the former with the centre feathers prolonged to four times the length of the others, which are graduated. In the female, the white is changed into an earthy-brown, and the tail is not prolonged.

In the non-breeding season, Le Vaillant says the males assume this livery ; that he found this species in Great Namaqualand, inhabiting trees overhanging rivers, and feeding on small flies. Their nests were placed on the end of branches overhanging the water; and the eggs, five in number, were pale-green, speckled with brown.

Sundevall shows that all this is utterly false, as the bird is composed chiefly of the American *Fluricola Bicolor*, with a false back let in, and the long tail-feathers of *Vidua Serena* attached.

Genus MUSCICAPA, Linn.

Bill rather short, broad at the base, and the sides compressed to the end, with the culmen much depressed, and slightly curved to the tip, which is emarginated ; the gonys long and ascending ; the gape furnished with bristles ; the nostrils basal, lateral, and partly concealed by the fronted plumes ; wings moderate, and rather pointed, with the first quill very short; the second rather shorter than the third and fourth, which are equal and longest; tail moderate and even; tarsi nearly the length of the middle toe, and covered in front; toes short, the lateral ones nearly equal, the inner the shortest, the hind toe long; the claws long, curved, and acute.

281. Muscicapa Grisola, Linn.; Hartl. Orn. W. Af., p. 97; Pl. Enl. 565, f. 1; Gould's B. of Eur., Pl. 65; Swains., Nat. Lib., Vol. 12, p. 52.

ABOVE grey, below whitish; the breast and sides of the neck striped with dusky lines. Length, 5″ 6‴; wing, 3″ 5‴; tail, 2″ 5‴.

The common European fly-catcher has been brought from Damaraland by Mr. Andersson in some abundance. I have never seen it from within the limits which I have laid down for my catalogue; but Hartlaub cites it, on M. Verreaux's authority, as from "the Cape," and Swainson (loc. cit.) alludes to it as from South Africa.

282. Muscicapa Cœrulescens; *Butalis Cœrulescens*, Hart, Ibis Vol. 1865, p. 267.

GENERAL colour, ashy; below paler; throat white; tail blackish; wing-feathers the same, margined with ashy. Length, 5″; wing, 2″ 9‴; tail, 2″ 5‴.

Found in Natal by Mr. Ayres (Ibis loc. cit.); but does not extend into the Cape Colony.

283. Muscicapa Fuscula, Sundev.; *Muscicapa Undulata*, Vieil.; *Butalis Adusta*, Boie.; *Alseonax Undulata*, Cab.; *L'Odulé*, Le Vail., Pl. 156.

ABOVE, ash-coloured; the head faintly spotted with the same, but darker; sides of wing and tail feathers lightish; chin and eyebrow whitish, with a rufous tinge; throat and belly white; chest and flank ashy. Length, 4″ 9‴; wing, 2″ 8‴; tail, 2″ 4‴.

Procured at Blanco by Mr. W. Atmore. Stated by Le Vaillant to be very common in Outeniqualand and Natal, and to construct their nests in the forks of branches, near to the trunks of the tree, of blades of grass and hair. Eggs, five; and reddish-grey.

284. Muscicapa Stellata, Vieil.; *Pogonocichla Stellata*; *Le Gobe Mouche Etoile*, Le Vail., Pl. 157; *Pog. Margaritata*, Sund.

HEAD, dull bluish-black, with a white spot in front of the eye; back yellowish-green; wings bluish-black; the outer edges of the quill-feathers pale'; centre tail-feathers black, the outer edges greenish, the rest of the tail-feathers bright yellow, broadly tipped with black; the outer pair margined exteriorly with the same; under parts all bright yellow; eye black. Length, 6"; wing, 3" 4'''; tail, 2" 9'''.

Sent from Blanco by Mr. Atmore, who writes that "it is very scarce." Le Vaillant states that it builds, in a fork of a low branch, a nest of woven grass, intermixed with lichens, the inside lined with very small rootlets. Eggs four; green, spotted with red.

285. Musicapa Ruficapillus; *Pindalus Ruficapillus,* Hartl. Ibis Vol. 1862, p. 153; *Pogonocichla Ruficapilla,* Sunde.; *Culicipeta Ruficapilla,* Grill.

Top of head rufous, blended into green on the back; a black stripe extends from the corner of the bill through the eye, above which is a yellow stripe, while beneath and under the chin, throat, and breast, it is all yellow; flanks cinereous; tail and wing feathers edged with green. Length, 4"; wing, 2" 1'''; tail, 1" 8'''.

From Swellendam (Cairncross), Traka (Atmore), Natal (Ayres); but it is one of our rarest species. Mr. Atmore says it resembles *Drymœca* in its habits.

[The following ten species have been reported by various authors to be natives of Southern Africa; but it is pretty well ascertained that not one of them really exists there. It is probable that the type specimens described may have come to Europe mingled with others from the Cape, and in this way acquired "a local habitation and (too often) a name" to which they were not in any way entitled.]

286. Muscicapa Afra, Gmel.; Cuvier. Vol. I. p. 330.

Dull-yellowish, black spotted; crown red, with black stripes; tail and wings rufous, the latter margined with fuscous; on the side of the neck two or three black marks; from the angle of the mouth a black stripe.

Habitat South Africa.—Cuvier. (loc. cit.)

287. Muscicapa Ochracea, Spar.; Cuv., Vol. 1, p. 363.

NECK and chest, ashy-ferrugineous; feathers lanceolate; wing and tail, ashy-black; head and back, brown; ears cil·iated with long feathers; belly yellow-brown.

A doubtful species. Supposed by Sundevall to be *Meliphagius Australiæ* of Australia.

288. Musicapa Tectes, Gmel.; Orn. 11, t. 39, f. 1 ; Cuv., Vol. 1, p. 343 ; *M. Borbonica*, Briss.

BROWN, dotted with red ; beneath reddish ; throat whitish ; quill and tail feathers, brown-edged ; the latter red-tipped.

Said by Cuvier (loc. cit.) to be a native of South Africa ; but is from the Island of Bourbon.

289. Todus Leucocephalus, Pallas ; *Le Capuchon Blanch*, Le Vail., Pl. 159.

BLACK, with white head and neck ; head crested. Length, about 4½″.

Le Vaillant says he found these birds only amongst the rocks and on the lofty mountains of the country of the Houzouanas. They are difficult of approach, being very suspicious. He learnt nothing of their habits, and could not find any nest. The stomachs of those examined contained nothing but insects. After all this account of Le Vaillant, the bird proves to be a South American species ! !

290. Muscipeta Azurea, Vieil. ; *L'Azurous*, Le Vail., No. 158 ; Cuv., Vol. 1., p. 331.

UPPER parts, shining azure-blue ; male, below orange, with white vent ; female, below all white.

Le Vaillant, who states he found this species in Namaqualand, says the natives told him these were " birds of passage." They feed only upon spiders and caterpillars. The nest was placed in a fork of a mimosa, strongly fastened to the neighbouring branches. It was beautifully rounded, and very deep, made of the stalks of creeping plants. It had no soft lining, not even moss. Eggs, five or six ; olive green, dotted with red, chiefly at the large end, where the spots form a complete circle. Sundevall denies that such a bird exists in South Africa.

291. Muscipeta Madagascariensis, Gmel. ; Cuv. Vol. 1, p. 330 ; Brisson, ij., t. 24, f. 5.

OLIVE ; throat yellow ; crop and chest, yellowish.

South Africa.—Cuvier (loc. cit.) ; but is a Madagascar bird.

292. Muscipeta Leucura, Lath. ; Cuvier, Vol.
1, p. 342 ; *Erythrosterna Leucura*, Bl.

ASHY-GREY ; beneath white ; middle tail-feathers black ; outer pair nearly quite white, the rest half obliquely white. Length, 4½".

Cited by Cuvier (loc. cit.) as a South African species ; but in reality comes from India.

293. Muscipeta Rufiventris, Gmel. ; Pl. En., t.
572, f. 3. ; Cuv., Vol. 1., p. 344.

BLACK ; vent red.

South Africa.—Cuvier (loc. cit.) ; but is in reality a Bourbon bird.

294. Muscicapa Albifrons, Sparm. ; Mus., t. 24 ;
Cuv., Vol. 1, p. 342.

BLACK-BROWN ; chest whitish ; belly, pale ferrugineous ; forehead whitish.

Supposed to be an Australian species, though cited by Cuvier as South African.

295. Muscicapa Cærulea, Gml., Pl. Enl., t. 666,
f. 1 ; Cuv., Vol. 1, p. 341 ; *Myiagra Cærulea*, Swainson ; *Mus. Caelestina*, Licht. ; *L'Azur à callotte et à collier noir*, Le Vail., Pl. 153.

GENERAL colour, blue ; nape and chest with a black bar ; belly and vent, bluish-white ; tail and quills, blue-black. Length, about 6".

I have not seen this bird in any collection made in this country. In Ceylon I was well acquainted with it. (See my notes on the Ornithology of Ceylon, No. 124, Ann. and Mag. Nat. His.)

Le Vaillant says he found them in the woods on the coast of Natal and Kaffraria ; they perched upon the tops of the large trees, in the thickest branches of which they built their nest, of long thin fibres, well lined with moss. Eggs, five ; of a grey russet. It is extremely improbable that this statement should be true. It is a well-known Indian species.

The Fourth Family, AMPELIDÆ, or Chatterers,

have the bill moderate, more or less broad at the base, and more or less depressed, with the sides gradually compressed to the tip, which is emarginated ; the wings long, and generally rounded ; the tail moderate, and usually even at its end ; the tarsi generally short and slender ; the toes moderate, with the outer more or less united to the base of the middle ; the claws short and curved.

The Sub-Family, CAMPEPHAGINÆ, or Cater-pillar-Catchers,

have the bill short, and rather depressed, with the culmen slightly curved, and the sides compressed to the tip, which is emarginated, and sometimes hooked ; the gape is furnished with a few short bristles ; the nostrils basal, rounded, and more or less exposed ; the wings moderate, with the third, fourth, and fifth quills the longest ; the tail long, and rounded at the sides ; the tarsi short, and covered with transverse scales ; the toes generally short, and the lateral ones unequal ; the claws moderate, compressed, and much curved.

Genus CAMPEPHAGA, Vieillot.

Bill short, and broad at the base, with the culmen rather depressed, slightly curved, and the sides gradually compressed to the tip, which is emarginated ; the gonys long and slightly ascending ; the gape furnished with a few short bristles ; the nostrils basal, lateral, rounded, and concealed by the frontal plumes ; wings moderate, with the first quill short, the second shorter than the third, and the third more or less shorter than the fourth, which is the longest ; tail long and broad, and rounded at the sides ; tarsi short, the length of the middle toe, and covered in front with broad scales ; toes moderate, the inner toe shorter than the outer, which is united at its base ; the hind toe moderate and broad, padded beneath ; the claws moderate, compressed, and curved.

296. Campephaga Niger, Vieil. ; Cuv., Vol. 1, p. 362 ; *Ceblephyris Ater*, Less. ; *Camp. Atrata*, Swain. ; *L'Echenilleur Noir*, Le Vail., Pl. 165, ♂ ; *C. Flava*, Vieil ; Cuv., Vol. 1, 362 ; *L'Echenilleur Jaune*, Le Vail., Pl. 164, ♀.

♂ —Shining metallic-black ; lower wing-coverts greenish ; ♀ greenish-grey, banded with black on the breast ; scapulars yellow ; wing and tail feathers edged with yellow. Length, 8½″ ; wing, 4″ ; tail, 3″ 9‴.

Not uncommon in the forests of the Knysna ; and I have seen it from Swellendam, Damaraland, and Natal. It feeds upon insects, which it captures among the upper branches of high forest trees. It is silent and retired in its habits, creeping about in search of its prey.

297. Campephaga Phœnicea. (Lath.) Swain.,

Nat. Lib., Vol. 11, p. 252 ; *Ampelis Phœnicia*, Lath. ; *Turdus Phœnicopterus*, Temm., Pl. Col. 71 ; Hartl., Orn. W. Af., p. 98 ; *Tanagra Dubia*, Shaw., Nat. Miscel., Pl. 252 ; *Ceblephrys Melanoxantha*, Licht.

♂ —Glossy blue-black ; shoulders crimson. ♀, above brown, with black bars tipped with white, and yellow-margined quill-feathers ; beneath white, with black spots. Length, 8″ ; wing, 4″.

Hartlaub, loc. cit., gives this as a Cape bird, on the authority of M. Verreaux. Mr. G. R. Gray, in his Genera of Birds, makes it equal to *C Flavus* of Temminck and *L'Echenilleur Jaune* of Le Vaillant, Pl. 164. This last I know to be the ♀ of *C. Niger* of Vieillot. A crimson-shouldered bird I have never seen in South Africa.

298. Campephaga Xanthornoides, Lesson ;

Ann. Des. Sci. Nat., 1858, p. 169.

GENERAL colour, shining blue-black ; shoulders bright yellow. Length, 8″ ; wing, 4″ 2‴ ; tail, 3″ 6‴.

Mr. Atmore procured a few of these beautiful birds in the neighbourhood of George. He describes their habits as similar to those of *C. Niger*.

299. Campephaga Levaillantii, Temminck ;

Ceblephrys Cæsia, Cab. ; *L'Echenilleur Gris*, Le Vail., Pl. 162, 163 ; *Ceb. Cana*, Cuv., Vol. 1, p. 362.

THE whole plumage of the male is a greyish-blue, rather darker on the head, the back of the neck, the scapularies, and the rump, with a black patch between the eye and the bill ; iris black. Length, 10″ ; wing, 5″ 9‴ ; tail, 4″ 4‴.

I found these birds in small flocks, frequenting the tops of lofty trees in the Knysna forests, searching after insects. Le Vaillant also seems to have procured them in the Outeniqua, and on the borders of the Sunday and Zwartkops Rivers.

300. Campephaga Ferruginea, Vieil. ; *Tanagra Capensis*, Sparm., Mus., t., 45 ; Cuv., Vol. 1, p. 363.

ABOVE, ferruginous-brown ; beneath, varied ferruginous and white ; tail blackish ; side-feathers reddish-brown ; bill yellow ; feet black.

This species is quoted by Cuvier loc. cit., as from the " Cape of Good Hope." It is, however, a New Zealand bird, and is only introduced here in accordance with the plan I proposed to myself of noticing *every* bird said to occur in South Africa.

T

The Sub-Family, DICRURINÆ, or Drongo Shrikes,

have the bill of various lengths, broad at the base, with the culmen more or less keeled, and curved to the tip; the sides compressed; the nostrils usually concealed by short compact plumes, and the gape furnished with strong bristles; wings long, with the fourth and fifth quills generally the longest; the tarsi and toes short, and strongly scutellated.

Genus DICRURUS, Vieillot.

Bill moderate, with the culmen more or less elevated, keeled, and curved to the tip, which is emarginated, the lateral margins curved, the gonys lengthened, and slightly advancing upwards; the nostrils basal, lateral, rounded, and concealed by projecting plumes and bristles; wings long, with the first three quills graduated, and the fourth and fifth equal and longest; tail long and forked—in some species with the outermost feathers prolonged, and webbed only at their ends; tarsi very short, longer than the middle toe, and covered with transverse scales; toes moderate, the lateral ones unequal, and both united at the base, especially the outer, which is united as far as the second joint; the hind toe as long as the middle one, strong, and armed with a strong curved claw.

301. Dicrurus Musicus, Vieil.; *Muscicapa Emarginata*, Licht.; *Le Drongear*, Le Vail., Pl. 167, 168.

ENTIRELY black, with forked tail, sometimes spotted with white on the breast and belly. Length, 10½"; wing, 5" 9"'; tail, 4" 9"'.

This drongo shrike is very abundant in the Knysna forests, and extends to Swellendam, Beaufort, and the Karroo. I have also seen it from Damaraland and Natal. It is a bold, fearless bird, darting about in the forest after insects, or chasing each other from tree to tree, uttering loud cries, which to some ears are anything but musical. Le Vaillant says they assemble in flocks, morning and evening, to catch the bees as they go out from, or return to, their hives. He says they make their nests in forks at the extremity of horizontal branches. They are composed of roots and flexible twigs, and of such a loose texture, that the eggs are plainly visible from below. The number of eggs is generally four, sprinkled with black dots, nearly all of a square form.

302. Dicrurus Ludwigii; *Edolius Ludwigii*, Smith, S. A. J., Vol. 2, p. 144; Zool. Z. A., Pl. 34.

BLACK, with a shining green gloss, most distinct on the upper parts and the breast ; tail slightly forked ; bill and legs black. Length, 7" 2''' ; wing, 4" ; tail, 3" 6'''.

Dr. A. Smith first observed this species in the forests about Natal in 1832. They were very shy, and only one specimen was procured. In habits they seemed to resemble others of the family, perching on dead trees, and performing rapid evolutions in the air in search of food.

303. Dicrurus Mystaceus, Vieil. ; *Le Drongo Moustache*, Le Vail., Pl. 169.

GENERAL colour black, with green reflections ; the wings and tail are umber-brown ; the latter slightly forked. The most distinguishing feature of this bird consists in tufts of stiff hairs which surmount the nostrils, not lying flat, but sticking up from the head.

Le Vaillant says he only once met with this bird in Kaffraria. He fell in with a flock of eight, of which he procured five. Like those of others of their family, their stomachs contained bees and caterpillars. Sundevall says it is a manufactured species.

304. Dicrurus Forficatus, Linn. ; *Dicrurus Longus*, Pl. Enl., t. 189 ; *Dicrurus Cristatus*, Vieil. ; *Lanius Drongo*, Sh. ; *Muscicapa Galeata*, Bodd. ; *Le Drongo*, Le Vail., Pl. 166 ; Cuv., Vol. 1, p. 371.

ENTIRELY black, shining, with a bluish lustre ; forehead with a crest, recurved, and pointed forward ; tail deeply forked.

Le Vaillant says he found this species in the country of the Kafirs, where it was extremely common, frequenting large forests, and living in small flocks, feeding on bees, which it seized on the wing in the same manner as fly-catchers. Sundevall says the bird is from Madagascar, not from South Africa. It seems strange that Le Vaillant should have got birds from an island which was more inaccessible in his day than in this ; but there is no doubt that he did do so, as in some instances he has actually adopted the *Malagash names* as his own.

Genus MELŒNORNIS,* Gray.

BILL short, with the culmen elevated, keeled, and curved to the tip, which is emarginated, the base broad, and the sides compressed towards the point ; the nostrils basal, lateral, rounded, and partly concealed by a few weak bristles ; wings moderate, with the fourth, fifth, and sixth quills equal and longest ; tail long, and rounded at the end ; tarsi longer than the middle toe ; toes moderate, with the outer as long as the inner toe ; the hind toe strong, shorter than the middle one, and armed with a strong claw.

* *Melasoma*, Swainson.

305. Melœnornis Ater ; *Melœnornis Atronitens,*
Licht. ; *Bradyornis Atra,* Sund.; Ofvers. Kongl. Vet. Ak. Forhandl., p. 105.

ALL black, with a blue gloss.   Length, about 8″.

Said to inhabit Kaffraria (Wahlberg).—Sed non vidi.

The Fifth Family, LANIIDÆ, or Butcher-Birds,

have the bill more or less long, strong, and straight, with the culmen curved, and the sides compressed to the tip, which is generally hooked and emarginated ; the gonys long and ascending ; the gape sometimes furnished with short bristles ; the wings moderate and rounded, or pointed ; the tail more or less lengthened, and usually rounded ; the tarsi strong, and more or less long ; the toes moderate ; with the hind toe long, and broadly padded beneath ; the claws long, curved, and very acute.

The Sub-Family, LANIINÆ, or Butcher-Birds,

have the bill moderate and strong, with the culmen curved, and the sides generally much compressed near the tip, which is mostly hooked or emarginated ; the gape sometimes furnished with short bristles ; the nostrils lateral, and rounded ; the wings more or less long, and generally rounded ; the tail of various lengths, and usually rounded ; the tarsi rather short, and strong ; the toes rather long, and strong, with the lateral ones generally unequal ; the outer toe the longest, and united at its base ; the hind toe long, and broadly padded.

Genus LANIUS, Linn.

Bill more or less long, and broad at the base, with the culmen curved, and the sides much compressed to the tip, which is hooked, and strongly emarginated ; the gonys long and ascending ; the gape furnished with a few short bristles ; the nostrils lateral : the opening rounded, and partly hidden by the projecting bristles ; wings moderate and rounded, with the fourth quill the longest ; tail more or less long, narrowed, and graduated ; tarsi rather longer than the middle toe , strong, and covered in front with broad scales ; toes moderate, with the lateral ones nearly equal ; the outer slightly united at the base ; the hind toe long, and broadly padded ; the claws moderate, curved, and acute.

306. Lanius Collaris, Gmel.; Cuv., Vol. 1, p. 265; *Le Fiscal*, Le Vail., Pl. 61 and 62.

GENERAL colour of the upper side, brown-black, with a white streak, forming a V-like mark ; under sides dirty-white ; in some specimens there is a rufous tinge on the thighs ; the four inner tail-feathers are black ; the outer feathers more or less white   Length, 9″; wing, 4¼″; tail, 4″ 9‴.

This shrike is more abundant in the neighbourhood of Cape Town than in any other part of the colony that I have visited. It appears to affect the vicinity of habitations at all times, as in the very wild country it is seldom seen.

It is bold, daring, and rapacious, a deadly foe to cage-birds, canaries in particular, entering into rooms regardless of the assembled family, and tearing them from their cages.

It preys upon all kinds of small living things ; and I have seen the spikes of an aloe—a favourite resort of these birds—garnished with snakes, locusts, small birds, hard-cased beetles, crabs, lizards, and sometimes even a fish. I have been assured on good authority that they seize gold-fish out of the fountains.

It builds in trees and bushes, and constructs a nest of grass, lined with fibres and hair. Eggs, four or five ; of a pale grey colour, blotched at the obtuse end, in the form of a ring, with greenish and reddish spots : axis, 12‴; diam., 9‴.

307. Lanius Subcoronatus, Smith, Illus. Zool. S. Af., Pl. 68.

CLOSELY resembling *L. Collaris*, but easily distinguishable from it by the white forehead and patch over the eye.

It also so closely resembles *L. Collaris* in all its habits, that it was not found to be distinct until one was accidentally killed by the members of Dr. Smith's expedition.

Dr. Smith saw it first at Latakoo ; but I have received it from Kuruman, where it was procured by Mr. R. Moffat.

308. Lanius Corvinus, Shaw ; *Corvinella Corvina ; Lanius Cissoides*, Vieil. ; Nat. Lib., Vol. IX., p. 233 ; *L. Mellivorus*, Licht. ; Griel. Zool., f. 2, p. 337 ; *Le Grande Pie-grieche*, Le Vail., Pl. 78.

ABOVE, light-brown, striped with blackish ; beneath dusky-white, with obscure spots ; tail long, curved ; flanks with a concealed spot.   Total length, 11″; wing, 4½″; tail, 6″ 9‴.

Le Vaillant does not appear to have met with this species in his travels ; but the editor of the "Naturalist's Library" states he has "received numerous specimens from Southern Africa. We have not seen it. Bonaparte (Conspectus) says from Senegal.

309. Lanius Cissoides, Licht.; *Basanistes Mela-noleucus*, Smith ; *B. Cissoides*, Licht.; Jard. and Selby, Ill. Orn., Pl. 117.

ENTIRELY brown-black, with the exception of a white stripe along the wings, some white feathers on the flanks, and a white rump; tail very prolonged; general aspect of the bird very robust. Length, $20\frac{1}{2}''$; wing, $5\frac{1}{2}''$; tail, 14".

We received several of these birds from Colesberg, from our valued correspondent, Mr. David Arnot. A Kafir in our employment stated it was very common in his country, and replaced the fiscal there. We have also received many specimens from Damaraland, and from the Victoria Falls.

Genus ENNEOCTORNIS, Boie.

Wings moderate and rather pointed, with the third quill the longest; tail moderate, and much rounded. The other characters like those of *Lanius*.

310. Enneoctornis Collurio. (Linn.) *Lanius Collurio*, Gmel., Cuv., Vol. 1, p. 264 ; *Lanius Œrugi-nosus*, Kl. ; *L'Écorcheur*, Le Vail., Pl. 64 ; *L. Spini-torques*, Becht. ; Gould., B. of Eur., Pl. 69.

HEAD, and lower part of the back, light grey ; upper part of back and wing-coverts, bright rust-red ; breast, belly, and sides, pale rose colour ; throat white ; a stripe of black passes from the bill through each eye ; two middle tail-feathers black, the rest white at the base. Length, $7\frac{3}{4}''$; wing, $3\frac{1}{2}''$; tail, 3" 2'''.

We have received this bird from Mr. Andersson, from Damaraland, and from Mr. J. J. O'Reilly, at Graaff-Reinet. Le Vaillant says it inhabits the country about the Sunday and Zwartkops Rivers (Port Elizabeth), Little and Great Namaqualand, and other places ; but not near Cape Town. Nests in trees, in the forks of the branches near the main trunk, and lays from three to six eggs.

311. Enneoctornis Rufus ; *Lanius Rufus*, Briss. ; *L. Pomeranus*, Gm.; *L. Ruficollis*, Shaw. ; *L. Rutilus*, Lath.; *La Pie-grieche Rousse*, Le Vaillant, Pl. 63 ; Cuvier, Vol. 1, p. 263 ; Shaw, Vol. 7, p. 316 ; *Lanius Ruficeps*, Becht.; Gould's Birds of Europe, Pl. 70 ; *The Woodchat*.

FORE part of the head, ears, shoulder-feathers, and side of neck, black ; a patch on the shoulder, throat, fore part of neck, and breast, white ; as are also the belly and vent, but slightly tinged with dull yellow; back of head and neck, bright chesnut; tail and rump dusky; the three outside feathers white at the base and tips. Length, 7"; wing, 3" 11"'; tail, 3" 5"'.

Le Vaillant says he obtained this bird in "l'intérieur de terres du Cap de Bonne Espérance;" but this is doubtful : it is probably not African.

Genus NILAUS, Swainson.

Bill lengthened, slender, with a prominent hook and tooth ; wings moderate ; the fourth and fifth quills longest ; tail short, nearly even ; the feathers narrow and obtuse ; tarsi moderate, slender ; inner toe shorter than the outer.

312. Nilaus Capensis, Shaw ; *Lanius Brubru*, Lath.; *Lanius Frontalis*, Forst ; *Le Brubru*, Le Vail., Pl. 71 ; Cuvier, Vol. 1, p. 265 ; Swain. Class. B., Vol. 2, p. 219.

Top of the head, and all the upper parts, black ; the latter mottled and marked with white ; tail black and white ; wings the same ; superciliary eye-streak, white ; throat, chest, and centre of belly, white ; sides of body, from the shoulders, rufous. Length, 6"; wing, 3" 4"'; tail, 2" 7"'.

Specimens of this bird have been brought from Damaraland by Mr. Andersson. I have not received them from other sources.

Le Vaillant states that they seek their food among the branches of high trees, hunting in small families. They make their nests in the forked branches of mimosas, using moss and small roots. Eggs five ; white, with brown blotches.

Genus PRIONOPS, Vieillot.

Bill moderate and straight, with the culmen and the sides much compressed near the tip, which is hooked and emarginated ; the gonys long, and curved upwards ; the lateral margins straight ; the nostrils lateral, with the opening rounded, and partly concealed by the projecting feathers ; wings long, and rather pointed, with the third and fourth quills equal and longest ; tail long and rounded ; tarsi as long as the middle toe, strong, and covered in front with transverse, broad scales ; toes moderate, with the outer longer than the inner one, and slightly united at the base ; the hind toe long, and broadly padded beneath ; the claws long, compressed, and much curved.

313. Prionops Talacoma, Smith; Zool. S. A., Pl. 5.

HEAD, light pearl-grey; cheeks white, margined behind by a black crescent; neck, and under parts, pure white; back and wings, glossy, greenish-black, with a white strip along the latter; two outer tail-feathers pure white; all the tail-feathers broadly tipped with white; bill black; legs pale yellow. Length, 8″; wing, 4¼″; tail, 3″ 10‴.

Dr. Smith first met with this species about 25° south latitude, in flocks of six or eight individuals, frequenting low bushes, or hunting the ground for insects. Termites seem to form a favourite prey.

Genus TELOPHONUS, Swainson.

Bill more lengthened than in *Lanius*, slightly hooked; the tooth smaller; wings very short and rounded; tail lengthened, graduated; lateral toes free; the inner very slightly shorter than the outer.

314. Telophonus Trivirgatus, Smith, Zool. S· A., Pl. 94.

COLOUR above, light hair-brown, tinged with yellow; below, lighter wood-brown; wings, reddish-orange; tail, dark-brown, inclining to black; the two middle-feathers approaching the colour of the back; three outermost more or less tipped with white. Length, 7½″; wing, 3″; tail, 4″.

Dr. Smith procured one specimen of this bird to the north-east of Kurrichane, between 24° to 25° south latitude. We have received it from Kuruman, from Mr. Moffat; and Mr. Andersson brought it from Damaraland, where it appears to be not uncommon.

315. Telophonus Erythropterus, Swainson, Nat. Lib., Vol. 11, p. 255; *L. Rutilus*, Lath., var. γ; *Le Tchagra*, Le Vail., Pl. 70; Pl. Enl., p. 479, f. 1; Cuvier, Vol. 1, p. 271; *Lanius Erythropterus*, Shaw, Genl. Zool., Vol. 8, 2, 307; *Lanius Coronatus*, Vieil.

ABOVE, brown; beneath whitish; wing-covers and quills (externally), rufous; crown, and stripe through the eye, black; sides of the head with a broad whitish stripe. Length, 8¼″; wing, 3″ 4‴; tail, 3″ 10‴.

Le Vaillant states that it lives only in the thickest brushwood, and densest foliage, such haunts being most productive of its favourite food, which consists of the larvæ and pupæ of different insects. Its eggs, five in number, are marked with brown.

I have received this bird from Damaraland, Swellendam, and Colesberg. It is very common at Nel's Poort and Zoetendals Vlei, creeping about the thick mimosa bushes, as described by Le Vaillant, uttering the cry which has obtained for it the name given by that traveller, whose description of its habits and food is correct.

316. Telophonus Longirostris, Swain., $2\frac{1}{4}$

Cent., p. 282.

ABOVE, brown; beneath cinereous; chin, ears, and stripe above the eye, whitish; ears margined above by a black line; bill much lengthened, and slightly curved. Length, $8\frac{1}{4}''$; wing, $3''$; tail, $4'''$.

Inhabits South Africa.—Dr. Burchell's collection. Swainson, "Two centenaries and a quarter." Swainson's type specimen is in the Cambridge Museum, and appears to me identical with the preceding species.

317. Telophonus Bacbakiri, Cab., Mus. Hein.,

p. 70; *Turdus Ceylonus*, Linn.; *Lanius Bacbakiri*, Sh.; *Laniarius Bacbakiri* Vieil.; *L. Ornatus*, Licht.; *Le Bacbakiri*, Le Vail., Pl. 67; *Telophonus Collaris*, Sw.; Cuv., Vol. 1, p. 271.

ABOVE, dull-green; below, bright chrome-yellow, with a shining black horse-shoe collar across the chest; a yellow stripe extends from the base of the bill over the eye; outer tail-feathers broadly tipped with yellow. Female resembles the male, except in being greenish underneath and wanting the black bar. Length, $9\frac{3}{4}''$; wing, $4''$; tail, $3''$ $10'''$.

Common throughout the colony, feeds on insects, and breeds in low bushes, making a nest of small twigs and bents of grass, coarsely lined with dry grass. Eggs, from three to five in number, of a beautiful verditer colour, spotted with brown, chiefly at the obtuse end : axis, $12'''$; diam., $9'''$. It is usually found in small families, except during the breeding season. Its loud call of "bacbakiri," its imitative powers, and bright plumage, render it one of the most conspicuous birds of the colony. I have not unfrequently heard two birds uttering their peculiar note for twenty or thirty minutes together—one bird giving out the harsh "*backback*," the other the shrill "*kiri*," the two performers being at a considerable distance from each other.

Genus EUROCEPHALUS, A. Smith.

Bill strong and short, with the culmen curved, and the sides compressed to the tip, which is slightly emarginated; the gonys long and ascending; the gape furnished with a few short bristles; the nostrils basal, with the opening oval, and covered by the projecting frontal plumes; wings long and pointed, with the third quill the longest; tail long and rounded; tarsi as long as the middle toe, strong, and covered in front with strong transverse scales; toes short and strong,

U

with the lateral toes nearly equal, and free at their bases; the hind toe moderate, and broadly padded; the claws moderate, curved, and acute.

318. Eurocephalus Anguitimens, A. Smith; Zool. S. Af.; *Chœtoblemma Leucocephala*, Swain., An. in Menag., p. 282.

FRONT and upper part of head, pure white; ears and sides brownish-black; body above, wings, and tail, brown immaculate; under parts white. Length, 9" 6'''; wing, 5" 7'''.

Inhabits South Africa.—Dr. Burchell's collection. Swainson, loc. cit. I have not seen it from any locality further south than Damaraland, whence it has been forwarded by Messrs. Andersson and Kisch.

The Sub-Family, THAMNOPHILINÆ, or Bush-Shrikes,

have the bill more or less lengthened, compressed on the sides, the culmen straight, but arched, hooked, and emarginated at the tip; the gape more or less bristled; the nostrils basal, lateral, and mostly rounded; the wings moderate, and more or less rounded; the tail usually long and rounded; the tarsi moderate, and covered with broad scales; the toes moderate; the lateral ones unequal; and the outer toe united to the middle one at the base.

Genus LANIARIUS, Vieillot.

Bill rather slender, straight, much compressed on the sides; the culmen slightly curved to the tip, which is slightly hooked and emarginated; the gonys moderate, and advancing upwards to the tip, which is emarginated; the gape furnished with a few short bristles; the nostrils basal, lateral, or rounded; wings moderate and rounded, with the fifth and sixth quills the longest; tail lengthened, and more or less rounded; tarsi longer than the middle toe, covered in front with transverse scales, and on the sides by an entire one; toes lengthened, the lateral ones unequal, and the outer united to the first joint; the hind toe strong, and longer than the outer toe; the claws moderate, arched, and acute.

319. Laniarius Silens, Shaw; Cuv., Vol. 1, p. 268; *La Pie-grieche Silencieuse*, Le Vail., Pl. 74.

UPPER parts, light brownish-black; wings with a white band, formed by the inner feathers being edged with white, exteriorly; tail-feathers of the same colour as the back, but with a white horse-shoe mark across the middle, broken by the two centre feathers, which have no white on them ; under parts, dull whitish-grey. Length, 7½" ; wing, 4" ; tail, 3" 8".

Le Vaillant describes the tail as having the lateral feathers bordered exteriorly with white, while he makes no mention of the centre white mark. He also states that it nests in trees, forming a beautiful structure of flexible twigs, lined with cotton gathered from cotton-producing plants. Lays three or four eggs, of a pale green colour, daubed (barbouillés) with light red.

Common about the Knysna, Swellendam, and Beaufort. It generally appears in pairs, and frequents one locality for a long period. While at Nel's Poort, in December, I procured several young birds in their first phase of plumage, being then spotted like a young chat ; but I never succeeded in finding a nest.

320. Laniarius Cubla, Lath. ; *Le Cubla*, Le Vail., Pl. 72 ; Cuvier, Vol. 1, p. 267 ; *Malaconotus Mollissimus*, Swain., Nat. Lib., Vol. 11, p. 260.

UPPER parts, all glossy black, except the rump, which is garnished with a tuft of soft dense white feathers ; under parts white, or greyish white ; iris bright yellow. Length, 7" ; wing, 3" 7''' ; tail, 3½".

Common in the forests of the Knysna, and generally found in pairs. I have also seen specimens from Damaraland and Natal. They utter a stridulous cry while in search of their insect prey, which they capture either at rest or in motion. Le Vaillant says they lay five or six eggs ; but does not inform us as to the colour.

321. Laniarius Atrococcineus. Burch., Zool. Journal, and Cuvier, Vol. 1, p. 272 ; *Malaconotus Atroccineus*, Swainson ; *Laniarius Atrococcineus*, Vieil. ; *Korokoba* of Natives, Burchell.

UPPER parts, thighs, and under the wings, shining-black ; under parts, from chin to vent-coverts, bright crimson ; a long white bar extends from the shoulder to the tip of the closed wing. Length, 9" ; wing, 4" ; tail, 4" 9'''.

This bird appears very common near Colesberg, Kuruman, and in Damaraland. I have also received numerous specimens from Mr. J. Chapman, procured near the Victoria Falls, on the Zambesi River.

322. Laniarius Similis, Smith ; Zool. S. Af., Pl. 46 ; *Malaconotus Chrysogaster*, Swain., *M. Aurantiopectus*, Less. ; *M. Affinis*, Less., Sup. Aux. Œuv. de Buf., Vol. XX., p. 330.

UPPER parts of the head and neck, blueish-grey ; back and rump greenish ; chin, throat, and under parts, light yellow ;

breast tinged with orange; stripe over the eye, whitish. Length, 7" 3"'; wing, 3" 6"'; tail, 3" 10"'.

Dr. Smith procured one specimen near Kurichane, searching for insects amongst the branches of an acacia-tree.

323. Laniarus Rubiginosus, Sundev.; Bp., Consp., p. 359; *Lanius Oleaginus*, Licht.; *L'Oliva*, Le Vail., Pl. 75, Fig. 2.

GENERAL colour above, dull green; head blueish-grey; a blackish, broad stripe extends from the eye over the ear-coverts; between the eye and the nostrils a white spot; under parts dirty white, tinged with rufous on the breast; flanks green; iris yellowish-red. Length, 7" 2"'; wing, 3" 5"'; tail, 3" 6"'.

Received from the Messrs. Atmore, who inform me that it is very scarce in the neighbourhood in which they discovered it, near George. It appears to be common at the Knysna, several specimens appearing in a small collection made there by Mr. G. Rex.

324. Laniarius Icterus; *Vanga Icterus*, Cuv.; *Lanius Olivaceus*, Vieil., Gal. Des. Ois., t. 139; *L. Poliocephalus*, Licht.; *Malacononotus Olivaceus*, Sw., West. Af., Vol. 11, p. 137, Pl. 22.

HEAD and neck above, greyish-green; back and wings green; tail-feathers black; the outermost, with the outer webs and tips, bright-yellow; the second pair the same, but less; the third pair merely tipped; a faint yellow line is also traceable on the edges of all the rest; cheeks black; eyebrow white; chin, throat, and breast, bright rufous; belly, vent, and flanks, greenish-grey; eyes yellowish red. Length, 7" 3"'; wing, 3" 4"'.; tail, 3" 8"'; tarsus, 1".

Inhabits the George forests (W. Atmore). Le Vaillant found it abundantly in the neighbourhood of Algoa Bay, the Gamtoos, Sundays, and Zwartkops Rivers, inhabiting forests, and nesting in trees and bushes.

325. Laniarius Ferrugineus, Cuv., Vol. 1, p. 265, and Shaw, Vol. 7, p. 310; *Lanius Boubou*, Lath.; *Le Boubou*, Le Vail., Pl. 68; *Malaconotus Rufiventris*, Swain., Class. B., Vol. 2, p. 220; *Zwarte Canaribyter*, or *Bonte Canaribyter*, of Colonists.

♂—Upper parts, glossy-black, with bar on the wing, white; under parts, from chin to middle of belly, pure white; the rest rufous.

♀—Upper parts, dull-brown; tail and wings darker, with

a faint white bar along the latter; under parts russet, paler on the chin and throat. Length, 8¾"; wing, 4¼"; tail, 4".

Not uncommon about wooded places in the neighbourhood of Cape Town and Rondebosch.

Le Vaillant says they nest in dense thickets, among thorns; but does not give a description of its eggs. I do not think they incubate here, but migrate and breed elsewhere, as I never heard of any one finding a nest. It occurs throughout the colony, and is usually found in pairs. By imitating the call of the male, any others, either male or female, may be attracted to the spot.

326. Laniarius Quadricolor, Cass. Proc. Ac. Phil. 1851, p. 245.

♂ Adult.—Entire superior surface of the head, body, and wings, olive-green, which is also the colour of the basal third of the external, and of three-fourths of the central tail-feathers; abdomen and ventral region yellow; throat fine scarlet; stripe through the eye, and a wide pectoral band, black; beneath the latter, a badly-defined band of yellowish-scarlet, fading into the rich greenish-yellow of the abdomen; under tail-coverts yellowish-scarlet; bill black.

The young Male resembles the adult, with the tail entirely green, and the throat, under parts, and inferior tail-coverts, pale greenish-yellow; a trace of black lines on the breast, and of scarlet on the throat. Length, 7" 3'''; wing, 3" 2'''; tail, 3" 3'''.

Inhabts the country about Port Natal.—Sed non vidi.

327. Laniarius Barbarus. (L.) *Malaconotus Barbarus*, Swain., Nat. Lib., Vol. 11, p. 243; *Lanius Barbarus*, Lin., Pl. Enl., p. 56; *Le Gonoleck*, Le Vail., Pl. 69.

Above, glossy black; beneath, crimson; crown fulvous yellow; vent and flanks buff.

A West African species. Does not extend to Namaqualand, though said by Le Vaillant to do so, and therefore included here.

328. Podobeus Fuscus, Lesson; Comp. Aux. Œuvres de Buff, Vol. XX, p. 302.

General colour, smoky-brown; the wings fuliginous; under parts smoky-grey; lighter on the belly and flanks; eyebrows white; the chin, and a longitudinal stripe on the neck,

whitish ; lesser wing-covers grey-brown ; tail, composed of
large fluffy feathers, is black, slightly tipped with white ;'
bill black ; tarsi brown.   Length, 17 centimeters French.

Quoted by Lesson as from the Cape of Good Hope.—Sed non vidi.
I believe this bird should be placed somewhere among the shrikes ;
but none of the authors that I have consulted know anything of it,
and I could not gain access to the work quoted.

The Fourth Tribe, CONIROSTRES, or Conical-billed Birds,

have the bill strong, more or less conical, with the tip slightly
emarginated, or entire ; the wings generally moderate, and
pointed ; the tarsi usually long, and strongly scutellated ;
the toes moderate, with the outer one sometimes united at
its base.

The First Family, CORVIDÆ, or Crows,

have the bill strong, with the base of the culmen sometimes
covered with projecting bristly plumes, arched, and the sides
compressed to the tip, which is more or less emarginated ;
the nostrils basal, and more or less concealed by the basal
plumes ; the wings generally long, and pointed ; the tail
moderate, or lengthened ; the tarsi and toes moderate.

The Sub-Family, CALLŒATINÆ, or Tree-Crows,

have the bill short, with the culmen elevated at the base, and
much curved to the tip ; the sides compressed, and the gonys
long and straight ; the wings short, and rounded ; the tail
lengthened, and graduated ; the tarsi more or less long, and
covered in front with broad scales ; the toes moderate, with
the lateral ones unequal.

Genus PTILOSTOMUS, Swainson.

Bill shorter than the head, much compressed ; the culmen
considerably arched and curved from the base ; rictus bristled ; wings moderate, slightly rounded ; the third, fourth,
and fifth quills longest ; tail long, cuneated : the feathers
lanceolate ; feet very strong and robust ; tarsus lengthened,
longer than the middle toe and claw ; lateral toes short, and
of equal length.

329. Ptilostomus Senegalensis. (Linn.) Pl.

Enl., p. 538; Swain. Nat. Lib., Vol. 11, Pl. 135; *Corvus Senegalensis*, Auct.; *Le Piapiac*, Le Vail., Pl. 54; *Coracias Nigra*, Lath.; *Corvus Afer*, Lin., Shaw, Vol. 7, Pl. 371; *Corvus Piapiac*, Daud.

GLOSSY black, with transverse linear shades on the tertial feathers; quills and tail light-brown, the latter with the feathers lanceolate. Length, $17\frac{1}{2}''$; wing, $6\frac{1}{2}''$.

Found, according to Le Vaillant, on the banks of the "Groot River." Builds a round nest, at the top of a high tree, composed entirely of thorny twigs, with a small round hole for entrance. Eggs six to eight, light blue, marked with brown patches. It has never occurred to any of my correspondents in this country. Sundevall states it is a Senegambian species; and I doubt if it is an inhabitant of South Africa.

The Sub-Family, CORVINÆ, or Crows,

have the bill of various lengths, the base broad, and the sides compressed, with the culmen more or less curved to the tip, which is usually entire; the nostrils entirely concealed by the projecting frontal plumes, except when the head or cheeks are denuded of feathers; the wings long, and slightly rounded; the tail long and graduated, or moderate, and nearly equal; the tarsi lengthened, and covered in front with broad scales; the toes moderate, and strong, the lateral ones of equal length.

Genus CORVUS, Linnæus.

Bill strong, generally long, with the culmen more or less arched, or nearly straight to the tip, which is entire; the sides gradually compressed from the base; the gonys long, more or less advancing upwards to the tip, the nostrils basal, lateral, rounded, or entirely covered by the projecting bristly plumes; wings long and pointed, with the first quill shorter than the second, and the third and fourth quills the longest; tail moderate, sometimes equal and rounded; tarsi longer than the middle toe, and broadly scutellated in front; toes moderate, strong, the outer slightly united at the base, and the lateral ones nearly equal; the claws strong, moderate, and curved.

330. Corvus Albicollis, Lath.; *C. Caffer*, Licht.;

Le Corbivau, Le Vail., Pl. 50; *C. Vulturinus*, Shaw., Vol. 7, p. 343; *Ringhals Kraai* of Colonists, lit. Ring-neck Crow.

GENERAL colour, black, with a bronze reflection; head, breast, and top of hind portion of neck, bronze: lower portion of latter white; a few white feathers occasionally divide the bronze of the breast from the black of the abdomen; bill large, curved, and white at the tip. Length, 1′ 8″; of wing, 1′ 5″.

This fine crow is abundant throughout the colony. It is very partial to the sea-side, where it preys largely on the rejectamenta of the waves, especially on the animal of the Paper Nautilus (*Argonauta Argo*); and I am informed that it is rare to get one of these shells perfect, unless it should happen to be thrown up at the very feet of the seeker, so keen is the appetite of these birds for them. It also frequents mountains not far from the sea. It is frequently seen about Table Mountain, descending from thence to the shores of the bay in the morning and returning at night, but always flying at a great height over the town.

It is a constant visitor at the "outspan" places along the "hardroad," and on the road itself, where it examines the droppings of the animals that pass along it, and picks the bones of those that die. It will attack weakly lambs and young antelopes, and it is said to do much mischief in this manner. Of course, it shares with the vultures in all their feasts, remaining to the last to pick up the small fragments that escape the notice of its larger confreres.

331. Corvus Scapulatus, Daud.; Tr. d'Orn. II.; p. 232, Pl. Enl. 327; *La Corneille à Scapulaire Blanc*, Vail., Pl. 53; *Bonte Kraai* of Colonists, lit. Particoloured Crow.

GENERAL colour, black, with purple reflections; head rather bronzed; a large patch between the shoulders, chest, and belly white; vent black. Length, 1′ 7″; wing, 1′ 2″; tail, 7″ 6‴.

This is the commonest crow of the Western country, and is sparsely distributed over the whole colony. It is usually found about the high roads, seeking its food in the same manner as *C. Albicollis*. In some districts, chiefly those where *C. Segetum* is most common, it is rarely seen. It makes its nest in trees, or on rocks; and its eggs, sometimes six in number, are light verditer blue, profusely spotted, chiefly at the obtuse end, with brown : axis, 1″ 9‴; diam., 1″.

332. Corvus Segetum, Temm.; Cuv., Vol. 7, p. 175; *La Corneille du Cap*, Le Vail., Pl. 52; *C. Macropterus*, Wagl.; *C. Le Vaillantii*, Less.; *Korenland Kraai* of Colonists, lit. Cornland-Crow.

ENTIRELY black, with purple reflections. Length, 1′ 7″; wing, 1′ 1′.

This species never approaches Cape Town; but appears to be otherwise generally distributed. I have seen it in considerable numbers near Caledon, and the Knysna, and have received it from Kuruman. A single specimen also occurs in a collection recently made in Damara-

land by Messrs. D. Kisch and C. Green. It keeps more to the cultivated land than do our other crows; but will not disdain to eat carrion
when its natural food (which consists of grubs) is scarce.

It breeds in trees, making a large nest of sticks, and laying from
three to five eggs, of a light pink colour, spotted with dark-brown
pink: axis, 2″ 2‴; diam., 1″ 2‴.

333. Corvus Montanus, Temm.; Cuv., Vol. 7, p. 175; *Le Grande Corbeau*, Le Vail., Pl. 51; *C. Major*, Vieil,; N. Dict. d'Hist. Nat. viii, p. 27.

ALL black, without any blue or green reflections, as in the
other species. Length, about 2 feet.

Inhabits, according to Le Vaillant, the mountains near Saldanha
Bay, dwelling in small parties, and never associating with the other
crows of the country. Lives upon worms, slugs, &c., and at times
attacks young antelopes.

I have never fallen in with any one acquainted with this bird, and
all my inquiries have failed to elicit any information concerning it.
Sundevall says it certainly is a fictitious species, and I incline to agree
with him.

334. Corvus Hottentottus, Linn.; Sh., Vol. 7., Pl. 1, p. 351.

"SIZE of a black-bird; feathers about the nostrils resembling
black velvet; bill and legs black; on each side of the bill three
very long black hairs; tail of moderate length, and nearly
even at the tip."

"Inhabits Kaffraria."—Shaw (loc. cit.); but is a well-known Indian
species of Drongo Shrike *(Chibia Hottentotta) !!!*

The Third Family, STURNIDÆ, or Starlings,

have the bill more or less long, with the sides compressed; the
culmen slightly curved, or straight to the tip, which is sometimes emarginated, and the gape often angulated; the wings
long and pointed; the tail generally long, even, or graduated;
the tarsi more or less long, strong, and covered in front with
broad scales; the toes long and strong, the hind toe very
long, and robust; the claws long, curved, and acute.

The Sub-Family, PTILONORHYNCHINÆ, or Glossy Starlings,

have the bill strong, and moderate, with the sides compressed,
and the culmen curved to the tip, which is emarginated; the
wings moderate and pointed; the tail of various lengths, and

V

graduated or even ; the tarsi strong, and covered with broad
scales ; the toes long, strong, the lateral one unequal, and the
hind toe long and strong.

Genus JUIDA, Less.

Bill moderate, more or less slender, the sides compressed,
and the culmen slightly curved to the tip, which is emarginated ; the gonys long and slightly advancing upwards ; the
nostrils basal, lateral, and sunk, with the feathers advancing
to the opening, which is exposed and oval ; wings moderate
and pointed, with the first quill short, and the second rather
shorter than the third, fourth, and fifth, which are the
longest ; tail more or less long, slightly rounded, or much
graduated ; tarsi about the length of the middle toe, robust,
and covered with broad scales ; toes rather long, strong, and
roughly scaled, the lateral toes unequal, the outer one longer
than the inner, the hind toe long and robust ; the claws
strong, short, slightly curved, and compressed.

335. Juida Australis. (A. Smith.) *Lamprotornis*
Burchellii, Smith, Zool. S. A., Pl. 47 ; *Megalopterus*
Australis, Smith, Rept. of Exp., p. 52.

PLUMAGE, a bright resplendent green ; ears violet-purple,
with a half collar of the same colour on the back of the neck ;
belly green-blue, with a steel gloss ; tail graduated, the
outermost feather of each side three inches shorter than the
two centre ones. Length, 13″ 6‴ ; wing, 7″ 6‴ ; tail, 6″ 9‴.

This fine bird was discovered by the expedition under Dr. Smith, in
lat. 25° south. It was rarely seen to leave high trees, but to seek its
food, which consisted of fruit and insects, exclusively upon them.

I have received it from the Transvaal Republic, through Mr. Arnot ;
and from Damaraland, through Messrs. Andersson, Kisch, Green, and
Chapman.

336. Juida Œnea. (Linn.) *Lamprotornis Longi-*
canda, Sw., Nat. Lib., Vol. II., p. 148 ; *Le Vert Doré*,
Le Vail., Pl. 87 ; *Turdus Œneus*, Auct.; Rüpp. Neue
Wirbelth. Voeg., p. 25.

HEAD metallic-green, glossed with violet-brown ; tail very
long, graduated, glossed with purple, and banded with
numerous dark lines ; body shining bluish-green ; purple on
the lower part of the back, rump, belly, and vent. Length,
20″ ; wing, 8″ ; tail beyond, 9½″.

According to Le Vaillant, this species is found in large flocks in
Great Namaqualand ; but are very difficult to procure, owing to their
shyness. They feed on berries and on worms. I have never seen it in
any of the collections formed there. It is a West African bird.

337. Juida Aurata. (Gmel.) Pl. En., 540; *Lamprotornis Lucida*, Norden; *Le Couigniop*, Le Vail., No. 90; *Lamprocolius Auratus*, Sund.; *Lam. Ptilonorhynchus*, Swain., Nat. Lib., Vol. 11, p. 140.

ABOVE, glossy metallic-green; head, tail, and plumage beneath, simple blue; *frontal feathers advancing and compressed forwards on the bill*; shoulders and tail-covers, scale-like, and greenish-blue. There is a row of black velvety dots at the tips of the lesser wing-covers, one on each feather, and the same on the greater covers; tail quite green; Length, 10″; wing, 6‴; tail, 3″ 9‴.

These birds, according to Le Vaillant, live in large flocks, and in their migrations extend as far as the south of Great Namaqualand. Sundevall denies that they come within these limits; but it is a common bird in Senegambia. I have received one specimen from Kuruman and two from Damaraland.

338. Juida Decorata. (Hart.) *Lamprocolius Decoratus*, Hart. Ibis., Vol. 1862, p. 148.

ABOVE shining-green, changing into blue, especially on the back and rump; ears purple; under part of throat and neck shining green; belly and vent black-brown, with blue reflections; tail and wing-feathers, dark black-brown, the latter not notched, the former nearly even, slightly exhibiting a barred appearance in certain lights; middle pair and outer webs of the rest blue-green, the blue predominating. Length, 7″ 9‴; wing, 4″; tail, 3″ 9‴; tarsus, 1″; lateral toes equal; bill to gape, 1″.

My specimens of this bird came from Natal. I know nothing of their habits.

339. Juida Phœnicoptera. (L.) *Lamprotornis Phœnicopterus*, Swain., An. in Menag., p. 360; *Le Nabirop*, Le Vail., Pl. 89; *Lamprotornis Nitens*, Pl. En., 561; *Green Spreo* of Colonists.

SEA-GREEN, glossed with blue on the head, rump, tail, and thighs, and with violet on the ears; shoulder covers, greenish-blue, margined by a flame-coloured and violet band; *inner webs of the primaries with a central notch*. Length, 9″ 6‴; wing, 5″ 4‴; tail, 4″ 2‴.

Abundant throughout the Eastern parts of the colony. Le Vaillant first met with it on the borders of the Gamtoos River in vast flocks. He says they are migratory, only visiting us during the dry season.

They subsist on berries and grubs of all kinds, as is the habits of other starlings, and breed in trees, or on the ground, laying five or six blueish-green eggs.

340. Juida Rufiventris. (Rüpp.) *Lamprotornis Rufiventris*, Rüppell, Faun., t. 11, f. 2.

HEAD brown, with a faint purple gloss; back, shoulders, throat, and breast, a shining coppery-green, with a brownish tint; wings brown, more or less glossed with green, inner vanes of feathers white towards quills; tail, deep shining green, belly and insides of shoulders ferruginous; bill and legs reddish brown. Length, 8".

Inhabits the Interior of South Africa.—Dr. A. Smith, S. Af. A. J., Vol. 2, p. 134. Swainson quotes it as a West African species.

341. Juida Erythrogaster, Bodd.; Cuv. Vol. 6, p. 394.; *Turdus Chrysogaster*, Gmel., Pl. En., t. 221; Cuv., Vol. 6, p. 373; *L'Orambleu*, Buff.

THE whole upper parts, greenish-blue; beneath orange; bill and feet black.

"Inhabits South Africa."—Cuvier (loc. cit.) sed non vidi.

342. Juida Bicolor, Gmel.; *Turdus Grillivorus*, Barrow; *Lamprotornis Bicolor*, Linnæus, Cuvier, Vol. 1, p. 393; *Turdus Bicolor*, Lath.; *Nautauges Bicolor*, Cab.; *Le Spreo*, Le Vail., Pl. 88; *Lam. Albiventris*, Swain.

GENERAL colour, brown, changing into shot-green on the neck and tail; lower part of belly and vent, white; base of lower mandible, yellow. Length, 11"; wing, 6"; tail, 4" 2'''.

The common spreo is found throughout the colony, frequenting places about which cattle are accustomed to graze, for the purpose of feeding on the insects which congregate upon their dung. They also constantly perch on the cattle, to rid them of the parasites with which they are infested.

They congregate during our winter season in small flocks, flying thickly together, and uttering loud and repeated chirps. At the breeding season they separate into pairs, and retire to rocks or houses to build, forming their nests of sticks, small roots, and fibres, in holes and crevices, and laying four or five lovely light-blue eggs, sometimes faintly spotted with brown at obtuse end : axis, 1" 2'''; diam., 10'''.

They also breed in the sides of the gullies so frequent in the surface of the country, and called *sluitjes*, digging holes into the clay. When walking about over the newly-ploughed land, hunting for grubs, they have much the manners of the European startling, and would be immediately recognised by even a casual observer as being allied to them.

343. Juida Morio. (Daud.) Pl. En. 199 ; *Corvus Rufipennis*, Sh. ; *Le Roupenne*, Le Vail., Nos. 83, 84. ; Cuv., Vol. 6, p. 393 ; *Rooivlerk Spreo* of Colonists, lit. Red-winged Starling.

GENERAL colour, dark steel-blue ; tail brownish ; wings deep rufous, the large feathers more or less tipped with brown-black. Length, 13″ ; wing, 6″ 3‴ ; tail, 6″.

The " red-wing spreo " is, equally with *J. Bicolor*, a resident in all parts of the colony. It is partially migratory, appearing in certain places at certain seasons, its visits being determined by the prevalence of some favourite food. Of this the farmer-gardener and wine-grower are but too cognizant. Woe to the crop of ripe luscious figs that are left undefended : but few of them will find their way to market ; but few would be the grapes that would reach the wine-press, was not the vineyard guarded by vigilant bipeds with loaded guns : the red-wing spreo, aided by finches, sparrows, &c., would soon pick the crop. They do not, however, confine themselves to the neighbourhood of gardens, being extremely partial to the sea coast, hopping about the rocks, even within the influence of the drenching spray, in search of small crustaceous or bruised shells ; nor do I think a bit of putrid fish would, in default of more cleanly fare, be rejected.

They breed among precipitous rocks, placing their nests in crevices, and laying four or five large blue eggs, slightly speckled with brown. I have seen eggs quite pure blue : axis, 1″ 5‴ ; diam., 10‴.

344. Juida Fulvipennis. (Swainson.) An. in Menag., p. 298 ; *Le Nabouroup*, Le Vail., Pl. 91 ; *Sturnus Nabouroup*, Daud. ; *Corocius Caffra*, Linn.

BLACK, glossed with purple ; greater quills externally rufous ; internally fulvous ; the shafts half white and half black. Length, 10″ ; wing, 5″ 6‴ ; tail, square, 4″ 8‴.

Stated by Le Vaillant to inhabit the country of both the Great and Little Namaquas, being found in flocks, with manners similar to those of the rest of the family.

I have received specimens from Damaraland, Colesberg, and the Free State. I also found it plentiful at Nel's Poort, flying about in small parties of from five to fifteen or twenty in number. Although *J. Morio* likewise inhabits that neighbourhood in smaller numbers, I never found the two species mixed together in one flock, and they can at once be distinguished by the square shape of the tail and the pale colour of the under side of the wing while in flight.

345. Juida Melanogaster, Sw. ; An. in Menag., p. 297 ; *L. Corrusca*, Licht. ; *L. Porphyropleuron*, Sundev.

SHINING sea-green, glossed with purple on the ears, scapulars, rump, and upper tail-covers, belly, and flanks, black ; the latter glossed with copper; quills and tail black, with obscure purplish edges; no spots on wing. Length, 8"; wing, 4" 5'''; tail, 3" 5'''.

I have received specimens from near Pietermaritzburg, and from Mr. Chapman.

346. Juida Leucogaster, Gm.; Pl. Enl. 648;

Cuv., Vol. 6, p. 394; *Pholidauges Leucogaster*, Swain., Nat. Lib., Vol. 11, Pl. 8.

♂—Violet; belly white; quills blackish. Length, 6"; wings, 3" 9'''; tail, 2" 5'''.

♀—Upper parts dark-brown, mottled with red-brown, chiefly on the head; under parts white, marked with longitudinal dark-brown spots.

This bird has frequently been obtained in Natal; and I have seen many specimens from Damaraland, brought thence by Messrs. Andersson, Kisch, Green, and Chapman. It does not appear to extend into the Cape Colony.

The Sub-Family, BUPHAGINÆ, or Beef-Eaters,

have the bill moderate, and broad, with the culmen slightly depressed, and curved to the tip, which is entire; the lateral sides of the lower mandible broad, and the gonys short and ascending; the nostrils basal, small, and partly closed by a membrane; the wings long and pointed; the tail long and graduated, with the end of each feather pointed; the tarsi short and strong; the toes moderate, strong, and armed with moderate, much compressed, curved, and acute claws.

Genus BUPHAGA, Linnæus.

Bill strong, and broad at the base, with the culmen slightly depressed, and curved to the tip, which is entire; the lateral sides of the lower mandible broad, projecting posteriorly under the eyes, and rounded; the gonys moderate, and curved upwards; the nostrils basal, and lateral, with the opening small, and partly closed by a membrane; wings long, with the first quill very short, and the second nearly as long as the third, which is longest; tail long, broad, and wedge-shaped, with the end of each feather pointed; tarsi as long as the middle toe, strong, and covered with broad scales; toes moderate and strong, with the lateral toes nearly equal, the outer united at the base, and the inner free; the claws much compressed, curved, and acute.

347. Buphaga Africana, Linn.; *B. Rufescens*,
Vieill., Gal. des Ois., t. 93 ; *Le Pique-bœuf*, Le Vail.,
Pl. 97.

GREYISH-BROWN; rump, breast, and body beneath, pale
fulvous ; lateral tail-feathers, ferruginous, shafts strong,
and somewhat rigid ; bill orange ; tip red. Length, $9\frac{1}{4}''$;
wing, $4\frac{1}{2}''$. .

Great Namaqualand and the neighbourhood of the tropic seem the
great habitats of this curious bird. In its manners it resembles the
starlings; but from the peculiar adaptation of its bill to extract " bots "
and other parasitic insects feeding upon cattle, it is constantly found
perching upon them. It must be getting rare, as I have never seen
a specimen in any collection that I have yet examined, except
Swainson's type in the Cambridge Museum.

348. Buphaga Erythrorhyncha, Stanley;
B. Habisinica, Ehrenb. Symb. Phys., t. 9 ; *B. Afri-
canoides*, Smith, Cont., Nat. His. S. Af., p. 12.

ABOVE, grey-brown ; tail brown ; the inner web of the external feathers, rufescent ; wing-feathers black ; beneath, pale-
fulvous ; throat grey ; bill red. Length, $8\frac{3}{4}''$; wing,
$4''\ 5'''$.

Inhabits Natal.—Teste Verreaux ; Hartlaub, Orn. W. Af., p. 121 ;
and Ayres, Ibis. Vol. 1863, p. 328, non vidi.

The Sub-Family, STURNINÆ, or Starlings,

have the bill more or less long, with the culmen curved, or
straight to the tip, which is obtuse, and somewhat flattened ;
the nostrils basal, placed in a membranous groove, and sometimes clothed with short feathers ; the wings moderate ; the
tail rather short, and even or rounded ; the tarsi rather long,
strong, and covered with broad scales ; the toes long and
strong ; and armed with strong, acute claws.

Genus PASTOR, Temminck.

Bill rather short, with the culmen curved from the base,
and the sides compressed to the tip, which is rather acute,
and only slightly emarginated, the lateral margins rather
curved, and the gonys long and ascending ; the nostrils
basal, lateral, and placed in a short broad groove, which is
membranous, and clothed with short feathers, with the
opening oval ; wings long and pointed, with the first quill
spurious, and the second the longest ; tail moderate and
even ; tarsi nearly as long as the middle toe, rather strong,
and covered with broad transverse scales ; toes long and

rather slender ; the outer toe rather longer than the inner, and united at the base; the hind toe long ; the claws long, curved, and very acute.

349. Pastor Roseus. (Linn.) Pl. Enl., 251 ; Le Vail., No. 96 ; *Turdus Seleucis*, Gmel., Gould's B. of Europe, 212 ; Cuv., Vol. 6, p. 392.

GENERAL colour of body, pale rose-colour ; head crested, and with the neck, throat, wings, and tail, shining black.

Le Vaillant states that this bird penetrates as far as 24° south latitude in South Africa, and that he killed specimens within that range. This is denied by all subsequent authors; and I believe the species never visits South Africa. It has certainly never appeared in any of the collections I have examined ; and some have been made up as far as the Zambezi.

Three other birds belonging to this family, and allied to this genus, have been, on various authorities, wrongly cited as South African. I include them in this list, without defining the generic characters, as they are well-known Indian forms.

350. Sturnus Capensis, Cuv., Vol. 7, p. 173 ; *Sturnopaster Contra.*

BLACK; white beneath ; round the eye, bare and orange-coloured ; patch of white on sides of face.

"Cape of Good Hope."—Cuv., loc. cit.; but is really an Indian species.

351. Hetœrornis Gingianus. (Lath.) Cuvier, Vol. 6, p. 423 ; *Acridotheres Gingianus*, Bp. Consp., 419 ; *Gracula Grisea*, Daud., Shaw., Vol. 7, p. 469 ; *Le Martin Gris-de-fer*, Le Vail., Pl, 95, f. 2.

ORBITAL spot, naked; behind, acute; above, iron-grey ; crown and cheeks, black ; beneath, reddish ; quills, purplish-black ; primaries. white based ; four wing-coverts on each side, reddish tipped. Length, 6″ 9‴.

Le Vaillant says he found a flock of these birds passing over the mountains of Bruintjes Hoogte, but observed none in other parts of South Africa. It is an Indian bird, and probably Le Vaillant's statement is false.

352. Hetœrornis Pagodarum, Gmel.; Cuv., Vol. 6, p. 423 ; *T. Melanocephalus*, Wahl. ; *T. Malabaricus*, Gm. ; *Le Martin Brame*, Le Vail., Pl. 95, f. 1.

GENERAL colour, creamy-grey ; head crested, black ; wings and tail black ; the two centre-feathers of the latter ruddy-grey. Length, 7″.

Le Vaillant states he found this species on the northern bank of the

Orange River, as it was migrating to the eastward. He only obtained one shot at the numerous flocks he observed, and killed one pair of males. This is probably another of Le Vaillant's inventions, the bird being a native of India.

Genus DILOPHUS, Vieil.

Bill long, straight, rather depressed, and broad at the base, with the culmen curved, and the sides compressed to the tip, which is emarginated ; the gonys curved, and ascending ; the nostrils lateral, and placed in a membranous groove, with the opening rounded and exposed; wings moderate, with the first quill spurious, the second rather shorter than the third and first, which are equal and longest ; tail even, short ; tarsi longer than the middle toe, strong, and covered in front with broad transverse scales ; toes moderate, the outer toe rather longer than the inner, and united at the base ; the hind toe long ; the claws moderate, slightly curved, and acute.

353. Dilophus Carunculatus. (Gmel.) *Sturnus Gallinaceus*, Lath. ; *E. Larvata*, Shaw, Vol. 7, p. 468 ; *Le Porte Lambeau*, Le Vail., Pls. 93 and 94 ; Cuv., Vol. 1, p. 425 ; *Creatophora Carunculata*, Lesson, Comp. Aux. Œuv. de Buf., Vol. XX., p. 308.

GENERAL colour, cinereous ; shoulder and upper part of wings, white; rest of wing and tail black, or dark-brown, with green reflections. Head in adult male naked, bright yellow, with black wattles. Female head not naked, and coloured as on back. Length, 8″ 6‴ ; wing, 4″ 6‴ ; tail, 2″ 10‴.

Frequents the same country as the common spreo, with which it is often found mingled ; though it not unfrequently flies in large flocks composed wholly of individuals of its own species.

Specimens with developed wattles are very scarce, and none that I have ever seen have these appendages as enlarged as those figured by Le Vaillant, loc. cit. Perhaps in the country where it breeds it may acquire such during the nesting season. Mr. Schwartz, of Zoetendals Vley, informs me that he once found this species breeding in his neighbourhood. A large company formed their nests in a dense bush, reared their young, and departed. He never saw them nest at any other time.

The Fourth Family, FRINGILLIDÆ, or Finches,

contains a numerous series of small, and for the most part strong birds, that have the bill short, thick, strong, and more or less conic, without emargination at the tip, but generally angular and thick at the base.

w

The Sub-Family, PLOCEINÆ, or Weavers,

have the bill strong and conic, with the culmen projecting on the forehead, and arched to the tip, which is entire; the wings somewhat rounded, with the first quill remarkably short; the legs and toes robust, and strongly scaled, and the hind toe strong, and nearly as long as the middle toe.

Genus TEXTOR, Temminck.

Bill more or less long, conic, broad at its base, laterally compressed, with the basal portion of the culmen advancing on the forehead (in some seasons swollen), and curved towards the tip; the lateral margins sinuated, and the gonys long and ascending; the nostrils basal, lateral, naked, and pierced in the substance of the bill; wings somewhat rounded, and reaching a little below the base of the tail, with the first quill very short, and the second nearly equalling the third and fourth, which are the longest; tail moderate, and rather rounded; tarsi equal in length, with the middle toe robust, and strongly scaled; toes short, and strongly scaled; the lateral toes equal, the hind toe equalling the inner, and the claws strong, and curved.

354. Textor Erythrorhynchus, Smith; Ill. S.

Af. Zool., Pl. 64; *Bubalornis Niger*, A. Smith.

GERERAL colour of adult male, black; the first half of each wing-feather white; bill orange-red; legs and toes, yellowish-brown. The young bird has some whitish patches on the neck and breast. Length, 9" 6'''; wing, 5"; tail, 4".

Dr. A. Smith procured this bird after passing the 25th degree of south latitude, frequenting herds of buffaloes, and perching on their backs in search of the parasitical insects which infest their hides. I have received specimens from Damaraland.

Genus HYPHANTORNIS, Gray.

Bill as long as, or shorter than, the head, broad at the base, laterally compressed at the tip, with the culmen broad, smooth, and rounded, advancing to a point on the forehead, and the lateral margins slightly angulated at the base, and straight towards the tip; the nostrils basal, exposed, oval, and pierced in the substance of the bill; wings reaching a little beyond the base of the tail, the first quill very short, the second equalling the sixth; the third, fourth, and fifth of nearly equal length, but the fourth the longest; tail rather short, even, or slightly rounded at the end; tarsi as long as the middle toe ; toes strong, the lateral toes equal in length, and the claws strong, and much curved.

355. Hyphantornis Capensis, Smith ; Zool. S. A., Pl. 66 ; *Ploceus Abyssinicus,* Cuv.

FRONT, chin, throat, and breast, saffron-yellow ; top of head, sides of neck, and vent, gamboge-yellow ; nape, back, and rump, lemon-yellow ; back of neck and shoulders, greenish-yellow ; the middle of each feather pale olive-brown ; wing feathers, dark purplish-brown, edged with sulphur-yellow ; tail olive-brown, tinted with yellow ; bill livid-brown. Length, 7″ ; wing, 3″ 6‴ ; tail, 2″ 6‴.

Generally diffused over the more southern districts of South Africa, in flocks of from 10 to 15 individuals, building on trees that overhang rivers and marshes. Nests constructed of coarse grass, somewhat kidney-shaped, with the entrance towards the upper end, and directed downwards.—Dr. A. Smith, loc. cit.

I have found this species to build indiscriminately over the driest spots, or over water. At this moment a large colony is established in some fir-trees opposite the house of a friend of mine, at the outskirts of the town, on the way to " Green-Point," close to the main road, and the constant stream of traffic. I should not describe the entrance of the nest as Dr. Smith has done ; but should rather say it was on the lowest or under side, opening downwards. The eggs are four or five in number, and of a beautiful spotless verditer : axis, 11‴ ; diam., 8‴.

I have kept these birds for a considerable time in a large cage, and have known several in the possession of others. They become very tame, and will readily answer to the call. If they are supplied with cotton, or thread, they will weave it most industriously into the bars of the cage, forming a dense mass, which it is impossible to unravel. This work they perform entirely with their bills, clinging the while to the side of the cage with their powerful claws.

They are most mischievous, cunning birds, and have more than once unfastened the bolts of my cage-door and got out ; such, however, is their attachment to each other (I have two now before me in my cage), that they have never left the room, though the windows, on each occasion, have been wide open.

A party have several times endeavoured to form a colony in the Governor's garden, but for some reason failed. This year (1861) a pair began to repair an old nest, in May, and I saw the male and several of his companions in full breeding dress. Cold, rainy, windy weather, however, blew away the structure, and though I hear the churring cry of the birds about, I see no signs of building up to the time I write (July 24).

356. Hyphantornis Aurifrons. (Tem.) Pl. Col. 175, 176 ; *Ploceus Icterocephalus,* Sw., Nat. His. of Birds, Vol. 1, p. 189 ; Cuv., Vol. 7, p. 132 ; Shaw, Vol. 14, Pl. 1, p. 35.

FOREHEAD, and top of head, fine gold colour ; cheeks and throat less brilliant ; sides of neck and beneath, citron-yellow ; nape (above), tail, and edges of wing, yellowish-green ; quills blackish, bordered with yellowish-green.

" Southern point of Africa."—Cuvier. (loc. cit.) Bonaparte, in his "Conspectus," says that this species resembles *P. Capensis*, but is larger, more yellow, and the forehead and throat "fere aurantiacis." The type specimen, with Swainson's own label still attached, is in the Cambridge Museum, and is nothing more than a fine ♂ of *H. Capensis*. I have shot many such, breeding in company, and pairing with the small race.

357. Hyphantornis Nigrifrons, Cabanis; Mus.
Hein. 1, p. 182 ; Hartlaub, Orn. W. Af., p. 126.

FOREHEAD, chin, and throat, black ; the head and remainder of the body below, yellow ; feathers of back and wing-coverts, greenish-yellow, with a fuscous medial stripe ; quill-feathers pale fuscous, with yellow margins ; tail-feathers greenish, margined with yellow ; bill black. Length, $6\frac{1}{4}$" ; wing, $3\frac{1}{4}$" ; tail, 2".

Kafirland.—Teste Cabanis. Hart., loc. cit.

358. Hyphantornis Nigriceps.

UNDER parts, bright yellow ; upper parts the same, variegated with black : this effect is caused by the bases of the feathers being black, and the edges yellow; wing and tail-feathers, dark-brown, the former with yellow margins to the outer webs, and the latter tinted with yellow ; collar round the neck, bright pure-yellow ; head, chin, and throat, black. Length, 6" 3''' ; wing, 3" 3''' ; tail, 2" 3'''.

This bird was forwarded from Kuruman by Mr. Moffat. I have not seen it from any other locality.

359. Hyphantornis Capitalis, Lath.; *Ploceus Velatus*, Vieil.

♂ —Top of head, breast, and under parts, bright-yellow, darkest on the head ; forehead, side of the head, beyond the eye, chin, and throat, black, the colour ending in a wedge-shaped mark on the chest; back greenish ; wings and tail brown, with a tint of yellow. The ♀ is a dull-green, mottled, and wants the black of the face and throat. Length, 6" ; wing, 3" 6''' ; tail, 2" 6'''.

Received from Kuruman, Colesberg, and Damaraland. I also found it abundantly at Nel's Poort, nesting on the trees overhanging the rivers and water-courses. Their nests are shaped just like those of *H. Capensis*, and suspended in the same manner. The eggs are subject to great variation : some are green, spotted with reddish-brown ; others are cream-coloured, minutely *spotted* with reddish-brown or light-purple, or heavily *blotched* with the same. I have taken eggs with green and cream-coloured grounds out of the same nest : axis, 11‴ ; diam., 7‴.

Mrs. Barber writes from " The Highlands," near Graham's Town : " I send herewith the nest of a kind of finch (a yellow bird, with a black head). They are common, and most likely you know both the bird and its nest, though I do not suppose that you know the material that the nest is made of ; for in our youthful, bird-nesting days it puzzled us amazingly, until at length we found out the secret, and then I do assure you we were very proud of it, and gave ourselves much credit for finding out what no one had been able to tell us, which was, that the nest of this bird was made of the fibres of the leaves of a species of *Sanseviera*, a plant belonging to the natural order *Asphodeleæ* ; but as our *Flora* has not yet been published up to that order, I cannot give you its specific name with any degree of certainty. It is not the tall aloe, like one that grows in our forests ; but the dwarf, thick-leaved, stemless *Sanseviera*, with the red edges to its leaves. The whole leaf is full of strong fibres, but from its tough nature the birds are only enabled to strip off the two marginal threads ; and to construct one of these nests many thousands of *Sanseviera* leaves are deprived of their red-edged fibres ; and in the neighbourhood of one or two of these nests you will not find a perfect leaf on any of these plants.

360. Hyphantornis Spilonotus, Vigors ; Proc.

Zool. Soc., 1830, p. 92 ; *Ploceus Flaviceps*, Swain., Nat. Lib., Vol. 12, p. 259, Pl. 32 ; *Ploceus Stictonotus*, A. Smith, S. Af. Q. Journal, No. 5, p. 11 ; Id., Zool. S. Af., Pl. 66, Fig. 1.

GENERAL colour of head, breast, belly, vent, and rump, bright gamboge-yellow ; back liver-brown, and yellowish-green variegated ; sides of head, chin, and front of throat, black ; wing-feathers, brown, edged with yellow ; tail, greenish-brown ; bill black ; feet flesh-colour. Length, 7" ; wing, 3" 6‴ ; tail, 2" 7‴.

Dr. A. Smith describes this species as inhabiting the districts upon the south-east coast, but not found to the westward of Kafirland ; frequenting rivers, and suspending their kidney-shaped nests of woven grass over the water. I have received the species from Kuruman. Eggs, according to Dr. Smith, blue ; but Mr. Ayers (Ibis., Vol., 1860, p. 212,) says they vary much in colour, being green, blue, white, and speckled white and brown.

361. Hyphantornis Subaureus. (Smith.) Zool.,

S. A., Pl., 30 ; F. 1 ; *Ploceus Tahatali*, Smith, Appendix to Report of Exped., p. 50, apud Gray.

FRONT, crown, sides of the head, and throat, bright saffron-yellow ; nape, back and sides of neck, back and shoulders, between sulphur and gamboge-yellow ; vent, light chrome-yellow; bill liver-brown.   Length, 6" 10'" ; wing, 3" 5'" ; tail, 2" 9'".

Rare. Dr. Smith only met with a few specimens; and those in the neighbourhood of Algoa Bay. I have not seen it.

362. Hyphantornis Tahatali. (Smith.) Appendix to Report of Exped., p. 50.

ABOVE, golden green ; feathers on top of head, back of neck, and interscapulars, dark-brown in the course of the shafts; beneath bright-yellow ; quills brown, edged with golden-green ; tail olive-brown, edged with greenish-yellow. Length, 5" 9'".

"Found between the Orange River and the Tropic."—Dr. A. Smith, loc. cit.

Mr. Gray, in his Genera of Birds, thinks this is identical with *H. Subaureus* ; but Dr. Smith would hardly have confused the two species with specimens of each before him.

363. Hyphantornis Ocularius. (Smith.) Zool. S. A., Pl. 30, F. 2.

GENERAL colour, golden-yellow ; back, rump, and shoulders, greenish-yellow ; chin, throat, and a transverse stripe enclosing the eye, deep-black ; bill black. In the Female the chin and throat are not black, and all the colours less vivid. Length, 7" ; wing, 3" 3'" ; tail, 3" 1'".

Found sparingly over South Africa, more especially in the vicinity of the south-east coast. Does not congregate. Suspends a nest constructed of delicate fibres of bark, closely interwoven into the shape of a retort, from the branches of trees. Eggs three, bluish-white, spotted with dusky-brown.—(Smith, loc. cit.)

I have not yet obtained this species ; but a nest precisely answering the above description was forwarded to me by Colonel Armstrong, of Bathurst.—Inhabits Natal. See "Ibis" Vol. 1860, p. 213.

364. Hyphantornis Mariquensis. (Smith.) Zool. S. A., Pl. 103.

UPPER and lateral parts of the head, and the back and sides of the neck, gamboge-yellow, faintly tinted with green, and variegated with umber-brown streaks; back and rump yellow, like the head ; wings brownish, edged with yellow ; breast, belly, and vent, between gamboge and lemon-yellow ; bill brownish.   Length, 3" 7'" ; wing, 3" 1'" ; tail, 2" 4'".

This species, like most of the others of the group, is gregarious, and inhabits the banks of rivers to the northward of Kurrichane, especially those near to the tropic of Capricorn.—Dr. Smith, loc. cit.

I have received it from Kuruman, Colesberg, and Damaraland.

365. Hyphantornis Auricapillus. (Swains.)
An. in Menag., p. 346.

HEAD above, upper tail-covers, and all the under parts, bright-yellow; back feathers have an olive tinge, with an almost imperceptible dusky stripe; frontal line, lores, ears, and chin, velvet-black; above the first, rufous. Length, 6"; wing, 3" 1"'.

Inhabits South Africa.—Dr. Burchell's Coll.; Swain., loc. cit., non. vidi.

Genus SYCOBIUS, Vieillot.

Bill as long as the head, with the culmen and the lateral margins curved, the latter very slightly sinuated; the nostrils basal, exposed, and rounded; wings moderate, and somewhat rounded, with the first quill very short, the second equalling the eighth, the third longer than the seventh, and the fourth and fifth of equal length; tail moderate, and even, or slightly rounded; tarsi robust, and strongly scaled; toes moderate, with the inner toe shorter than the outer, and the claws strong, and curved, especially that of the hind toe.

366. Sycobius Bicolor, Vieil.; Ency. Meth, p. 698; *Symplectes Chrysomus*, Swain., Nat. Lib., Vol. 11, p. 170; *Fringilla Gregalis*, Licht.; *Pl. Chrysogaster*, Vieil.; *Eupodes Xanthosomus*, Jard. and Selby, Ill. Orn., N. S., Pl. 10; *Syncobrotus Bicolor*, Caban., *Pyranga Icteromelas*, Vieil.; Hartlaub., Orn. W. Af. p. 135.

ABOVE, dark-brown, tinged with green, including wings, tail, sides of head, ears, and neck; chin dirty-white, tinged with pale yellow, mottled with brown; rest of under parts bright yellow. Length, 6" 3"'; wing, 3" 2"'; tail, 2" 3"'.

A single specimen of this bird was procured by Henry Bowker, Esq., of the Frontier Armed and Mounted Police, in the country beyond the Kei, and kindly presented to the Museum by Lady Wodehouse, the wife of the present Governor of the Cape Colony. Another specimen appears in a collection formed near Algoa Bay, and now in the possession of Professor Newton, of Cambridge.

Genus PLOCEUS, Cuvier.

Bill short, thick at the base, pointed at the tip, laterally compressed, with the culmen smooth, rounded, and advancing on the forehead in a point; the lateral margins slightly sinuated, and angulated at the base, and the gonys lengthened and ascending upwards; the nostrils basal, rounded, and partly hidden by the frontal plumes; wings moderate, with the first quill very short, the second shorter than the third, the third to the fifth longest, and the tertials nearly the length of the quills; tail moderate, with the end even, or slightly rounded; tarsi as long as, or longer than, the middle toe; toes moderate, the inner toe shorter than the outer, and the claws long, slender, slightly curved, and compressed.

367. Ploceus Taha. (Linn.) Z. Ill. p. 7; *Ploceus Dubius*, Smith, App. to Report of Exp., p. 50 (non-breeding plumage); *Ploceus Melanocephalus*, Vieil.

♂—In summer plumage, black; the crown of the head, back, rump, vent, and a narrow oblique stripe on each side of the breast, immediately in front of the shoulders, bright-yellow; wings and tail, grey-brown; thighs, pale-yellow, speckled with brown.

♀—In winter plumage, above, pale yellowish-brown; head, and neck freely dashed with longitudinal, brownish-black stripes; back and rump the same; under parts greyish-white; throat and breast tinged with sienna-yellow, and with the flanks variegated by longitudinal brown streaks. Length, 4″ 9‴; wing, 2″ 7‴; tail, 1″ 1‴.

The species does not extend south of 26°, and is said to breed in the reeds which skirt the rivers north of that line. They congregate in large flocks, and are very destructive to the corn-fields.—-Dr. Smith, loc. cit.

368. Ploceus Capensis; *Euplectus Capensis*, Swain., B. of W. Af., Vol. 1, p. 180; *Loxia Capensis*, Linn.; Cuv., Vol. 2, p. 137; *Pyromelana Capensis*, Blyth; *Icterus Flavescens*, Daud.'; *Yellow Kafir Fink*, or *Finch* of Colonists.

GENERAL colour of male in breeding plumage, deep black; the feathers of the head and neck short, and so closely set as to resemble the richest velvet; rump and shoulders brilliant yellow; wings brown. Length, 5¾″; wing, 3″ 3‴; tail, 2″ 3‴.

This is a very common species throughout the colony, affecting indiscriminately the solitary vley (morass) in the midst of the veldt (open, uncultivated country), or the homestead of the farmer. During the breeding season it is seen about reeds, among which it breeds, placing its nest with great art, so as to include in its structure three or four of the firmest and most upright stems, which support it like pillars. The male generally perches on the topmost twigs of the bushes mingled with the reeds, or on the heads of the reeds themselves; and his brilliant yellow rump, contrasting with his otherwise intensely black plumage, render him peculiarly conspicuous. The female resembles the male in the winter dress, being a light-brown, with a dark stripe down each feather.

Having kept several in cages, I am enabled to describe the moult which takes place. The first thing which changes is the bill: the black commences to show at the tip of the lower mandible, which gradually darkens, from its usual horn-colour, to the deepest black; before this is accomplished the upper mandible has begun to blacken, and a few black feathers are visible round the neck; the breast next shows a change, and a feather or two on the back darkens, and thus by degrees the whole bird changes to its splendid black. This transformation commences about July, and is complete in September, which is the nesting season. The moult back again, begins in the same manner.

This species lives well in cages. I had one which had been about six years in confinement, and effected his escape after all, from the warping of his cage-bottom, owing to its being exposed to the sun after my pet had enjoyed his bath. He would come to the cage-bars whenever my wife approached them, his long legs drawn out to their full extent, his bright black eye glittering even in the midst of his deep black plumage, watching for his morsel of bread. On his being addressed as "*Fink*," he instantly replied, stretching his neck to the utmost, and uttering the most piercing, discordant shriek. We used to laugh, and say he would some day dislocate his neck in his attempts to sing, in rivalry to the canaries. After his bath, or when enraged, which was pretty often, he would erect the feathers of the back at right angles to his body, and then looked beautiful, the light shining through the brilliant yellow, setting it off to great advantage.

The nest is constructed of grasses, and is domed, with an entrance near the centre. The eggs, four or five in number, are very pale verditer, thickly marked everywhere with dark, greenish-brown blotches and spots: axis, 10‴; diam., 7‴.

369. Ploceus Oryx. (L.) *Euplectes Oryx*, Swain., *Pyromelana Oryx*; *Red Grenadier Grosbeak*, Cuv., Vol. 2, p. 136; small var., *Eu. Sundevalli*, Bp.; *Emberiza Oryx*, Gmel.; *Red Fink* of Colonists.

♂, in breeding plumage: upper parts, brilliant scarlet; lower parts, with the exception of throat, and vent, which are scarlet, deep velvetty black; forehead, cheeks, and chin, black; wings and tail brown. Length, 5″; wing, 2″ 9‴; tail, 1″ 10‴. Female and male in non-breeding plumage, brown, the centre of each feather with a dark stripe.

The "Red Caffre Fink," though not an uncommon bird, is certainly a very local one, that is, though distributed over the whole colony, it seems to be confined to narrow limits. About Cape Town, I only know of one place where it is to be found, and that is the swampy ground near the Royal Observatory. It breeds among the reeds growing in the river, supporting its nest on three or four stems, like its congenor, *Ploceus Capensis*. In the winter time it congregates in flocks, and does much damage to the grain fields in the neighbour-

x

hood of its nesting-place, but never strays away from that locality. Nest like that of *Loxia Capensis*. Eggs four or five; pure, light blue: axis, 10'''; diam., 7'''. Builds in thousands about the swamps and river running out of Zoetendals Vley.

370. Ploceus Lathamii; *Loxia Lathamii*, Smith, App. Rept, of Exped., p. 51.

ABOVE, brown; interscapulars and wing-covers edged, and tipped with pale rufous or rusty white; chin, middle of belly, and vent, white; breast and sides of belly, rusty-white, faintly clouded with brown. Length, 4".

"Inhabits the country near to and beyond Kurrichane."—Dr. A. Smith, loc. cit.

This name was probably sunk in some other by Dr. Smith, as I cannot find reference to it in any of the works consulted by me. Another species quoted by Shaw, Vol. 9, Pl. 2, p. 320, as from the Cape of Good Hope, under the name of *Loxia Aurantia*, the "Orange Grosbeak," is a South American bird, viz., *Spermophila Aurantia*, Vieil., Ois. Chant., p. 44.

Genus PHILETÆRUS, A. Smith.

Bill more compressed than in *Ploceus*, with the culmen slightly arched from the base, and the lateral margins sinuated; wings moderate, reaching to about the middle of the tail, with the first quill very short, almost rudimentary, the second, third, and fourth nearly of equal length, and longest; tail somewhat rounded at the end; tarsi strong, shielded by transverse scales; toes moderate, the outer and inner toes nearly of equal length; claws much curved and acute.

371. Philetærus Socius. (Lath.) *Euplectes Lepidus*, Swain., An. in Menag., p. 309; *Philetærus Lepidus*, Smith, Zool. S. A., p. 8; *Loxia Socia*, Lath.; Cuv., Vol. 2, p. 133; *Social Grosbeak*, Paterson's Voyage, Pl. 19.

FLANKS with a blackish patch, the feathers of which are margined with whitish; chin black; plumage above, drab-brown; the margins of all the feathers, and the whole of the under parts, pale isabella. Length, 5"; wing, 3"; tail, 2".

Inhabits South Africa.—Swainson, loc. cit.

This bird is most abundant in the Interior. Dr. Smith looks upon Latakoo as its metropolis; but I have heard of it in great numbers all over Damaraland, and even as low down as Namaqualand. Being the builder of the huge nests so often mentioned by travellers, it is well known. Eggs three or four, blueish-white, mottled at the large end with small brown dots. Feeds on seeds and insects.

Genus PLOCEPASSER, Smith.

Bill large, conical, pointed, and laterally compressed; the culmen slightly arched, advanced on the forehead in a point; wings moderately long and rounded, with the first quill very short, the second and third equal and longest; tail square at the end, or slightly emarginated; tarsi strong, with distinct shields in front, and entire behind; toes moderate, the outer and inner toes equal, the hinder toe slightly shorter, but stronger, the middle one much the longest; claws strong and curved.

372. Plocepasser Mahali, Sm.; Z. S. A., Pl. 65;

Agrophilus Mahali, Swain.; *Pl. Melanorhynchus.* Rüpp.

UPPER part of head, and stripe on each side of neck, black-brown; eyebrows white; side of head, brown; side of neck and back, light-brown; rump, vent, chin, and throat, pure white; breast and belly, dull-white; wings with two white stripes across the shoulders. Length, 6″ 6‴; wing, 4″; tail, 2″ 9‴.

Congregates in large flocks between the Orange River and the tropic; builds in companies. The nests are composed of the stalks of grasses, the thickest extremities being placed so as to protrude externally, and offer a defence against snakes, &c.—Dr. A. Smith, loc. cit.

I have received a few specimens of this bird from Kuruman and Damaraland.

Dr. Smith's statement that the protruding sticks of the nest are meant as a defence against snakes, appears to me about as well founded as the idea that other members of this family construct their bottle-shaped nests at the extremity of branches, so as to be out of the way of monkeys and snakes. Why should these birds, beyond all others, be endowed with such prescient wisdom? Why should they depart from their custom sometimes, and build their nests on reeds? Or why take these precautions in places where neither snakes nor monkeys exist?

373. Plocepasser Pileatus; *Leucophrys Pileatus,* Swain., An. in Menag., p. 347.

ABOVE, light earthen-brown; beneath white; front, crown, and sides of the chin, black; stripe above the eye, rump, and upper tail-covers, pure white; the white stripes over the eye are very broad behind, and narrow in front, and almost meet at the nape; wings with two broad cream-white bands. Length, 7″; wing, 4″ 2‴.

" Inhabits South Africa."—Swain., loc. cit., non vidi.

Genus VIDUA, Cuvier.

Bill more or less lengthened, conic, laterally compressed, with the culmen much arched, or nearly straight from the base to the tip, and advancing on the forehead in a point; the lateral margins sinuated or straight, and the gonys lengthened and ascending; the nostrils basal, lateral, mostly hidden by the frontal feathers, and rounded; wings moderate, with the first quill spurious, the second nearly as long as the third, the third, fourth, and fifth nearly equal, and longest; tail varying in length; with some of the coverts, and tail-feathers considerably lengthened, and of various forms; tarsi slender, shorter than the middle toe, covered in front with conspicuously-divided scales; toes long and slender, the lateral ones nearly equal, the hind toe as long as the inner, and all armed with long claws.

374. Vidua Regia. (L.) Hartlaub, Orn. W. Af., p. 136; Cuv., Vol. 2, p. 149; Shaw., Vol. IX, Pl. 2, p. 426; *Shaft-tailed Bunting; Emberiza Regia*, L.; *Vidua Riparia Africana*, Briss., Orn., Vol. 3, p. 129.

SIDES of head, under parts of body, and round the neck, rufous; hind part of neck spotted with black; plumage above, lower part of thighs, and vent, black; four middle tail-feathers nearly 10" long, and webbed only for about two inches at the end, the rest simple shafts; other feathers, even, short, and black; bill and legs red. Female brown, and without the long tail-feathers. Length, 5"; tail (without long feathers), 2"; wing, 2" 10"'.

South Africa teste Hartlaub, loc. cit.—I have received it from Damaraland only.

375. Vidua Principalis. (Linn.) Cuv., Vol. IX; Pl. 2, p. 423; *Emberiza Principalis*, Vieil., Ois Chant., t. 36; *Em. Vidua*, L.; *Em. Serena*, L.; *V. Minor et Major*, Briss.; *V. Angolensis*, Briss., Orn. App., p. 80; *V. Erythrorhyncha*, Swain., B. of W. Af, Vol. XI, p. 176; *Koning-Roodebec* of the Colonists, lit. King of the Red-bills.

♀—General colour, deep glossy black: this colour covers the crown and the back; collar pure white; wings and scapulars the same; lower parts of back and rump, white; on the wings a large patch of white; ears, side of head, and under parts, pure white; but the black colour of the back advances on to the sides of breast, so as to form a half collar, open in front; tail-feathers externally black, internally white, this latter colour predominantly on the outermost feathers; four middle tail-feathers greatly elongated, two convex and two (one within the other) concave. Bill, brilliant rose-red. Length, 10"; wing, 2" 9"'; tail, 8".

These birds are found in small flocks throughout the colony. The male only assumes the long tail-feathers during the breeding season. In the winter he closely resembles the female, who is clad in a sober livery of mottled brown, and black on the upper parts, whitish on the centre of the belly, and rusty on the flanks; on the head are six dark black-brown stripes, divided by brown stripes, those which pass through the eye immaculate, the others dotted with dark-brown.

They feed on grass seeds, and are only to be found in open grassy country; they rarely perch on trees, though a pair in my aviary always roost on the perches, the ♀ having taken possession of the highest in the whole cage.

The ♀ amuses me by the way in which she scratches on the sandy cage-bottom. How she does it I cannot tell, her motions are so quick; but she apparently springs forward and spurs backwards, and the sand and seed fly in showers at each repetition of the manœuvre. I have frequently observed the hovering motion described by Mr. Ayres (Ibis., Vol. 2, p. 212), but attributed it to a design of the ♂ to take care of his long tail.

Swainson's type specimen of *V. Erythrorhyncha* is in the Cambridge Museum, and is certainly nothing more than than ♂ of our *Principalis*; the bill is now faded to the usual yellow assumed by all old specimens.

376. Vidua Axillaris, Sm.; Z. S. A., Pl. 17.

GENERAL colour, velvet-black; shoulders reddish-orange; feathers of wings and tail black, with a greenish gloss, narrowly edged with pale-brown. Length, 6"; wing, 3"'; tail, 2" 6"'.

Dr. Smith procured but one specimen of this bird, the male here described. When shot it was perched upon some bushes growing out of some marshy ground in Kafirland. Mr. Gurney has received it from Natal (Vide Ibis., Vol. 1860, p. 205).

377. Vidua Albonotata, Cassin.; Proc. Acad. Philad. 1841, p. 65; *Urobrachya Albonotata*, Hartl., Orn. W. Af., p. 137.

GENERAL colour, shining black; shoulders yellow; base of the tail-feathers, white; wing-feathers tipped with white. Length, 6" 6"'; wings, 3".

Natal (Verreaux)—non vidi.

378. Vidua Ardens. (Bodd.) Pl. Enl., 647 ; *Emberiza Panayensis*, Gmel. ; *E. Signata*, Scop., Sonn. Voy., t. 75 ; *Vidua Lenocinia*, Less., Tr. d' Orn., p. 437 ; *V. Torquata*, Less. ; *V. Rubritorques*, Swain., West. Af., XI, p. 174 ; *Pentheria Rubritorques*, Bp. Consp., 1, p. 448.

MALE, in full breeding plumage, generally velvet-black, with a broad half collar beneath of brilliant scarlet; wings, thighs, under tail-coverts, and rump striped with grey, that colour being on the edges of the feathers; tail elongated. Total length, 10″ 6‴ ; tail, 7″ ; wing, 3″ ; tarsus, 11‴.

Inhabits the country beyond the Kei (from whence it has been sent by Mr. H. Bowker) and Natal. It does not appear within the colony.

Genus CHERA, J. R. Gray.

Characters in common with *Vidua ;* but the wings long, with the first quill spurious, the second, third, and fourth slightly graduated, and the fifth and sixth nearly equal, but the fifth the longest; tail and its coverts much lengthened, compressed, and arched.

379. Chera Progne. (Bodd.) Pl. Enl., 635 ; *Orange-shouldered Bunting*, or *Kaffrarian Grosbeak ;* Cuv., Vol. 2, p. 150 ; *Vidua Phœnicoptera*, Swain. ; *Loxia Caffra,* Gmel. ; Shaw, Vol. 24, Pl. 1, p. 17 ; Shaw, Vol. IX., Pt. 2, p. 286 ; *Emberiza Longicauda*, Gmel. ; Shaw, Vol. IX., Pt. 2, p. 422.

♂—General colour, glossy-black ; shoulders fulvous and brilliant crimson; tail enormously developed. ♀, or young ♂ : general colour, pale yellowish-brown ; the centre of each feather being darker, approaching in some instances to black, give the bird a mottled appearance; wing-feathers black, with pale, yellowish-brown edges; tail-feathers umber-brown, with light edges; shoulders showing a bright orange patch. Length, 21″ ; wing, 6″ ; tail, 16″.

I have only procured this species from the Eastern side of the colony. It is said to be very common in Kaffraria ; and Mr. Arnot has sent several specimens (the ♀ or young ♂ here described, among the number) from Colesberg.

I am informed that in the breeding season, when the male has assumed his nuptial livery and long tail-feathers, his flight is so laboured that the children constantly run them down. They are quite unable to fly against the wind, and in rainy weather can hardly be got to move out of the thick bushes in which, knowing their helplessness, they conceal themselves. The Kafir children stretch bird-limed lines across the fields of millet and Kafir corn, and snare great numbers of the males by their tails becoming entangled in the lines. I am informed they breed among rushes and reeds, like *Ploceus Oryx*.

The Sub-Family, COCCOTHRAUSTINÆ, or Hawfinches,

have the bill large, short, strong, conic, and very broad at the base; the culmen and gonys more or less equally curved at the tip; the lateral margins more or less angulated at their base; the wings lengthened, and more or less pointed; the tail generally short; the tarsi as long as, or shorter than, the middle toe, robust, and strongly scaled, and the hind toe as long as, or rather shorter than, the inner one, and strong.

Genus PYRENESTES, Swainson.

Bill enormous, perfectly conic; the two mandibles equal, or the lower somewhat thicker; upper mandible with an obsolete tooth at its base; tip entire; commissure straight; wings and tail rounded, first quill very small, spurious.

380. Pyrenestes Frontalis; Zool. S. A., Pl. 61, 62; *Pyrrhula Albifrons*, Vigors.

GENERAL colour, orange-coloured brown, with the feathers on each side of the upper mandible white, and a speculum of the same colour on the wings, formed by all the quill-feathers being white at their base. Length, 7″ 4‴; wing, 3″ 9‴; tail, 3″.

Inhabits the forests on the Eastern frontier and Natal, feeding upon berries and small fruits.—Dr. A. Smith, loc. cit.

The Sub-Family, FRINGILLINÆ, or Finches,

have the bill short, conic, and more or less broad at the base, with the culmen sloping, and the sides compressed to the tip, which is generally entire; the gonys long, and advancing upwards; the wings more or less lengthened, and pointed; the tail varying in length, sometimes with the ends of the feathers acuminated; the tarsi as long as, or shorter than, the middle toe, slender, and transversely scaled; the toes more or less long and slender; the hind toe moderate; the claws generally long, curved, and acute.

Genus ESTRELDA, Swainson.

Bill small, moderate; tail lengthened, graduated, or rounded; feet moderate; tarsus longer than hind toe.

381. Estrelda Astrild. (Linn.) Pl. En., p. 157,

f. 2 ; Cuv., Vol. 2, Pl. 153 ; *Fringilla Undulata*, Pall. ; *Loxia Astrilda*, Shaw, Vol. IX, Part 2, p. 254 ; Edw. Birds, Pl. 354, f. 2 ; *F. Tricolor*, Vieil., Ois. Chant., t. 12 ; *Senegalus Striatus*, Briss. ; *Wax-billed Grosbeak* ; *Roodebec* of Colonists.

UPPER parts, brown, minutely waved with darker brown lines ; under parts, lighter brown, also waved with brown lines ; over the whole of these parts, in certain lights, there is spread a lovely rose colour, changing into purple ; bill, stripe through the eye, and down the centre of the breast, and stomach, brilliant scarlet ; vent, intense black ; tail graduated, brown ; legs black. Length, 4″ 9‴ ; wing, 2″ ; tail, 2″ 2‴.

These elegant little birds congregate in prodigious flocks throughout the colony, frequenting the fields of grain, ploughed lands, and vineyards. They are not unfrequently close to habitations, and their little shrill, piping notes may be heard even in towns.

They breed in communities, and it is said that several inhabit the same nest, laying and sitting in it promiscuously, and sometimes three or four together. The nest is a large structure, composed of straw, grasses, feathers, wool, paper, rags, &c. It is often as large as a stable bucket, round, and with an entrance in the side. The interior is a mass of feathers ; and the eggs, from 8 to 14 in number, are pure white, oval, and about 7‴ long, by 4‴ broad.

These birds live well in confinement, and add much to the appearance of an aviary, by their sprightliness, and the beauty of their plumage when many are together. It is no uncommon thing to see a dozen or more on one perch, sitting close up to each other, for company's sake. I have four which have separated themselves into two pairs, though agreeing well all together. The birds of each pair are always together, cleaning and caressing each other. When I put in the bathing vessels, they crowd into them in defiance of all the bigger and stronger birds ; and it is no unusual thing to see two canaries, or one canary, the " king of the roodebecs," and three or four " roodebecs," all in the vessel at once, splashing, ducking, and jostling one another in the narrow bounds to which their cage restricts them.

Mr. Ayres, of Natal, states (Ibis Vol. 4, p. 156,) that he has found several nests of these birds built upon the ground, generally in some convenient indentation, sheltered by a clump of grass.

382. Estrelda Nitidula ; Hartl., Ibis, Vol. 1865,

p. 269.

GENERAL colour, olive-green ; rump and upper tail-coverts, yellowish-green ; an orange mark in front of the eye ; body spotted with white dots, surrounded with black. Length, 3″ 2‴ ; wing, 1″ 9‴ ; tail, 1″ 2‴.

Natal.--(Ayres.)

To prevent misconception, it must be noted that there is an error in the paging. The text proceeds regularly.

–

383. Estrelda Rubricata. (Licht.) Vieil., Ois.
Chant., t. 9 ; *E. Rubricata et Rufo-pectus*, Gr. ; Bp. Cons. Av., p. 459.

ABOVE, olive; rump purple; beneath purple; vent black; bill black. Length, 4".

Natal. (Ayres.)—Ibis, 1864, p. 352.

384. Estrelda Melanogenys, Sundev.; Oivers.
K. Vet. Ak., Forhandl.

BACK, olivaceous, slightly undulated ; head cinereous ; rump red ; throat, cheeks, and tail, black ; breast cinereous ; belly dull whitish-yellow ; upper mandible black ; lower red.

Inhabits Port Natal.—Sund., loc. cit.

385. Estrelda Incana, Sundevall ; Ofvers. K. Ak.
Forhandl., 1850, p. 98 ; *E. Natalensis*, Caban. ; Hartl., Orn. W. Af., p. 143.

ASH-COLOURED, with red rump; throat whitish ; chin, stripe between eye, and bill, black ; tail black, graduated ; vent dark-fuscous ; bill and legs black.

Inhabits South Africa.—(Wahlberg)—sed non vidi.

386. Estrelda Dufresnii. (Vieil.) Hartl. Orn.
W. Af., p. 142 ; *Fringilla Melanotis*, Temm., Pl. Col. 221, f. 1 ; *Fringilla Neisna*, Licht.

♂—Top of head and back of neck, ashy ; back and shoulders, green ; wings rufous ; rump red ; tail black ; ears, chin, and upper part of throat, deep black ; lower part of the latter, white ; chest ashy-white ; belly and vent yellowish-white ; bill, upper mandible, black, lower scarlet. ♀ wants the black on the head and throat, and all her colours are less vivid. Length, 3" 9''' ; wing, 1" 9''' ; tail, 1" 8'''.

Swellendam and Knysna are the only places whence I have received this pretty species. Mr. Ayres has found it in Natal ; and Hartlaub quotes it from Kafirland.

387. Estrelda Minima ; *Lagonosticla Minima,*
Vieil., Ois. Chant., t. 10.

PALE-RED ; the lower part of the back pale-brown, with a ruddy tint ; wings pale-brown ; scapulars and lesser wing-

Y

coverts ruddy ; rump more red ; vent pale-brown ; tail-feathers
fuscous ; the webs near their base ruddy.

South Africa.—Hartlaub.

388. Estrelda Margaritata, Strick. ; Ann. and Mag., Nat. His., Vol. 13, 1844, Pl. 10 ; *E. Verreauxii*, Massena ; Hartlaub, Orn. W. Af., p. 143 ; P. des Murs., Pl., p. 64.

UPPER parts, rich ferruginous-brown, except the quills, which
are dusky within ; upper tail-coverts and outer margins of
retices dull vinous-red, and their inner webs and apical
portions black; the circuit of the eyes, cheeks, throat, and
breast, pale claret-red ; rest of lower parts deep black, spotted
next the chest and on the sides with large pearl-like spots,
the colour of peach-blossom, of which two are placed transversely and subterminally on each feather. Length, 4" 9''' ;
wing, 2" 1''' ; tail, 2".

Hartlaub says from South Africa. Mr. Strickland, whose description I copy, says his specimen "was purchased at Cape Town, and
was said to have been brought from Madagascar."

389. Estrelda Erythronota. (Vieil.) Ois. Chant., t. 14 ; *Estrelda Lipiniani*, Smith, Append. to Report of Exped., p. 49.

TOP of head, neck, and interscapulars, tinted with crimson ;
back, rump, and upper tail-coverts, crimson ; hinder part of
breast, and belly, dull crimson ; sides of head, chin, under
tail-coverts, and tail, black ; quills, ashy brown ; wing-
coverts and scapulars, grey, closely banded with brown.
Length, 4" 6'''.

Inhabits the country to the north and east of Kurrichane.—Dr.
Smith, loc. cit.

390. Estrelda Polysona ; *Ortygospiza Polyzona*, Sund. ; *Fringilla Atricollis*, Vieil., Ency., p. 980 ; *Frin. Polyzona*, Temm., Pl., Col. 221, f. 3.

ABOVE, fuscous-grey ; below, banded black and white ; chin
and eyebrows white ; throat, belly, and front, black ; breast
rufous ; vent white ; tail tipped with white ; bill red.
Length, 3" 6''' ; wing, 1''' ; tail, 1" 9'''.

Kafirland (Wahlberg).—Hartlaub, Orn. W. Af., p. 148.

391. Estrelda Benghala. (Linn.) Pl., Enl. 115,

f. 1 ; *Uraeginthus Phœnicotis*, Swain., Nat. Lib., Vol. 11, p. 192 ; *Fringilla Benghala*, L., S. N., Vol. 1, p. 323 ; *F. Angolensis*, L., Bonap. Cons., Vol. 1, p. 458.

ABOVE, grey-brown ; rump, tail, and under parts, with the exception of the centre of the body, which is white, verditer-blue; the ears of the ♂ are covered with a deep purplish-red spot, which is wanting in the female. Length, 4″ 6‴ ; wing, 1″ 11‴ ; tail, 1″ 10‴.

Dr. Smith procured this species between Kurrichane and the Tropic of Capricorn. Mr. Andersson and others have brought it from Damaraland ; and I saw flocks of them feeding in the fields of ground-nuts as far as 1° 30″ south.

392. Estrelda Melba. (Linn.) Edw. Birds, Pl.

278, f. 2, p. 273 ; *Tringilla Elegans.* ·

♂ —Forehead, chin, and throat, scarlet; back of head and neck ashy ; back and chest golden-green, the latter spotted at the junction of the belly with white ; vent white; rump and centre tail-feathers, dull-scarlet ; outer tail-feathers black-brown. ♂ —Ashy, barred on the hinder parts with white ; vent white ; rump dull-scarlet. Length, 4″ 9‴ ; wing, 2½″ ; tail, 2″.

Inhabits South Africa.—Hartlaub, Orn. W. Af., p. 145.

393. Estrelda Squamifrons, Sm., Z. S. A., Pl.

95 ; *Amadina Squamifrons*, Smith.

GENERAL colour above, light liver-brown ; feathers of wings and tail, darker, edged with white ; tip of head and moustache, black; the feathers of the first being edged with white, giving a scaley appearance ; under parts dusky white. Length, 2″ 6‴ ; wing, 2″ 5‴ ; tail, 1″ 9‴.

Dr. Smith states that this species is frequently found to the northward of Latakoo, but rarely to the southward. I have received it from Kuruman ; and it has appeared in all the collections I have examined from Damaraland.

394. Estrelda Granatina, Linn. ; Pl., Enl. p. 109,

f. 3 ; *Uraeginthus Granatina* ; *Fringilla Granatina*, L., Briss., Orn., Vol. 3, p. 216 ; Vieil., Ois. Chant., t. 17, 18 ; Edw. Birds, Pl. 191.

CHESNUT ; cheeks, a well-defined and beautiful violet ; front and rump, blue; throat, vent, and tail, black ; bill red. Length, 6″ ; wings, 2″ 2‴ ; tail, 3″.

Found by the expedition under Dr. Smith, between the Kiegariep and Kurrichaue, and by Mr. Andersson in Damaraland.

395. Estrelda Sabflava. (Vieil.) N. Dict. d'Hist.

Nat. XXX, p. 575 ; *Sporaeginthus Subflava*, Hart. Orn. W. Af., p, 144; *Fringilla Subflava*, Vieil.; *F. Sanguinolenta*, Temm., Col. 221, fig. 2 ; *Estrelda Sanguinolenta*, Swain., Nat. Lib., Vol. II, p. 190.

ABOVE, light-brown; beneath, varied with yellow and orange ; flanks with transverse brown lines ; eyebrows and rump, crimson; tail blackish; two lateral feathers on each side edged with white. Length, 3″ 3‴ ; wing, 1″ 8‴.

South Africa.—Hartl., loc. cit.

Genus AMADINA, Swainson.

Bill short, very thick, and broad at the base ; tail short, rounded, or even ; tarsus longer than hallux.

396. Amadina Alario. (Lin.) Pl. Enl. 204;

Crithologus Alario ; *Crithagra Bistrigata*, Swain., 2 Cent., p. 318, adult. ; *C. Ruficauda*, Swain., 2 Cent., p. 317, juv. ; *Spermatophila Daubentoni*, Gray, Gen. B. ; *Fringilla Personata*, Licht. Bp. Consp. ; *Berg-canarie* of Colonists, lit. Mountain Canary.

♂—Upper parts, rufous; under parts, white ; tail rufous ; head, neck, and throat, black : this colour extends from the throat, and divides into two broad lines on each side of the breast; quill-feathers of wing, black ; shoulders and coverts rufous, a white colour extends from the back of the throat round to the back of the neck ; bill black. The ♀ is generally of a dull brown colour, lightest on the under parts. Length, 5″ ; wing, 2″ 8‴; tail, 2″ 2‴.

I have received this species from the Knysna, Malmesbury, Swellendam, and Colesberg. It is not uncommon in the neighbourhood of Nel's Poort (Beaufort), ranging in small flocks among the thornbushes, and feeding on grass seeds. It often intermingles with flocks of the common *A. Astrild*. Its song is very sweet and sustained, for which reason it is kept in confinement, which it endures very well. It seems to vary somewhat in the distribution of its colours : but that described above is the general character.

397. Amadina Erythrocephala, Hart., Orn.

W. Af., Pl. 146 ; *Loxia Erythrocephala*, Lin. ; Smith, Z. S. Af., Pl. 69 ; *Cardinalis Angolensis*, Briss. ; Orn. App., p. 78 ; *Fringilla Reticulata*, Vogt. ; *Loxia Maculosa*, Burch. ; *Sporothlastes Erythrocephalus*, Cab.

ABOVE, brown ; beneath, grey-brown, with a tinge of red, and barred with dark-brown, which imparts to it a scaled appearance; top and sides of head, and chin, pink; tail-feathers brown, all, with the exception of the two centre ones, tipped with white. The female resembles the male, but wants the crimson head. Length, 5″ 6‴; wing, 3″; tail, 2″ 3‴.

Dr. A. Smith found this species only in the neighbourhood of Latakoo, feeding in the native gardens, and congregated together in considerable numbers. I have received it from Kuruman and Damaraland.

398. Amadina Nitens. (Linn.) Pl. Enl. 291, f.

1, 2; *Hypochera Ultramarina; Frin. Ultramarina,* Gmel.; Vieil., Ois. Chant., t. 21; *F. Funerea,* De Turr.; Edw. Birds, pl. 362, f. 1.

SHINING blueish-black; wings and tail rather fuscous, with pale edges; under wing white; bill red. Length, 4″ 6‴; wing, 2″ 4‴; tail, 1″ 2‴.

Natal, &c.—Hartl., Orn. W. Af., p. 149. (Ibis Vol. 2, p. 205, &c.)

Genus FRINGILLA, Linn.

Bill short, conical, and broad at the base, with the culmen rounded, sloping, and the sides compressed to the tip, which is acute and entire, the lateral margins straight; the gonys long and ascending; the nostrils basal, lateral, and sunk in a small groove, with the opening generally concealed by the projecting small plumes; wings mostly long and pointed, with the second and third quills equal, and rather longer than the first; tail more or less long, and generally slightly forked at the end; tarsi shorter than the middle toe, and covered in front with broad scales; toes moderate and slender, the lateral toes nearly equal, the outer toe united at its base, the hind toe long, and armed with a long curved claw.

399. Fringilla Canicollis; *Serinus Canicollis,* Sw.; *Cape Canary.*

FOREHEAD, back, and under parts, greenish-yellow, mottled with brown on the two first; nuchal collar and flanks grey; vent albescent; wings and tail dark-brown; the outer webs narrowly margined with yellow; the inner webs of the latter broadly margined with brown-yellow. Length, 5″ 3‴; wing, 3‴; tail, 2″ 2‴.

The Cape canary is a common bird throughout the colony, congregating in flocks on the open and ploughed lands, and feeding on grains and seeds of all kinds. It sings very sweetly, and breeds in captivity with the tame canary, the mule bird being very handsome—and, I am told, again producing with either the yellow bird or one of its own species. I propose trying this experiment. The nest of the Cape canary is generally constructed in a low thick bush, and is of a cup-shape, formed of hair, moss, and feathers. The eggs, three to four in number, are white, streaked and spotted at the obtuse end with purplish-brown : axis, 9′′′; diam., 6½′′′.

400. Fringilla Capensis ; *Citrinella Capensis*
Less ; Sup. Aux. Œuv. de Buf, Vol. XX., p. 335.

HEAD, olive-green ; back and wing-covers, reddish-brown ; rump yellowish-green ; and greenish-yellow on the chin, throat, chest, and belly ; the lesser coverts are grey ; wing-feathers black, each feather terminated by an oblong white spot ; tail-feathers black, with a white border at the tip, the border wider on the lateral feathers, beneath lighter ; bill and legs reddish.

Quoted by Lesson as from the Cape of Good Hope. I cannot identify it ; but sometimes fancy it may be our " Pietje Canarie," the next species.

401. Fringilla Totta ; *Citrinella Totta*, Bp. ;
Loxia Totta, Gmel. ; *Pietje Canarie* of Dutch Colonists.

HEAD, greenish-brown ; rump, yellowish-green, mottled with brown ; eyebrows yellow ; under parts, greenish-yellow, the yellow predominating on the chin and down the centre of the belly ; wing and tail feathers blackish-brown, tipped with white ; tail-coverts grey. Length, 5′′ 3′′′ ; wing, 2′′ 9′′′ ; tail, 2′′ 2′′′.

This canary is found in similar situations to the Cape canary. It is said that the cock will breed with the hen yellow-canary in captivity ; but that the hen pietje-canary will not breed with the male of the yellow species. This may be from an unwillingness of the wild hen to build in a cage.

402. Fringilla Gularis ; *Linaria Gularis*, Sm.,
Append. to Rept. of Exped., p. 49.

TOP of head, dark-brown, feathers edged with white ; eyebrows white ; neck and upper parts, grey-brown, clouded with darker tints ; chin, pure white ; breast, pale grey-brown ; belly tawny-white, darkest on the sides ; quills and tail brown, the feathers edged with greenish-white. Length, 5′′ 6′′′.

Inhabits the colony, and country as far as Latakoo. Rare.—Dr. Smith, loc. cit.

M. Victorin procured this species at the Knysna. See J. W. Grill. Zool. Antic. Af.—Victorin.

403. Fringilla Striaticeps; *Poliospiza Striaticeps*, N. Sp., Hart.

GENERAL colour above, dark reddish-brown, the centre of the feathers being darkest; head variegated with white streaks, which, coalescing over the eye, form themselves into an eyebrow; chin and throat white, mottled on the former with brown; under parts ruddy-brown, the colour clearest on the belly and vent. Length, 5" 9'''; wing, 2" 9'''; tail, 2¼".

Received from Mr. Atmore, Swellendam; and found also in Damaraland by Mr. Andersson.

404. Fringilla Angolensis, Hartl., Orn. W. Af., p. 150 ; *Linaria Atrogularis*, Smith, Append. to Rept. of Expd., p. 49 ; *L. Angolensis*, Briss., Orn. App., p. 71 ; *Frin. Angolensis*, Gmel. ; *Frin. Tobaca*, Vieil. ; *Fringil. Uropygialis*, Licht., Bp. Consp., Vol. 1. p. 520.

ABOVE, grey-brown, the centre of the feathers dark-brown ; rump sulphur-yellow; chin and throat black ; breast, belly, and vent, rusty-white; wings brown, the tips of the quills and the inner edges of inner vanes, towards bases, white ; the outer edges golden-green ; tail slightly forked, dark-brown, with the tips, and more or less of the inner vanes, white. Length, 4½".

Inhabits the country about and beyond Kurrichane.—Dr. A. Smith, loc. cit. Kafirland, Krebs.—Hartl., loc. cit.

405. Fringilla Nœvia, Gmel. ; Sys. Nat. 1, p. 11 ; *White-cheeked Finch*, Shaw, Vol. IX, Pt. 2, p. 496 ; Lath. Gen. Syn. 3, 278, 44.

HEAD and neck, pale cinereous ; upper parts of latter marked with dusky streaks; sides of head white, with a reddish streak passing through the eye, which is bounded on the under part with black, and joins another black streak arising at the base of the under mandible ; back and wings, pale rufous, with dusky streaks ; under parts of body, plain cinereous ; tail dusky. Length, 9".

"Native of Cape of Good Hope."—Shaw, loc. cit.

This species has not, that I can learn, been recognised by any subsequent author. It is uncertain what bird is meant.

Genus PASSER, Brisson.

Bill strong, sub-conical, and broad at the base, with the culmen rounded, slightly arched, and the sides compressed to the tip, which is slightly emarginated; the lateral margins straight; the gonys long, and curved upwards; the nostrils basal, lateral, with the opening partly covered by the frontal plumes; wings moderate, with the second and third quills rather longer than the first; tail moderate and even, or slightly forked; tarsi strong, nearly the length of the middle toe, and covered with transverse scales; toes moderate, the lateral toes nearly equal; the claws moderate and acute.

406. Passer Motitensis, Smith; Zool. S. Af., Pl. 114.

Top of head, and back of neck, grey; eyebrows, sides of neck, and back, chesnut; the latter dashed with dark-brown; streak through the eye, chin, and throat, black; cheeks, breast, and belly, rusty-white; wing and tail feathers, light brownish-red. Length, 7"; wing, 3" 5'''; tail, 3".

Dr. Smith only procured two specimens of this fine sparrow, "both about sixty miles north of the Orange River. When they were killed they were on the ground actively employed picking up seeds which had fallen from the shrubs among which they were moving."

407. Passer Diffusus; *Pyrgita Diffusa*, Smith; Appendix to Report of Exped., p. 50.

HEAD and neck, dull rusty-grey; interscapulars brown, washed with grey; wing-coverts, back, and rump, dark chesnut-brown, some of the coverts tipped with white, which form one or two oblique bands on the wings; chin, throat, breast, and flanks, pale grey-brown; belly and vent, rusty-white; quills and tail, dark-brown. Length, 5" 9'''.

"Inhabits the country between the Orange River and the Tropic."— Dr. Smith, loc. cit.

408. Passer Arcuatus, Gmel.; Pl. En., 230, f. 1; Cuv., Vol. 2, p. 135; *Cape Sparrow.*

♂—Top of head, cheeks, chin, throat, and breast, black; back of head and neck, brown; back, rump, and shoulders, rufous; wing and tail feathers brown, with light margins; a white stripe extends from over the eye, round the back of the cheeks, and nearly unites on the throat; under parts, dirty white. ♀ less brightly coloured. Length, 6" 6'''; wing, 3" 4'''; tail, 2" 7'''.

This Cape sparrow, like its cousin, the English bird, is essentially a "cit." In the country you certainly find him, but never away from human habitations. He seems to think man only builds houses for him to dwell in; only grows corn for him to eat; only plants trees for him to roost in. The airs he gives himself are amusing; and you feel inclined to forgive his peculations, out of sheer admiration for the boldness with which he executes the theft.

With the earliest dawn he is up and doing, and his chirrup arouses you from your slumbers; but as he has not got to dress, and you have, he is off to inspect your farm-produce before you are. As he has wings, he visits all your property (not to count your neighbours'), and levies toll where he likes; and you find him in the evening, when you reach home, tired and footsore, there before you, and with unabated vigour fighting for the snuggest and warmest berth under the eaves, or the cosiest branch upon your pet oak-tree. Well, don't be hard on him! He will in his season rid you of thousands of caterpillars and grubs; and if your " eldest hope " is old enough to begin to shoot, he will do no great harm in thinning their numbers in the autumn, and manufacturing puddings for his brothers and sisters with the bodies of the slain.

Sparrows build in holes, in walls, or in trees, indiscriminately. If they select the former, they accumulate a lot of sticks as a ground-work, and fill up with straw and feathers. If they build in trees, they construct a large ball of straw, and line it with feathers. Their eggs, three to five in number, are light verditer, with brown blotches; but they vary much in shape and colour.

409. Passer Simplex, Swain., Nat. Lib., Vol. 11, p. 208; *P. Gularis*, Less.; Rev. Zool. 1839, p. 45; *P. Swainsonii*, Rupp. Faun., t. 33, f. 2; *P. Grisea*, Lafren.; *Pyrgita Gularis*, Less.; *P. Spadicea*, Licht., Bp. Consp., Vol. 1, p. 510.

HEAD and neck, grey; plumage above, rufous; beneath whitish; wings and tail, brown; stripe under chin, white. Length, 6"; wing, 3"; tail, 2".

Hartlaub (Orn., W. Af., p. 150) gives this species as from South Africa. I have only seen it from Damaraland.

The Sub-Family, EMBERIZINÆ, or Buntings,

have the bill conical, acute, compressed, with the culmen more or less straight to the tip; the lateral margins sinuated, and more or less inflected; the interior of the upper mandible furnished with a palatine knob, which is more or less apparent; the wings moderate, and somewhat pointed; the tarsi about the length of the middle toe, and scutellated; the hind toe longer than the inner, and robust; and the claws slender, and more or less curved.

Genus FRINGILLARIA, Swainson.

Bill with upper mandible entire; wings short; the primary quills not much longer than the tertials; tail moderate, either

z

quite-even or very slightly rounded; lateral toes small, equal; all the claws small and short.

410. Fringillaria Flaviventris, Vieil.; Ency.
Meth., p. 929; *Emberiza Capensis*, Gml., var. β, Pl., Enl., 664, f. 2; *E. Xanthogaster*, Steph.; *E. Flavigastra*, Rupp. Atlas., t. 25; *E. Quinquevittata*, Licht.; *Frin. Bicincta*, Forst; *Frin. Capensis*, Swain., Nat. Lib., Vol. 11, Pl. 18, p. 211.

HEAD black, with seven white stripes; plumage above varied with rufous and grey; beneath bright-yellow; chin, vent, and tips of lateral tail-feathers, pure white. Length, 5″ 6‴; wing, 2″ 8‴.

"Natal,"—Sir W. Jardine and Mr. Gurney (Ibis Vol. 1); "Damaraland,"—Andersson, Chapman, and Kisch.

411. Fringillaria Capensis. (L.) Hartl., Orn., W. Af., p. 152; *Emberiza Capensis*, Linn., Pl. Enl. 158, f. 2; *E. Erythroptera*, Temm., Pl., Enl. 664, f. 1; *Cape Bunting*, Shaw, Vol. 9, Pt. 2, p. 372; *Em. Caffrariensis*, Steph.; *Passerina Capensis*, Vieil.

TOP of head and neck, dirty-grey, variegated with black; sides of head and chin, dirty-white, marked with two black streaks, one passing through the eye, the other beneath; under parts, dirty yellowish-white; lesser wing-covers rufous; greater covers, quills, and tail dusky, with the margins rufous; iris hazel. Length, 5½″; wing, 2″ 10‴; tail, 2¼″.

Mr. Atmore procured this pretty species at Meiring's Poort, where he met with it "in small flocks of four or five individuals," probably family parties. With this exception, I have not seen it from any other place in the colony. Mr. Andersson has it from Damaraland.

412. Fringillaria Impetuani, Smith, Append.
to Rept. of Expd., p. 48; *Fringillaria Anthoides*, Swains. An. in Menag., p. 316.

COLOURED like a lark; above isabella, striped with brown; beneath paler and immaculate; above the eye, and beneath the ears, a pale stripe; outer edge of the secondaries, and inner edge of all the quills, rufous; a dark stripe behind the eye; throat inclining to dusky-white; belly and vent almost white; wings and tail brown. Length, 5¾″; wing, 2″ 8½‴; tail, 2″.

"Inhabits South Africa."—Swains. loc. cit.; also Hartl., Orn. W. Af., p. 152.

413. Fringillaria Vittata, Swain. ; Ad. in Menag., p. 315 ; *Strœpkopje* of Colonists, lit. Striped Head.

ABOVE, grey, striped with black ; beneath, cinereous-grey ; sides of the head with two white and two black stripes ; wing-covers rufous ; quills and tail, blackish ; chin, and under tail-covers, whitish. Length, 6″ ; wing, 2″ 9‴ ; tail, 2″ 6‴.

This little bird has very much the habits of the European " Hedge-Sparrow." It frequents low bushes and rocks, creeping about the former, and running up the latter, and sitting on their summits, opening and shutting its wings, and uttering a little short, piping note. It breeds about bushes and rocks, placing its nest in crevices or among roots, or even on the ground. The eggs, three to five in number, are of a dirty-white ground, profusely speckled with minute rufous and light purple markings : axis, 10‴ ; diam., 6‴.

It is universally distributed, and feeds on seeds and insects.

414. Fringillaria (?) Africana; *Zonotrichia Africana,* Smith, Append. to Rept. of Exped., p. 48.

ABOVE, black-brown; feathers broadly margined with rufous ; chin white ; throat and breast, pale rufous, marked with stripes of dark-brown ; middle of belly white ; quills dark-brown ; outer edges towards base, dark-chesnut ; inner edges pale rufous ; tail rounded, the two outermost feathers of each side white, the second with a brown stripe in the course of the shaft near the point, the intermediate ones dark-brown, the two middle ones edged with light rufous. Length, 5″.

" Lives amongst the grass, both to the north and south of Kurrichane."—Dr. A. Smith, loc. cit.

I have failed to identify this species, and place it here with a mark of doubt. The genus in which Dr. Smith placed it is an American one.

415. Fringillaria Tahapisi. (S.) *Emberiza Tahapisi,* Smith, Append. to Rept. of Exped., p. 48.

HEAD and throat black, the former with seven longitudinal white stripes, one from base of mandible to centre of nape, and three on each side ; body above light-brown, dashed with dark-brown, beneath light cinnamon-brown ; tail dark-brown, the outermost feather on each side tipped and edged with pale rufous. Length, 5″ 3‴.

Inhabits the country towards the sources of the Vaal River.—Dr. A. Smith, loc. cit.

The Sub-Family, ALAUDINÆ, or Lark,

have the bill short, more or less lengthened, conical; the wings with the tertials generally as long as the primaries; the claws lengthened, more or less curved, and the hind one very long, and generally straight.

Genus ALAUDA, Linn.

Bill moderate, conical, with the culmen slightly arched to the tip, and compressed on the sides; the nostrils basal, lateral, and hidden by the frontal plumes; wings lengthened, with or without a short spurious quill, and with the third quill generally the longest; tail moderate, and slightly emar-ginated; tarsi rather longer than the middle toe, and slender; toes moderate and divided, and the hind claw lengthened, and more or less straight.

416. Alauda Crassirostris, Vieil.; Nat. Dict.
D'Hist. Nat., 1, p. 373; *Brachonyx Crassirostris*, *L'Alouette à gros bec*, Le Vail., Pl. 193; Cuv., Vol. 2, p. 119; *Alauda Magnirostris*, Steph.; *Dubbelde Leeuwirk* of Colonists.

FEATHERS of upper surface thoroughout, dark burnt-umber-brown, margined with a lighter shade; outer tail-feather, on each side broadly margined; the inner less so; the neck and breast marked in the same manner, but the dark parts of each feather are narrow, and the margins broader, and with a decided yellow tinge, which prevails over the whole belly and chin. Length, 6½"; wing, 4" 3'''; tail, 3.'''

This fine lark is abundant throughout the colony, frequenting equally the cultivated and bush-covered lands. It feeds on insects and seeds; and makes a rough nest in a depression of the soil, under the shelter of a low bush, or large clod of earth, and lays four or five eggs, of a very pale cream colour, profusely dotted throughout with small light-brown and purple spots: axis, 11'''; diam., 8'''.

417. Alauda Ferruginea, Sm., Zool. S. Af., Pl. 29.
HEAD, back, and sides of neck, back, rump, shoulders, and two centre tail-feathers, bright ferruginous-red; under parts, dull white; chin, breast, throat, and flanks mottled with dark-brown spots; a blackish line commences at the base of the lower mandible, and terminates below the points of the ears; eyebrows, dirty-yellow. Length, 7" 5'''; wing, 4" 1'''; tail, 3" 2'''.

Habitat: Arid plains to the southward of the Orange River, called the Bushman Flats. Is very wary, shy, and difficult of approach.

Soars very frequently, particularly in the early part of the day, and at sun-rise whistles delightfully, generally at such times perching on the summit of the highest bush in the neighbourhood.—Dr. A. Smith, loc. cit.

418. Alauda Codea, Smith, Zool. S. Af., Pl. 87, f. 1.

UPPER surface of body, head, and neck, rusty yellowish-grey; the two latter and interscapulars variegated with streaks of brownish-red, washed with yellowish-grey; chin and throat, pure-white; breast, white, spotted with brownish red; belly rusty-white; flanks tinted with pale-brown; wing-feathers brownish-red, edged and tipped with white; two middle tail-feathers, reddish-brown, glossed with grey; the rest umber-brown, edged and tipped with rusty-white; eyebrows white; ear-covers, light orange-coloured brown; iris hazel. Length, 6″ 6‴; wing, 3″ 6‴; tail, 2″ 9‴.

Generally, though sparingly, scattered over the Western Province, particularly in the Karroo. Perch readily on bushes.—Dr. A. Smith, loc. cit.

Mr. C. A. Fairbridge procured a single specimen on the Cape Flats, near Rondebosch. I found it plentiful at Nel's Poort, in the Karroo. It first appeared in the neighbourhood of Beaufort. Nearer Cape Town we did not observe it. A single nest fell under our observation in December. The eggs, four in number, and of a mottled brown, were deposited in a cap-shaped nest, on the side of a low bush, at the edge of a foot-path. When we approached it the bird crept away to a little distance, then rose, flew a few yards, and perched on an ant-heap to watch us. As we were proceeding to kill a *Hyæna*, which had been caught in one of our traps, we deferred taking the eggs until our return; but coming home by a different route, we had to leave them till next morning, when on proceeding to the spot we found the young excluded. The hen bird, on this and several subsequent occasions, executed the same manœuvres to escape detection. "Common at Traka."—Mr. Atmore.

419. Alauda Tartarica, Pall.; *Alauda Mutabilis,* Forst.; *Tanagra Sibirica,* Mus. Carls., t. 19; *Le Tracal,* Le Vail., No. 191; Cuv., Vol. 2, p. 118.

THIS lark cannot be mistaken for any other of the family. Its entire plumage is black, transversely and irregularly striated above and below with white.

Le Vaillant states he procured it breeding in Great Namaqualand, on arid plains, at the foot of craggy mountains. This is another of his inventions. The bird is a native of Tartary.

Genus PYRRHULAUDA, A. Smith.

Bill short, with the culmen arched, the sides much compressed, the tip entire, and the lateral margins sinuated; the nostrils basal, and concealed by the frontal plumes; wings long, with the first quill spurious, and the second rather shorter than the third or fourth, which are nearly equal and longest; tail moderate, and slightly emarginated; tarsi moderate, longer than the middle toe, and slender; toes very small, the lateral ones equal, and strongly scutellated above; the claws short, slightly curved, and the hind claw rather long, and nearly straight.

420. Pyrrhulauda Leucotis, Stanley; Smith, Zool. S. A., Pl. 26; *Loxia Leucotis*, Salt.; *Alauda Melanocephala*, Licht.; *Fringilla Otoleucus*, Temm., Pl. Col., 269, f. 2, 3.

♂—Head, neck, shoulders, breast, and abdomen, black; ears, half-collar on back of neck, and a blotch on each side of the breast, hinder portion of flanks, and rump, white, the two latter tinged with brown; back and shoulders chesnut; tail umber-brown, except the outer feathers, which are rusty-white. The female is dusky-brown, with a speckled breast. Length, 5"; wing, 3" 3'''; tail, 2'''.

Inhabits plains thinly covered with dwarf shrubs, near to the Tropic of Capricorn.—Dr. A. Smith, loc. cit.

421. Pyrrhulauda Verticalis, Smith, Append. to Rept. of Exped., p. 48, and Z. S. A., Pl. 25.

HEAD, neck, and under parts, black-brown; ears, top of head, nuchal collar, and spot in front of each wing, white; back grey-brown; tail slightly forked, the outermost feather of each side brown towards the base of the inner vane, elsewhere white. Length, 4" 9'''; wing, 3" 2'''; tail, 2" 3'''.

"Lives amongst grass, both to the north and south of Kurrichane." —Dr. Smith, loc. cit.

422. Pyrrhulauda Australis, Smith, Zool. S. A., Pl. 24, and Append. to Rept. of Exped., p. 49; *P. Melanosoma*, Swain.

HEAD, above, black-brown; back ferruginous, dashed with dark-brown, beneath black; quills and tail, black-brown, the two centre-feathers of the latter edged and tipped with light rufous. Length, 5"; wing, 3" 2'''; tail, 2" 2''.

"Inhabits the country along the Orange River, but extends farther south than *P. Verticalis.*--Dr. A. Smith, loc. cit.

Dr. Smith states that all the African species are gregarious. This accords with what I have observed respecting the Ceylonese *P. Grisea.*

P. Australis congregates in flocks of several hundreds, and builds on the ground, under the shelter of a tuft of grass or some dwarf shrub. It has been sent to me from Colesberg by Mr. Arnot; and I found it plentiful on a limited tract of country near Nel's Poort, Beaufort. It seemed to keep in small parties of four or five individuals, and feed on small grass seeds.

Genus MEGALOPHONUS, G. R. Gray.

Bill moderate, slender, compressed, with the culmen somewhat curved to the tip ; the nostrils basal, lateral, and covered only by a membrane; wings very short, with the first quill short, and the second nearly as long as the third, which is the longest ; tail moderate; tarsi longer than the middle toe, and slender ; toes moderate, the lateral ones equal; and the hind claw short, and somewhat curved.

423. Megalophonus Rufipileus. (Vieil.) Cuv.,
Vol. 2, p. 118 ; *Alauda Ruficapilla,* Steph. ; *L'Alou-ette à Callotte Rousse,* Le Vail., Vol. 198.

Top of head, marroon-red, with a black line down the centre of each feather; back of neck, back, shoulders, and rump, earthy-brown, with blackish marks across the feathers ; throat, chest, belly, and flanks, dirty greyish-white, with brown markings ; belly and vent inclining to yellow ; wing-feathers reddish grey-brown, waved with brownish-black ; tail dirty-white on the edge.

Le Vaillant states he found but three specimens of the lark, all in the country of the Houswaanas.

Sundevall doubts its existence, as it has not been seen since. I cannot help thinking that Le Vaillant's figure and description have been taken from the bird identified as *Brachonyx Erythronothus* of Stephens.

424. Megalophonus Pyrrhonothus. (Vieil.)
N. Dict. d'His. Nat., 1, 361 ; Cuv., Vol. 2, p. 117 ; *L'Aloutte à dos Roux,* Le Vail., No. 197 ; *Enkelde Leeuwerk* of Dutch Colonists.

ALL the upper parts, with the exception of the rump, which is reddish, are of a grey-brown, shading into brown-black ; underneath dirty-white, with some brown lines on the chest.

Le Vaillant states that this bird frequents bushy plains, perching readily on bushes and trees at the edges of woods, singing sweetly. I feel convinced that this is identical with *Anthus Leucophrys,* Vieil., which goes by the name of " Enkelde Leeuwerk " among the colonists.

425. Alauda Erythronotus, Steph.; Zool. S.
Af., Pl. 110, fig. 2; *Brachonyx Pyrrhonotha*, Smith; Z. S. A., Pl. 110, f. 2.

BACK of the neck and back, rufescent, variegated with brown; throat and breast white, the latter streaked with brown; belly, flanks, and vent, rusty buff-orange; wing-feathers brown, margined with rufous; scapulars rusty orange-brown, edged with white, and barred with liver-brown; tail-feathers umber-brown, the outer vanes and tips edged with pale rufous, the two middle feathers rusty reddish-orange, barred with liver-brown.

I very much doubt the identity of the bird described by Dr. Smith, loc. cit., with the " Alouette à dos Roux " of Le Vaillant, as Dr. Smith believes; and my doubts are confirmed by Sundevall. See his review of Le Vaillant's birds.

426. Megalophonus Cinereus, Vieil.; *Calandrella Ruficeps*, Brehm; *La Petite Alouette à Tête Rousse*, Le Vail., No. 199.

GENERAL colour above, brown, the edge of each feather being ashy; top of head and sides of breast, bright rufous; throat, chest, belly, and under tail-coverts, white; wing-feathers brown, the outer one with the outer web white; tail brown, with the edges of the two central feathers the colour of the back. In some specimens the throat and breast is sparsely freckled with brown. In a variety received from Colesberg, the whole bird is brighter coloured, the red of the head and breast more vivid, the whites purer, and the ash colour of the back and centre tail-feathers charged with ruddy tinge; iris hair-brown. Length, 6″; wing, 4″; tail, 3″ 2‴.

The red-headed lark is common throughout the colony, affecting, however, some spots in preference to others; for instance, just on the top of the little hill before reaching Mowbray, on the high-road from Cape Town, a few pair may always be found dusting themselves; yet a mile on each side the species is not often found.

It feeds on insects and seeds, builds a nest on the ground by a tuft of grass, and lays three to five eggs, of a cream colour, profusely speckled with brown and purple : axis, 11‴; diam., 8‴.

427. Megalophonus Rostratus, Hartl.; Ibis Vol., 1863, p. 326.

ABOVE, fulvous-brown, with blackish markings; back and neck pale rufous, striped with black; top of head rufous, marked in the same manner; below fulvous; throat longitudinally marked with black; chin white; under side of wings, lively red; base of primaries red; tail-feathers fuscus, the outer ones, with the external web, white. Length, 6″ 6‴; wing, 3″ 6‴; tail, 2″ 3″.

Found by Mr. Ayres at Natal, running with great swiftness amongst the grass. Mr. Arnot has also sent a single specimen, killed near Griquatown.

428. Megalophonus Africanus; *Mirafra Africana*, Sm.; Z. S. A., Pl. 88, f. 1; *Megalophonus Planicola*, Licht.

UPPER surface of head, streaked brown and reddish-orange; back and sides of neck brown, the edges of the feathers being reddish-orange or sienna-yellow; back brown, the feathers edged with rusty-white or grey; chin and fore part of throat white; middle of throat and sides of neck white, mottled with triangular brown spots; breast, anteriorly, sienna-yellow: posteriorly, light reddish-orange, marked with narrow brown stripes, expanding into angular spots; belly yellowish; wing-feathers towards the base, reddish-orange, towards points, brown; tail brown, outer webs of two lateral feathers on each side, cream-yellow; eyebrows sienna-yellow. Length, 7″ 6‴; wing, 3″ 9‴; tail, 2″ 4‴.

Dr. Smith states, loc. cit., that this bird is confined to the Eastern districts, frequenting the grassy plains between Algoa Bay and Graham's Town. Generally seen in pairs, and feeds on insects and seeds. A single specimen of this lark was forwarded to the Museum by the late Mr. R. Moffat, from Kuruman.

429. Megalophonus Sabota; *Mirafra Sabota*, Smith, Z. S. A., Pl. 89, f. 1.

FEATHERS of the upper surface of the head, of the back and sides of the neck towards the body, of the back, and rump, umber-brown, broadly edged with clear reddish-orange, fading at their margins into rusty-white; chin, throat, breast, and belly, rusty-white, the former, and the flanks, tinted more or less strongly with reddish-orange, and the breast and lower part of the throat mottled with umber-brown stripes; wings reddish-brown, narrowly edged with light-brown; two middle tail-feathers reddish-brown, tipped and edged with rusty-orange, the rest umber-brown, edged with wood-brown; eyebrows and under eyelids, pale cream-yellow; ear-coverts, dull reddish-orange. Length, 6″ 1‴; wing, 3″ 3‴; tail, 2″ 3‴.

Found on the arid plains north of Latakoo. Perched frequently on low shrubs.—Dr. A. Smith, loc. cit.

*A

430. Megalophonus Africanoides; *Mirafra Africanoides*, Smith, Zool. S. A., Pl. 88, f. 2.

HEAD, superiorly, umber-brown, the feathers edged with a reddish-orange; back and sides of neck, dirty pale-brown, with light edges to the feathers; back brown, feathers edged with light reddish-orange; rump and upper tail-coverts greyish-brown, feathers edged with rusty-white; chin, throat, breast, and belly, rusty-white; sides of breast and flanks tinted with reddish-orange, and the middle of breast sparingly mottled with triangular pale, umber-brown spots; vent pure white; wing-feathers brown, more or less edged with wood-brown, most distinct on the outer vane of the outermost feather of each side; the two middle feathers between broccoli and umber brown; eyebrows white. Length, 6″; wing, 3″ 6‴; tail, 2″ 6‴.

Specimens of this bird are occasionally to be procured to the south of the Orange River, but most readily on the arid open plains which lie between that river and the Tropic of Capricorn.—Dr. Smith, loc. cit.

431. Megalophonus Chenianus; *Mirafra Cheniana*, Sm., Z. S. A., Pl. 89, f. 2.

UPPER surface of head, brown, the feathers edged with rufous-white; back, and sides of neck, dull umber-brown, each feather edged with rusty-white or wood-brown; back and rump, umber-brown, feathers edged with greyish-white, tinted with pale reddish-orange; chin and throat white, speckled with umber-brown; breast and flanks, pale reddish-orange, the former variegated with umber-brown stripes; middle of belly and vent, cream-yellow; under tail-coverts white; wing-feathers margined with reddish-orange, two middle tail-feathers narrowly edged with rusty-white, the two outermost of each side white, with more or less of the inner edges of inner vanes umber-brown; eyebrows sienna-yellow; ear-coverts, pale reddish-orange. Length, 5″ 4‴; wing, 2″ 10‴; tail, 1″ 11¼‴.

Rare. Found on the extensive grassy plains to the northward and eastward of Latakoo. Prefers situations abounding in long rank grass, among which it runs. Feeds on seeds and insects.—Dr. A. Smith, loc. cit.

432. Megalophonus Lagepa; *Mirafra Lagepa*, Smith; Z. S. A., Pl. 27., f. 2.

UPPER surface of head, neck, and back, reddish-orange, variegated with short streaks of umber-brown; under parts white, streaked on the throat, breast, and flanks, with umber-brown ; wing-feathers brown, edged with reddish-orange ; tail the same ; vent white, streaked with clear reddish-orange ; eyebrows rusty-white. Length, 6″ 6‴ ; wings, 3″ 7‴ ; tail, 2″ 10‴.

Thinly distributed between the Berg and Orange Rivers, close to the Western Coast. Found also on karroo sparingly covered with brush-wood. It whistles in the mornings, soars like a true lark, and on descending commonly perches on a shrub. Feeds on seeds and small insects.—Dr. A. Smith, loc. cit.

Mr. Atmore found it at Traka ; and forwards the eggs, which are cream-coloured, profusely spotted, especially at the obtuse end, with purple and brown : axis, 11‴ ; diam. 8‴. He states that the birds are never gregarious, but always in pairs.

433. Megalophonus Guttatus. (Lafren.) Rev. Zool., 1839, 259.

ABOVE rufous, with a broad dark-brown stripe down the centre of each feather, those of the wings and tail edged with dirty nankin ; under parts all dirty-white, much mottled longitudinally with brown lines. Length, 6″ ; wing, 3″ 4‴ ; tail, 2″ 9‴.

This species was procured in the Karroo by Mr. Atmore. I found it abundantly as far as Nel's Poort, and observed it in the Karroo between Worcester and Robertson. It constructs a snug cup-shaped nest of hair and grasses, in a hole under a low bush ; and the eggs, three to five, are of a dirty-white, spotted with brown and purple, chiefly in the form of a ring round the obtuse end : axis, 9‴; diam., 7‴

434. Megalophonus Apiatus, Vieil., N. Dict. d'Hist. Nat. 1, p. 342 ; *Brachonyx Apiata*, Smith, Zool. S. A., Pl. 110 ; *L'Al. Bateleuse*, Le Vail., No. 194 ; Cuvier, Vol. 2, p. 117 ; *A. Clamosa*, Steph. ; *Clapert Leeuwerk* of Colonists.

UPPER parts, ashy-grey, variegated with reddish, dark-brown, black, and white markings, very elegantly blended ; on the wings these colours appear to form small bars ; on the head longitudinal stripes ; chin white ; sides of head, neck, breast, and belly, dirty-nankin, the latter with a deep rufous tinge, the former all speckled with dark-brown ; tail-feathers brown, margined with nankin, which colour prevails over the outer web of the outer pair, two inner pairs ashy-grey, faintly barred like the wings ; iris rich brown-chesnut. Length, 6″ 3‴ ; wing, 3″ 2‴ ; tail, 2″ 3‴.

This beautiful lark, of whose elegant colours it is impossible to convey any idea by mere words, is common in all the Western districts of the colony, and is well known from its singular habit of rising fifteen or thirty feet into the air, perpendicularly, making a sharp cracking sound with its wings as it rises, uttering a long shrill " phew," and then falling as abruptly to the earth. This action it will repeat at intervals of a minute or two, for an hour or more, chiefly during bright sunny mornings. It delights in warm sandy soils; but I met with it on the high table-land of the Cold Bokkeveld in considerable abundance.

Genus CERTHILAUDA, Swainson.

Bill slender, lengthened, more or less curved; nostrils round, naked; wings very long; the first quill spurious; the three next nearly equal; tail moderate, even; feet lengthened; the lateral toes equal; length of hinder claw variable, although typically short and straight

435. Certhilauda Garrula, Smith; Zool. S. A., Pl. 106; *Certhilauda Albofasciata*, Lafresnaye, Mag. de Zool., 1836, Ois. 59.

Top of the head and back, blackish-brown; the feathers margined with rufous; tail-feathers, with the exception of the two centre, blackish-brown; broadly tipped with white; throat white; breast and belly ruddy. Length, 6" 6"'; wings, 3" 3"'; tail, 2" 8"'.

"Inhabits the northern parts of the colony, is very noisy, and generally found in small flocks. It moves rapidly, and when at rest is found either on the ground or perched on ant-heaps or small shrubs. It constructs its nest in a hole or excavation in the ground; the foundation is small twigs; the inner part grass or hair. It lays four or five eggs, of an olive colour, with darker spots of olive-brown."—Dr. A. Smith, loc. cit. Several specimens received from Mr. Atmore, who writes, 24th October, 1864.—" I have just stumbled upon a nest of *C. Garrula* with two eggs, hard set; they are brown spotted, much like those of the " Kalkoentje," but smaller. Nest open, and nicely secreted under a bush. The whole family keep together till pairing time : they run very swiftly on the ground. Eggs—three, four, or five. The male is very attentive to his sitting hen, spreads his wings, sets up his feathers, and sings.

I obtained a few specimens of this species at Nel's Poort; and observed it at the "River Zonder End," on Mr. Vigne's farm. It appears to be widely distributed. Nests seen by me resembled those described by Mr. Atmore.

436. Certhilauda Semitorquata, Smith; Zool. S. A., Pl. 106; *C. Rufo-palliata*, Lafres.

Upper surface of the head and back, rufous, variegated with longitudinal brown lines; back of neck with a half-collar of pale ash-grey; throat white, variegated with brown; breast and belly, pale rufous, the breast marked with longitudinal brown blotches; eyebrows grey; tail-feathers ruddy-brown, margined with reddish-orange. Length, 7″ 9‴; wing, 4″ 2‴; tail, 5″ 6‴.

"Inhabits arid plains in the interior of Southern Africa; and when disturbed in one place flies to another not very distant. It has but few of the habits of the larks."—Dr. A. Smith, loc. cit. I have received a few specimens from Mr. Arnot, killed near Colesberg.

437. Certhilauda Africana, Sm.; Z. S. A., Pl. 90; *Alauda Africana*, Gmel.; Cuv., Vol. 2, p. 119; *Le Sirli*, Le Vail., No. 192; *Certhilauda Longirostris*, Swain.; *A. Africana*, Pl. Enl. 712; Cuv., Vol. 1, p.479.

ABOVE, grey-brown, the feathers being margined with dirty white; beneath white; the lower part of the throat, breast, and flanks, variegated with short, broad, brown streaks; wing-feathers yellowish-brown, margined externally with yellow; tail-feathers grey-brown, margined externally with ruddy-white. Length, 8″ 3‴; wings, 4″ 3‴; tail, 3″ 3‴.

Common in most parts of the colony, as I have received it from all my correspondents. It frequents the sandy Cape Flats, the cornlands of Malmesbury, the uplands of Caledon, and the grassy plateaux of the Knysna. It never congregates in flocks, rarely more than two being found within a certain range. Feeds on insects and seeds. It constructs a cup-shaped nest of hair and grasses, lined with feathers, under the shelter of a bush or stone; the eggs, generally three in number, are dirty-white, faintly and minutely speckled with light brown: axis, 11‴; diam., 8‴. Some specimens run much darker than others.

438. Certhilauda Subcoronata, Smith; Zool. S. Af., Pl. 90, f. 2.

ABOVE, rufous, striped with brown; below, ruddy-white; chin, throat, breast, anterior part of belly and flanks, whitish, striped with brown; wing and tail-feathers, grey-brown, margined with yellowish-brown; eyebrows rusty-white, prolonged so as nearly to meet on the nape; iris hazel. Length, 8″ 6‴; wing, 4″ 6‴; tail, 3″ 2‴.

But few specimens fell under Dr. Smith's notice, and these were procured in the arid plains of the Middle and Eastern districts. I have received a few specimens from Mr. Atmore, procured in the neighbourhood of Traka, in the Albert division, in August.

The Sub-Family, PYRRHULINÆ, or Bullfinches,

have the bill very short, strong, more or less compressed, and entire, with the culmen arched and convex; the wings moderate, and somewhat rounded; the tail moderate, and slightly emarginated; the tarsi short and scutellated; the toes moderate, the lateral ones generally unequal.

Genus CRITHAGRA, Swainson.

Bill short, swelled on every side; the upper mandible smaller than the under, and considerably wider above than it is high; culmen arched from its base; commissure curved; wings moderate, the three first quills nearly equal; tail moderate, slightly forked; tarsus shorter than the middle toe; lateral toes equal; claws lengthened, slender, acute, and but slightly curved, the hinder as long as its toe.

439. Crithagra Scotops, Sundev.; Ofvers. Kong. Vat. Akad. Forhandl.

GENERAL colour, yellowish-green, with brown centres to many of the feathers of all the upper parts, except the rump; under parts less marked, except on the flanks; belly and vent immaculate; forehead and chin almost black, a narrow eyebrow extending from each nostril over the eye, and a half-collar above the neck, bright yellow; tail and wing primaries brown, faintly edged with green. Length, 5″; wing, 2″ 10‴; tail, 2″.

Sent from Swellendam by Mr. Cairncross. Found also at the Knysna by M. Victorin; and in Kaffraria by Prof. Wahlberg.

440. Crithagra Sulphurata. (Linn.) Swain., Jardine and Selby's Orn., t. 109, f. 1; *Buprinus Sulphuratus; Loxia Sulphurata*, L.; *Brimstone Grosbeak*, Shaw., Vol. 9, p. 313; *Crithagra Albogularis*, Smith, S. Af. Q. J., p. 48, the ♀.

GENERAL colour, yellow, very faintly striped with rufous below, and much variegated with dark-brown above, a yellow band extends from the nostrils over each eye; the outer edges of the wing and tail feathers, bright-yellow; rump rather free from the brown markings of the back. Length, 6″ 2‴; wing, 3″ 2‴; tail, 2″ 5‴.

Received from Mr. Cairncross, Swellendam. It is apparently not uncommon in that district, extending as far westward as Caledon. It is also plentiful at Nel's Poort, where I took the eggs in some abundance. The nest is usually placed in a low bush, often only a few inches from the ground: it is very compact, and cup-shaped; the eggs, generally four in number, are white, slightly tinged with green, and dotted at the obtuse end, mostly in an annular form, with intensely dark or light purple spots, with here and there a wavy streak: axis, 10‴; diam., 7‴.

441. Crithagra Selbyi, Smith ; Swain., An. in Menage., p. 319 ; Dr. Smith, Appendix to Report of Exp., p. 50 ; *C. Cinerea*, Swain.

CINEREOUS grey, striped above, and immaculate beneath ; those on the crown, and the spots on the back, dark-brown ; wing and tail feathers, blackish-brown, edged with light-grey ; stripe above the eye, a small spot beneath it, and another divided into two ; beneath the ears, white ; rump, yellow ; middle of throat, belly, and vent, white. Length, 6″ ; wing, 3″ ; 2‴.

Inhabits South Africa.—Dr. Burchell.

442. Crithagra Butyracea ; *Fringilla Buty-racea*, Linn. ; *Chloris Indica*, Bris. ; *Loxia Buty-racea*, L., Sh., Vol. 9, p. 274 ; *Yellow Finch*, Shaw., Vol. IX, p. 477 ; *C. Flava*, Swain., An. in Meng., p. 318 ; *Kleine Seisje* of Dutch Colonists.

PLUMAGE above, olive-green ; quills the same, with whitish edges ; from the forehead over the eyes, a yellow streak, and one of the same colour beneath, from the angle of the bill ; under parts of body and rump, yellow ; tail yellowish-green, and slightly forked. Length, 5″ 2‴ ; wing, 2″ 10‴ ; tail, 2″ 4‴.

Abundant throughout the colony, congregating during the autumn in large flocks, and doing extensive mischief to the grain crops It sings very sweetly, and thrives well in confinement. The nest is a cup-shaped structure, found in a low bush ; the eggs, three to five in number, are precisely like those of No. 440, only rather less spotted, and smaller : axis, 8‴ ; diam., 5‴.

The type of Swainson's *C. Flava* is in the Cambridge Museum, and is nothing more than a ♀ or a ♂ in the non-breeding plumage.

443. Crithagra Strigilata, Swainson ; An. in Menag., p. 317.

ABOVE, greyish-olive, striped with dusty ; rump and tail-covers, greenish-yellow ; sides of the head, grey-brown, with two whitish stripes, one above the eyes, the other beneath the ears ; body beneath with dusky stripes ; belly white ; inner wing-covers, yellow.

Inhabits South Africa.—Swainson, loc. cit.

444. Crithagra Chrysopyga, Swains., Nat. Lib., Vol. 11, p. 206 ; Hartlaub, Orn. W. Af., p. 154.

UPPER plumage, including the crown, ears, wings, and tail, olive-gray, with a dark stripe down each feather ; a broad stripe of bright-yellow is above each eye, meeting on the front; another below the ear; upper tail-coverts, and the whole of the under plumage, from chin to vent, also yellow. Length, 4''; wing, 2'' 4'''; tail, 1'' 5'''.

I have seen this from Damaraland; but have not met with it within my limits. Hartlaub, however, gives it as a South African species.

445. Crithagra Flaviventris, Gmel.; *Loxia Flaviventris*, Shaw, Zool., Vol. IX., p. 271; *Cocco-thraustes Lutea*; Briss., Orn. III., p. 227, t. 11, f. 2; *Yellow-bellied Grosbeak*, Lath., Gen. Syn., 3, 138, 42.

On each side of head, passing over the eye, is a stripe of yellow; head, hind part of neck, and back, olive-green, sprinkled with brown; quills and tail brown, with olive-green edges; under parts of body plain yellow; rump olive-green; tail slightly forked. Length, 5''.

Inhabits the Cape of Good Hope.—Shaw, loc. cit.

446. Crithagra (?) Africana; *Loxia Africana*, Gmel.; *African Grosbeak*, Shaw, Vol. IX., Part 2, p. 272; *Le Verdier Sans Vert*, Buff., His. Nat., Vol. 4, p. 364.

Upper parts, olive-brown and grey; upper wing-coverts rufous; greater quills edged with rufous-white; under parts of body white, varied on the breast with brown; tail olive-brown, outer feathers edged with rufous-white, the outermost with a white spot near the tip. Length, 6½''.

"Comes from the Cape of Good Hope."—Shaw, loc. cit. This species has not been recognised by any subsequent author: it is probably some species of *Crithagra* in non-breeding plumage.

The Fifth Family, COLIDÆ, or Colies,

have the bill moderate, with the culmen elevated at the base, and the sides compressed to the tip, which is acute; the nostrils placed in the middle of a large membranous groove, with the opening small and exposed; the wings short; tail much lengthened, and graduated; the tarsi robust; and the toes lengthened, the outer as long as the inner; the hind toe moderate, united at the base of the inner toe and directed forward.

The only Sub-Family, COLINÆ, or Colies,

have the characters given above.

Genus COLIUS, Brisson.

Bill moderate, broad, and elevated at the base, with the culmen arched, and the sides compressed to the tip, which is acute ; the lateral margins sinuated ; the gonys short, and slightly ascending ; the nostrils placed in a short, broad membranous groove, with the opening small and exposed ; wings moderate, with the first quill long, the second to the fourth nearly equal and longest ; tail much lengthened, gradu ted, with the feathers narrowed throughout ; tarsi as long as the middle toe, robust, and covered with broad scales ; toes rather long, the lateral toes equal, the hind moderate, united at the base of the inner toe, and directed forward ; the claws long, curved, and very acute.

447. Colius Erythropus, Gml. ; Lath., Gen. Syu, Pl. 41 ; *C. Leuconotus*, Lath. ; *C. Erythropygius*, Vieil., *C. Capensis*, Gmel. (apud Sundevall) ; *Le Coliou a dos Blanc*, Le Vail., No. 257.

UPPER parts, generally ash-coloured ; a white line, bordered on each side by a broad black one, extends from between the shoulders to the rump, which is purple, glossed with red ; throat and chin cinereous, changing into vinaceous on the breast, and a dirty-yellow on the belly ; head crested ; tail very long, and graduated ; wings short and rounded ; legs in a fresh specimen bright coral-red ; bill bluish ; plumage of body singularly short and hair-like. Total length, 13″ or 14″ ; wing, 3″ 7‴ ; tail, 10″ ; bill, 6‴ ; tarsus, 8‴.

Of the three species of this genus found in South Africa, and known by the trivial name of *Muisvogel*, or *Mousebird*, this is the only one that is found in the neighbourhood of Cape Town. It is not uncommon in gardens during the fruit season, ranging about in small families of six or eight individuals. They fly with a rapid, though laboured flight, generally at a lower level than the object at which they aim, and on nearing it rise upward with a sudden abrupt curve. They creep among the branches like parrots and hang suspended, head downwards, without inconvenience; indeed, it is said that they invariably sleep in this position, many of them congregated together in a ball. They are said to breed in holes of trees, laying three or four eggs, somewhat rounded at each end, of a dull white colour : axis, 10‴ ; diameter, 8‴.

In habits, the three species closely resemble each other ; and at the Knysna, where they are all to be found, we frequently shot them, and could not distinguish which we had obtained until we picke l them up.

*B

448. Colius Capensis, Gmel. ; Pl. Enl. 283, f. 1 ;
Colius Carunculatus, Steph.; *C. Erythromelon*, Vieil. ;
N. Dict. d'His. Nat. VII., p. 378 ; *Coliou Quiriwa*,
Le Vail., No. 258 ; *Colius Quiriwa*, Less.; *Colius
Indicus*, Lath.

Upper parts, shining bluish-green, inclining to grey ; forehead reddish-brown ; under surface reddish-brown, with a grey tinge on the breast ; vent, and under side of elongated tail-feathers, reddish-brown ; base of bill, and cere round the eye, a fine coral-red, fading to dirty-yellow in dried specimens. Length, 13″ 6‴; wing, 3″ 9‴; tail, 9″ 9‴.

Generally distributed in the George district, and to the Eastward. It does not appear to frequent the sparsely wooded country west of Swellendam. I saw it abundantly at the Knysna; and Mr. Atmore obtained nest and eggs at Traka. The eggs resemble those of *C. Erythropus* in every respect.

449. Colius Striatus, Gmel. ; *C. Panayensis*,
Gmel. ; *Tanagra Macroura*, Scop. ; *Le Coliou Rayé*,
Le Vail., No. 256.

Above, cinereous, with a tinge of glossy greenish-brown on the wings and tail, lightest on the head, which is crested ; under part of throat and chest cinereous, minutely striated with brown ; centre of belly dirty-yellow, inclining to rufous on the vent ; inner surface of wing rufous ; tail long, and graduated ; upper mandible black, and much curved ; lower, light horn-colour ; feet horn-colour. Length, 13″ ; wing, 4″ 10‴; tail, 9″ ; tarsus, 11‴; bill, 6‴.

Not uncommon in the Karroo. In habits resembling the preceding. Eggs received from Mr. Atmore resemble those of preceding species, but are rather larger : axis, 11‴ ; diam., 8½‴. Its nest is said to be an open, cup-shaped structure, built in a mass of small branches. Several birds are reported to lay in one nest.

It is not uncommon about the Knysna, Caledon, and Worcester, and extends this side of the mountain as far as Eerste River, as I found by shooting them in Mr. Lawrence van der Byl's garden.

The Sixth Family, MUSOPHAGIDÆ, or Plantain-Eaters,

have the bill moderate, broad at the base, the culmen much elevated, curved, and the sides much compressed towards the tip, which is strongly emarginated ; the wings more or less lengthened and rounded ; the tail long and broad ; the tarsi moderate, strong, and covered in front with broad transverse scales ; the toes long, and the outer one sometimes versatile.

The Sub-Family, MUSOPHAGINÆ, or Plantain-Eaters,

have the outer toe capable of being placed posteriorly.

Genus TURACUS, Cuvier.

Bill short, with the culmen elevated, and arched to the tip, the sides much compressed, and the lateral margins finely serrated; the nostrils placed in the middle of the upper mandible, large and open, but in most, more or less covered by the recumbent plumes; wings short, and rounded, with the fourth to the seventh quills the longest; tail lengthened, broad and rounded; tarsi shorter than the middle toe, and covered in front with broad transverse scales; toes long, united at their base, and the sides margined by a membrane; the lateral toes nearly equal, the outer one versatile; the claws moderate, thick, and curved.

450. Turacus Persa. (Linn.) Edw. Birds, Pl. 7;

Corythaix Persa, Vieill.; *Opœthus Africanus; Cuculus Persa,* Linn.; Shaw, Vol. IX., P. 1, p. 63; *Corythaix Albocristatus,* Strickl.; *Common Louri.*

GENERAL colour, green, iridescent, with steel-blue on the wings and tail, and approaching to dull black, or rifle-green, on the thigh and vent; head much crested, each feather being tipped with white; two white lines also extend from the corner of the bill, and passing one on each side of the eye, reunite behind it. The wings when expanded display the most brilliant carmine, glossed with purple; bill reddish, but black at the base. Length, 19"; wing, 8"; tail, 9".

Very common throughout the forest districts. It feeds on fruits, and frequents the highest trees, rarely if ever descending to the ground, over which it can, however, travel with great rapidity if brought down by a shot. The motions of this bird are very graceful and light, and performed with an ease and rapidity that delight the eye of the beholder. Strange to say, though I inquired carefully, I never could obtain any information respecting the nidification of this beautiful and common bird. Mr. Atmore, however, states that the eggs are white; but this must be from hearsay, as he writes: "How difficult it is to find these forest birds' nests. The Lories are breeding now: but for the life of me I cannot find a nest. The young ones go in troops, and are delicious eating: the old ones in pairs. We never shoot specimens out of a troop, except for the pot."

451. Turacus Porphyreolophus; *Corythaix*

Porphyreolopha, Vigors, Pro. Zool. Society, 1831 ; *C. Burchelli*, Smith, S. Af. Qu. Journal, 1831 ; *Vley-Lourie* of Colonists.

GENERAL colour, dark, iridescent green, with steel-blue reflections on the back, wings, and tail ; head slightly crested, shining green and blue ; under parts a dull green, with a reddish tinge on the breast ; wings, when expanded, display the same brilliant carmine, glossed with purple, as seen in *C. Persa.* Length, 16″ ; tail, 8″ ; wing, 7″.

This very beautiful bird has been found at the Knysna ; but is there extremely scarce. It increases in numbers as the collector travels East-ward, until at Natal, as I am informed, it is not uncommon. Dr. Smith states that it inhabits thickets near the coast, feeding upon fruits, &c.

Genus SCHIZORHIS, Wagler.

Bill short, with the culmen scarcely higher than broad at the base, much arched to the tip, and the lateral margins sinuated, and nearly smooth ; the nostrils lateral, exposed, and placed near the basal part of the culmen, with the opening longitudinal ; wings moderate, and pointed, with the fourth, fifth, and sixth quills the longest ; tail long, and nearly equal, with the feathers broad ; tarsi shorter than the middle toe, robust, and covered in front with broad trans-verse scales ; toes long ; the lateral toes nearly equal ; the outer one scarcely versatile, the base united, and the margins bordered by a membrane ; the claws rather long, and curved.

452. Schizorhis Concolor; *Corythaix Concolor*,

Smith, S. A. Zool., Vol. 2, p. 48 ; *Chizaerhis Felicia*, Lesson.

FOREHEAD with a long crest, which, with the entire plumage, is greyish ; quill-feathers dark-brownish ; tail very long, hoary grey, with a deep greenish gloss, particularly towards the tip ; bill, legs, and toes black. Length, 19″ ; tail, 10″ ; wing, 9″.

Inhabits South Africa, inland of Port Natal, according to Dr. A. Smith ; but I have only seen it as yet from Damaraland, where it seems not uncommon, from the number of specimens procured by Messrs. Andersson, Chapman, and others.

The Seventh Family, BUCEROTIDÆ, or Hornbills,

have the bill more or less.lengthened, curved, broad at the base, and compressed to the end with the culmen furnished with a singularly-formed helmet, or only curved to the tip, which is acute; the nostrils basal, and usually rounded; the wings moderate; the tail generally long, broad, and more or less graduated; the tarsi in most, short and strong; toes moderate, strong, and more or less united at their base, especially the outer toe; claws short, and rather obtuse

The only Sub-Family, BUCEROTINÆ, or Hornbills,

have similar characters to those given above.

Genus BUCEROS, Linn.

Bill large, more or less long, broad at the base, and more or less surmounted by a helmet of various shapes, with the culmen curved, and the sides much compressed to the tip, which is pointed; the lateral margins smooth in the young, and more or less jagged in the adult; the nostrils basal and lateral, with the opening small, exposed, and rounded; wings rather short, with the third quill nearly as long as the fourth or fifth, which are equal and longest; tail generally long, broad, and more or less graduated; tarsi rather short, robust, and covered in front with broad transverse strong scales; toes rather long, broad; the outer toe longer than the inner, both united at the base, especially the outer, the union of which is continued to the second joint; the hind toe long, strong, and strongly scutellated above; the claws strong, long, curved, and acute; the face and throat more or less naked, the latter sometimes furnished with a gular pouch.

453. Buceros Coronatus, Shaw; Enc. Meth.
Ois, t. 240; *B. Melanoleucus*, Licht. Cat.; Le Vail, Pl. 234, 235; Swain., Zool. Illus., Pl. 178.

HEAD grey, some of the feathers at the back rather elongated, and centred with white, giving the bird when in some positions the appearance of having a white half-circle at the back; back dark-brown, each feather edged with lighter brown; wing-feathers very dark brown, edged with white;

tail same as wings, and broadly tipped with white, with the exception of the two centre feathers; under side, with the exception of the throat and chin, dull white; bill large, slightly casqued, red, with a narrow yellow band along the base; eye-lashes very stiff, and projecting. Total length, including bill, 22″; bill, 3″; tail, 11″; wing, 10″ 6‴.

This Hornbill has a wide range, extending from the Knysna along the East Coast as far as Kisiludini, 5° south of the Line. How much farther I do not know; but at this place I found it very abundant, flying in small flocks or families, and feeding on berries and fruits. At the Knysna it is also common. Le Vaillant states that the nest is made in a large hole in a tree; and that the eggs, four in number, are white.

My friend Mrs. Barber thus writes of the singular habit which this bird exhibits, in common with other species of the genus, of blocking up the sitting female in her nest :—" By the bye, do you know that our hornbills build their nests of mud and sticks in old broken and hollow trees, or between the crowded stems of the tall Euphorbia in our forests; and while the process of incubation is going on the ♂ bird *builds the* ♀ *into the nest,* closing up the entrance in such a manner that it is impossible for her to escape, leaving only a small hole for the purpose of feeding her during her long imprisonment. I do not know how long she is kept in durance vile; but we have sometimes taken them out, and found them so cramped and weak that they were unable to fly. This peculiar habit may be a precautionary measure to protect the ♀ during the season of incubation; for it may be that during that time she is too weak and dull to fly away from any approaching danger. Depend upon it, that it is not done in vain, and that there is a sufficient reason for this strange and curious habit. We self-willed and presumptuous human beings often act without reason or reflection; but the 'birds of the air' and the 'lilies of the field' are guided by a higher Power! And if we do not understand the wisdom of all their ways, it is because we have not studied them sufficiently."

454. Buceros Buccinator, Tem., Pl. Col. 284.; Cuv., Vol. 2, Pl. 416.

" GREEN above; belly and vent white; tail tipped with white; prominence on bill, trumpet-shaped."

Procured by Mr. Ayres at Natal, and also included in Victorin's list. It has been forwarded to me by Mr. Henry Bowker, who procured it in the Transkei country. His sister, Mrs. Barber, informs me that it " is a very wild and shy bird, and very difficult to shoot."

455. Buceros Pœcilorhynchus ; *Tokus Pœcilorhynchus,* Lafren., Rev. Zool., 1859, p. 257 ; Hart., O. W. Af., p. 164.

ABOVE, pale brown; wing-coverts with pale margins; tail-feathers brown, with the exception of the tips, which are white; head and neck cinereous; back of head crested; eyebrows, breast, and belly white, as are also the shafts of the middle tail-feathers; legs brown; bill slightly arched, notched; maxilla slightly arched with a white keel; the region of the nostrils white, with black margins; tip and denticulations reddish; mandibles black, with five elevated ribs, white spotted. Length, 17″; wing, 8″; tail, 7‴.

Inhabits South Africa, according to Hartlaub.—Sed non vidi.

456. Buceros Erythrorhynchus; *Tockus*
Erythrorhynchus, Temm., Pl. En. 260; Hartlaub, O. W. Af., 165; *Hydrocorax Senegalensis Erythrorhynchus*, Briss., Orn. IV., p. 575; *Le Toc*, Le Vaillant, Pl. 238.

Top of head grey; forehead, cheeks, a line half-way down the middle of back, some spots upon the shoulders, some of the inner wing-feathers, more or less of the three outer tail-feathers, and all the under parts, white; outer wing-feathers black, with some white markings; four inner tail-feathers and rump black; bill deep-red, elongated, curved, sharply keeled above, not casqued, 3½ inches long. Total length, 19″; tail, 7″; wing, 7″ 6‴.

Sundevall gives this species as an inhabitant of Kaffraria. I have not heard of it there; but have received it in considerable numbers from Mr. J. Chapman, who procured it towards the Zambesi. Mr. Andersson obtained it in Damaraland.

457. Buceros Nasutus. (L.) Hartlaub, O. W.
Af., p. 164; *Hydrocorax Senegalensis Melanorhynchus*, Briss., Orn. IV., p. 573, t. 46, fig. 1; *Calao Nasique*, Le Vail., O. Af., Pl. 236, 237; *Tockus Hastatus*, Cuv., Pl. Enl. 890.

Above, pale-brown, the margins of the feathers lightest; head and neck, dusky-cinereous; a line over each eye, patch on the back, and the nuchal half-collar white; under side whitish; breast brown; the quill-feathers of wing with light margins; base of tail white; all tail-feathers black, with the exception of the centre ones, which are of the colour of the back, tip white, shafts black, centre ones white; bill curved, casqued; upper mandible white, or pale yellow, at the basal half; tip and edge red; lower mandible, basal half, black, with five whitish raised striæ; tip and edge red. Length, 20"; wing, 8" 6'''; tail, 7".

Hartlaub and Sundevall quote this as an inhabitant of Kaffraria. All the specimens I have seen came from Mr. Chapman, shot towards the Zambesi.

Genus BUCORVUS, Lesson.

The characters similar to those of the former genus, except that the tarsi are very long, strong, and covered in front with large scales, those near the toes being hexagonal; toes short, thick, the inner toe rather shorter than the outer, both united at the base, especially that of the outer; the hind toe long and strong; the claws moderate, curved, and acute.

458. Bucorvus Abyssinicus. (Gm.) Pl. Enl. 779; *B. Carunculatus*, Wagl. Syst. Av. Spec. 6; *B. Leadbeateri*, Vig. Av. Juv.; *Le Calao Caronculé*, Le Vail., Pls. 230 and 231; *Brom-Vogel* of Colonists.

ALL black, with the exception of the webs of the primaries, which are white; bill very large, and much casqued, with a large patch of bare red skin at the base; size very large.

Common on the Eastern frontier; but in consequence of their feed-ing on carrion, and emitting a dreadful stench, I have not succeeded in inducing any of my correspondents to send me one. I am told that they associate in large flocks, and devour vast quantities of grubs and locusts. They get their name from the droning cry they utter. The Fingoes seem to attach some superstitious veneration to them, and object to their being shot in the neighbourhood of their dwellings, lest they should lose their cattle by disease.

Le Vaillant figures loc. cit. a head of this bird, in which the bare space round the eye and the lower portion of that on the neck are blue. In a single *dried* head which we possess, these parts are deep orange-red.

Order III.   SCANSORES.

THE Third Order, SCANSORES, or Climbers, are at once distinguished by the position of their toes, which are placed two anteriorly, and two posteriorly.

The First Family, RAMPHASTIDÆ, or Toucans,

have the bill much prolonged, broad at the base, with the culmen curved ; the sides compressed to the tip, the lateral margins more or less serrated.   We have no birds of this family inhabiting South Africa : they are confined to the New World.

The Second Family, PSITTACIDÆ, or Parrots,

have the bill more or less large and strong, with the culmen arched to the tip, which is prolonged, and acute, the lateral margins sometimes dentated, and the base covered by a cere, of a greater or less size, in which the nostrils are placed.

Genus PSITTACUS, Linn.

Bill large, and rather compressed, with the culmen triangular, and much arched to the tip, near which the lateral margin is strongly emarginated, that of the under mandible much sinuated, and the anterior part sharply edged ; the gonys advancing upwards, and angular ; the nostrils basal and lateral, with the opening small and rounded ; wings mostly reaching to the end of the tail, with the first quill nearly as long as the second and third, which are longest ; tail short and even ; tarsi very short, and covered with small scales ; toes long, the lateral ones equal, and all covered with small scales ; the claws short, and slightly curved ; the wings and the tail generally long ; and the tarsi usually very short and robust.

*c

The Sub-Family, PSITTACINÆ, or Parrots,

have the bill more or less large, broad at the base, and the sides compressed, with the culmen much arched to the tip, which is acute and prolonged, the lateral margins dentated or festooned; the nostrils basal, lateral, and rounded; the wings more or less long and pointed; the tail usually short and squared; the tarsi short, and covered with small scales; the toes moderate.*

459. Psittacus Levaillantii, Latham; *P. Flaminiceps*, Bechst.; *P. Fuscicollis*, Kuhl; *P. Infuscatus*, Shaw, Vol. VIII., p. 523; *Pionus Levaillantii*, Wagl. Mon. Psitt.; *Psittacus Robustus*, Gmel.; *P. Caffer*, Licht.; *Piroquet à Franges Souci*, Le Vail., Tab. 130 and 131.

HEAD, neck, and throat, brownish green; wings, and back between the shoulders, dark-green; back and rump, bright grass-green; belly, thighs, and under tail-coverts the same; upper portion of outer edge of wing, and knees, brilliant orange-red; between the eye and the bill a black spot. Length, 12″; wing, 8″; tail, 3″ 6‴.

Inhabits the forests of the Eastern frontier, the Knysna, and the Zuurberg. Le Vaillant states that they breed in hollow trees, and lay four white eggs, about the size of those of pigeons. It is apparently a scarce bird, and rather difficult to obtain, as but few specimens have reached my hands: these have all been from the Knysna and the forests of the Eastern frontier seaboard.

Genus PSITTACULA, Brisson.

Bill large, rather compressed at the sides, with the culmen much arched to the tip, which is prolonged and acute; the lateral margins festooned; the nostrils basal, lateral, rounded; wings reaching to the end of the tail, and pointed; with the first and second quills nearly equal and longest; tail short and even, with the ends of the feathers truncated or pointed; tarsi very short, and covered with small scales; toes long, with the lateral outer toes equal, and all covered with small scales.

---

* Many parrots have been accidentally introduced into South Africa in the neighbourhood of Cape Town, having escaped from confinement. Some of these are now breeding here: among them may be mentioned *Palæornis Torquatus*.

460. Psittacula Passerina. (Linn.) Kuhl. Ed.

Birds, Pl. 235 ; *Psittacula Capensis*, Gmel. ; *Agapornis Cyanopterus*, Swain. ; *Perruche à Ailes Bleues*, Buff. Ois. ; *Petite Perruche de Cap de Bonne Esperance ; Ps. Gregarius*, Spix., Pl. Enl. 445, f. 1 ; *Blue-winged Parrakeet*, Lath. Syn. ; *P. Cyanopterus*, Bodd. ; *Ag. Guianensis*, Swain.

PLUMAGE green, some of the wing-feathers blue; under wing-coverts blue; bill and legs reddish. It is sometimes said to have a blue spot on the crown, and yellow legs. Length about 4″ 6‴.

Native, according to Buffon, of the Cape of Good Hope (Shaw's Zoology, Vol. VIII., p. 556) ; but I much doubt this being a Cape species. It was probably introduced from the Eastward.

461. Psittacula Roseicollis. (Vieil.) Shaw, Zool.,

Vol. XIV., Pl. 1, p. 143 ; Kuhl., N. Dist. d'His. Nat. XXV., p. 377 ; *P. Pullarius*, var. β, Linn.

PLUMAGE pale green ; forehead and eyebrows red ; face and neck in front, rosy ; lower wing-coverts, blue-green ; rump and upper tail-coverts, azure ; tail scarlet, with tip bluish ; a black band at the tip ; outer webs edged with green. Length, 6″ ; wing, 3″ 10‴ ; tail, 2″.

"Inhabits South Africa."—Shaw, loc. cit. I have received several specimens from Damaraland, but have not yet seen it from any locality actually within my limits.

462. Psittacula Swinderiana, Wagler ; *Psittacus Swinderianus*, Kuhl., Consp. Psitt., P. 62, t. 2 ;

Agapornis Swinderianus, Nat. Lib., Vol. VI., Parrots, P. 118 ; *Swindern's Lovebird*.

HEAD and nape, beautiful lively green, bounded by a black nuchal collar; neck and breast yellowish green ; mantel and wings green ; lower back, and upper tail-coverts, deep azure blue ; the tail has the two intermediate feathers green ; the rest on each side have their basal half vermilion-red, bounded by a bar of black, the tips green. Length about 6″.

Said by Swainson to be a native of South Africa.—Sed non vidi.

The Third Family, PICIDÆ, or Woodpeckers,

have the bill generally lengthened, acute, and straight, with the base more or less broad, and the sides much compressed towards the tip, which is usually obtuse.

The Sub-Family, CAPITONINÆ, or Barbets,

have the bill large, broad at the base, usually furnished with bristles, and compressed towards the tip; the tail generally short, even, and the feathers rounded at the ends.

Genus LAIMODON, Illiger.

Bill large, elevated, and broad at the base, with the culmen arched, and the sides compressed towards the tip, which is acute; the lateral margins strongly and irregularly dentated; the nostrils basal, lateral, and concealed, and the base of both mandibles furnished with long bristles; wings moderate, with the third to the sixth quills nearly equal and longest; tail moderate, even; tarsi as long as the outer toe, and covered in front with broad scales; toes unequal, the outer pair equal, and the two anterior ones united to the first joint; claws short, compressed, and curved.

463. Laimodon Leucomelas. (Bodd.) Pl. Enl., 688, f. 1; *Bucco Rufifrons*, Steph.; *Red-fronted Barbet; P. Stephenii*, Leach; *Le Barbu à Plastron Noir*, Buff.; *Bucco Niger*, Gmel.; Shaw, Vol. IX., Pt. 1, P. 30; Le Vail. Barb., t. 29, 30, 31.

FOREHEAD crimson: from whence a black stripe passes over the head, and down the back part of the neck to the back; sides of head, neck, and breast white; the white on the side of the head is diversified first by a streak of yellow, which is placed over the eyes, and secondly by an irregular one of black, that begins at the base of the upper mandible, and divides the white into two parts, ending on the shoulders; chin and fore parts of neck black; upper parts of body, and wings, brown and yellow mixed, the edges of the feathers being generally fringed with yellow; rump pale bright yellow; tail brown, with yellow margins.

South Africa (Shaw).—Sed non vidi.

464. Laimodon Unidentatus. (Licht.) Verz

Sud arc, Thiere, p. 17.

GENERAL colour above, black, pencilled with yellow; fore-head deep crimson; a stripe extends from the nostril, over the eye, to the back of the head; the first part of this is yellow, the last part white; a broad white line also extends from the corner of the bill down the side of the throat; chin and gorget black; belly dirty white, mixed with grey; wing-secondaries and tail-feathers margined with yellow; bill black, strong, and with a tooth in the upper mandible. Length, 6″; wing, 3″ 6‴; tail, 2″ 3‴.

Common in mimosa bush throughout the Karroo, and has apparently a very wide range, having been received from Beaufort, Swellendam, Knysna, Kaffraria, Colesberg, Hopetown, Kuruman, and Damara-land. It also appears in Mr. Chapman's collection.

It is a solitary bird, never more than one pair being seen together, and that but seldom. Its call is similar to that of the Indian *Bucco Indicus*, viz., " poo-poo-poop," three syllables, constantly repeated. While uttering this cry, it remains stationary in some tree, but evidently expends much labour on its monotonous call, its body being jerked up and down the while. It feeds on fruits and berries, and is said to build in hollow trees; but I was not fortunate enough to obtain its nest. Mr. Atmore writes: "Feeds on seeds and insects. I see them frequently feeding on the ground. I know of a tame one which roved about unconfined. It eat meat, bread, sugar, corn, in fact anything—appeared very much attached to its mistress, and came regularly into the house to be fed. It disappeared at the pairing season, so I suppose it got married! They nest in holes of trees, but don't make the holes. Eggs spotted."

465. Laimodon Nigrithorax. (Cuv.) *Pogonias*

Personatus, Tem.; Pl. Col. 201; Cuv., Vol. 2, p. 467;
Pogonorhynchus Torquatus; Barbican Masqué, Le
Vail. Barb., t. 28; Lesson, Vol. 2, p. 137.

TOP of head and front, sides of head, chin, throat, and gorget, brilliant scarlet, all bounded by a band of glossy black, more or less broad; back and shoulders grey-green; quill-feathers of wings and tail dark-brown, margined with bright-yellow, the latter very slightly so; under parts green-yellow, tinged here and there with scarlet; base of bill sparsely covered with long bristles. Length, 6½″; wing, 3″ 7‴; tail, 2″ 6‴.

Inhabits Kaffraria and Natal, from whence came the specimen here described.

Genus MEGALAIMA, Gray.

Bill more or less long, broad at the base, and compressed on the sides; the culmen generally arched to the tip, which is acute; the gonys long, and advancing upwards; the lateral margins curved; the nostrils basal, lateral, and rounded, and the base of the upper mandible furnished with very long and strong bristles; wings moderate, with the fourth, fifth, and sixth quills nearly equal and longest; tail short, and rounded on the sides; tarsi shorter than the outer, anterior toe, and covered in front with broad scales; toes long, and the outer pair of equal length and longest; claws moderate, compressed, and acute.

466. Megalaima Bilineata, Sundev.; Ofvers.
Kongl. Vet. Ak. Forhandl., p. 109; *Barbatula Leucolaima*, Verr., Hartl. W. Af., p. 173.

ABOVE, black; forehead and eyebrows white; cheeks black, with black ring; chin and throat whitish; breast grey; belly yellow; rump deep-yellow; wing-feathers black, those on the outside bordered with yellow; tail-coverts above, black, below, olive-grey; wings black, variegated with greenish-yellow; underneath whitish; webs of secondaries narrowly edged with yellow; primaries all black; bill black. Length about 3″ 9‴; wing, 3″; tail, 1″.

Kaffraria (Wahlberg)—loc. cit; sed non vidi.

467. Megalaima Leucotis, Sundev.; Ofvers.
Kongl. Vet. Ak. Forhandl., p. 109.

BLACKISH, with a band behind the eye, and the belly white; head black; back reddish. Length, $6\frac{1}{3}″$.

Kaffraria (Wahlberg).—Sed non vidi.

468. Megalaima Barbatula. (Tem.) Pl. Enl.,
746, f. 2; *Barbatula Minuta*, Tem., Hartl. W. Af., p. 173; *Bucco Chrysoptera*, Swain., An. in Menag., p. 322; *B. Parvus*, Cuv., Le Vail. Barb., t. 32; *Capito Rubrifrons*, Vieil.; *B. Pusillus*, Dum.; *B. Nanus*, Vigors, Proc. Z. S., 1831, p. 93; *B. Chrysozonicus*, Rupp.

ABOVE black, striped with yellow; beneath, yellowish white; frontlet crimson; chin yellow; ears black, margined with white ; wing-coverts and margins of the lesser quills, golden yellow. Length, 4" 6'''; wing, 2" 4'''; tail, 1" 2'''.

Inhabits South Africa.—Dr. Burchell's Coll. Swain.: loc. cit.—Sed non vidi.

Genus CAPITO, Vieillot.

Bill long, broad at the base, with the sides suddenly compressed, and the culmen arched towards the tip ; the gonys long and advancing upwards ; the nostrils basal, lateral, and rounded ; the base of the upper mandible furnished with a few short bristles ; wings moderate, and pointed, with the fourth, fifth, and sixth quills equal and longest ; tail long, and rounded ; tarsi as long as, or longer than, the anterior outer toe, and covered with broad scales ; toes unequal, the outer pair the longest and equal, the inner pair short and unequal ; the claws moderate, compressed, and curved.

469. Capito Vaillantii ; *Picus Cafer*, Gml., Hartl.

W. Af., p. 176 ; *Trachyphonus Vaillantii*, Ranz. ; *Micropogon Sulphuratus*, Lafren. ; *M. Occipitalis*, Rupp. ; *Polysticte Quopapa*, Smith ; *Picus Lathamii*, Wahl., Le Vail. Promerops, t. 52 ; Bp. Cons. Av., p. 142, Mag. Zool. 1836, t. 60.

HEAD crested, black ; forehead, cheeks, throat, rump, and belly, yellow ; the latter spotted with red-brown ; the cheeks imbricated, and the rump just above the tail barred with the same colour ; tail black, barred and tipped with white ; wings and back black, barred with white ; chest with a black band. Length, 8" ; tail, 3" 3'''.

From Mosilikatzi's country, procured by MM. Verreaux—non vidi.

The Sub-Family, PICINÆ, or Woodpeckers,

have the bill more or less long, broad at the base, much compressed towards the tip, which is truncated ; the sides of the upper mandible sloping, and furnished with a lateral ridge that springs from the middle at the base, gradually bends towards the lateral margin, and then extends above it to the tip ; the outer posterior toe generally longer than the outer anterior one.

Genus DENDROBATES, Swainson.

Bill as long as, or shorter than, the head, broad at the base ; the nostrils lateral, basal, and hidden by projecting plumes ;

wings long and pointed, with the second, third, and fourth quills nearly equal and longest; tail moderate, graduated, and rigid; tarsi shorter than the outer anterior toe; toes unequal, and the outer posterior one generally longer than the anterior; the hind toe short and slender; the claws long, compressed, and much arched.

470. Dendrobates Griseocephalus, Bodd.,

Pl. Enl. 786, f. 2 ; *Picus Capensis*, Gmel. ; *P. Caniceps*, Wagl. ; Swain., Nat. Lib., Vol. 8, p. 154 ; *Le Pic Olive*, Le Vail., Nos. 248, 249.

GENERAL colour above, rich olive-yellow ; crown of head, rump, and upper tail-coverts, brilliant crimson; sides of head and under parts cinereous. The female resembles the male in all except the crimson crown, which in the former is cinereous. Length, 7″ 6‴; wing, 4″ 6‴; tail, 3″.

Not uncommon in all wooded parts of the colony. Many have been sent me from Constantia and Rondebosch. Le Vaillant states that it breeds in holes of trees, and lays four white eggs. Found also in Natal.—(Ayres.)

471. Dendrobates Namaquus, Licht., Cat.

Dupl. Berl. Mus., p. 17 ; *Picus Mystaceus*, Vieil. ; *P. Diophrys*, Steph. ; *P. Biarmicus*, Cuv.; *Le Pic à Double Moustache*, Le Vail., Nos. 251, 252.

UPPER parts, dull-green, barred with dirty-yellow; forehead black, each feather tipped with white ; top of head crimson in the male (black in female); back, black; cheek, chin, and throat, white, crossed by two black lines, one passing over the ear from the corner of the eye, the other extending from the base of the lower mandible to the shoulders ; under parts greenish-grey, faintly barred with dirty-yellow ; shafts of wing and tail feathers, yellow, as are also the tips and underside of the latter, and the upper coverts. Length, 8″ 6‴ ; wing, 5″; tail, 3″ 6‴.

Le Vaillant states he found this species in Kaffraria, and that it lays four white eggs. I have never received it from any of my correspondents, except Mr. F. Green, who procured it in Damaraland. Mr. Andersson also obtained it there in some plenty.

472. Dendrobates Fulviscapus. (Illig.) Swain.;

Picus Fulviscapus, Ill. ; *P. Fuscescens*, Vieil. ; *Le Petit Pic à Baquettes D'Or*, Le Vail., No. 253 ; *P. Chrysopterus*, Cuv.

GENERAL colour above, sepia-brown, barred all over with dirty-yellow or white ; forehead brown ; top and back of head crimson ; under parts grey, striped on the breast, and barred on the belly and vent with the colour of the back ; shafts of the wing and tail feathers, and under side of the latter, golden-yellow. Female resembles the male, but has the crimson of head replaced by dark-brown, approaching to black. Length, 6″ ; wing, 3″ 9‴ ; tail, 2″ 6‴.

This little woodpecker has a wide range over the colony. I have received it from Rondebosch, Beaufort, Colesberg, and the Knysna. It appears in Chapman's collection, and also in Mr. Andersson's. It affects the dead stumps of euphorbias and aloes ; and reminded me, from this peculiarity, of the Indian *P. Maharattensis*, which constantly keeps to the "Candelabra Euphorbia" of Ceylon. Le Vaillant states that they lay from five to seven white eggs.

The Sub-Family, GECININÆ, or Green Woodpeckers,

have the bill more or less long, strong, and straight, with the base broad, and the sides compressed towards the tip, which is truncated and acute ; the sides sloping, and furnished with a lateral ridge, which springs from above the nostrils, and runs along near the culmen towards the end of the upper mandible for two-thirds of its length.

Genus CAMPETHERA, Gray.

Bill long and strong ; with the base broad, the culmen slightly curved, and the sides compressed towards the tip, which is acute ; the lateral ridge running near the culmen from the base towards the tip, and the gonys moderate, angulous, and ascending ; the nostrils basal, lateral, and covered by the projecting plumes ; wings long, with the first quill nearly as long as the second, which is the longest ; tail moderate and graduated, with the tips of the feathers pointed and rigid ; tarsi short, and covered in front with broad scales ; toes long, the outer pair unequal, the anterior one the longest ; the claws long, strong, curved, and acute.

*D

473. Campethera Nubica, Bodd ; Pl. Enl. 667 ;
Picus Cafer, Lath. ; Picus Punctatus, Cuv., Vol. 2, p. 451 ; Le Pic Tigré, Le Vail., No. 250 ; Picus Notatus, Licht., Nat. Lib., Vol. 18, p. 158 ; Tachepetes Cafer, G. R. Gray.

UPPER parts, dull-green, sparsely mottled with yellow; approaching to yellow on the rump, where it is barred with yellowish-white; under parts dirty-yellow, profusely spotted with large bean-shaped blotches of the colour of the back; sides of the head whitish, mottled with dull-black; moustache and back of head bright crimson; forehead in the male, dark-brown, each feather tipped with crimson; in the female, with yellow: she also wants the moustache. Length, 7"; wing, 4" 9'''; tail, 3" 6'''.

Received from Mr. Henry Bowker, from the Transkei, and from Mr. Arnot at Colesberg. Appears in Chapman's collection in considerable number. Le Vaillant cites the George forests and the neighbourhood of the Gamtoos River, and Kaffraria, as habitats, and states that it lays four eggs, marked with brown, on a bluish ground.

474. Campethera Chrysura. (Swain.) Reich. ;
Dendromus Chrysurus, Swains. B. of W. Af. ; Dendrobates Chrysurus, p. 158, Bonap. ; Picus Smithii, Malh.

ABOVE, olive-grey, spotted and banded with whitish; beneath fulvous-white, striped on the body, and spotted on the throat with black. Male with the upper part of the head, and stripe from the end of gape, crimson. Female, with the front and crown blackish and unspotted; shafts of tail-feathers golden-yellow. Length, 8"; wings, 4" 3'''; tail, 3".

Hartlaub, in his Birds of Western Africa, gives South Africa as a habitat of this species, on the authority of M. Malherb. It has also been procured by Mr. Ayres at Natal.

The Sub-Family, COLAPTINÆ, or Ground Woodpeckers,

have the bill broad at the base, and the sides compressed to the end, with the culmen much curved to the tip, which is acute; the sides of the upper mandible sloping, and the lateral ridge scarcely visible, or entirely wanting.

Genus COLAPTES, Swainson.

Bill long, rather slender, broad at the base, and narrowed towards the end, with the culmen elevated at the base, and curved to the tip, which is rather acute, the lateral margins slightly curved, and the gonys angulated and curved upwards to the end of the lower mandible; the nostrils basal, lateral, and covered by projecting plumes; wings long, with the first quill short, and the fourth and fifth quills the longest; tail rather long, graduated, and the ends of the feathers narrow and rigid; tarsi short, robust, and covered with broad scales; toes unequal, the outer anterior toe longer than the outer posterior one; the claws moderate, compressed, and acute.

475. Colaptes Olivaceus. (Lath.) *Picus Arator*, Cuv. Vol., p. 450; *Picus Olivaceus*, Lath.; *Le Pic Laboureur*, Le Vaillant, No. 254, 255; *Geocolaptes Terrestris*, Burchell.

GENERAL colour, brown, mottled with dirty-yellow; rump crimson; breast and belly, pale-crimson, in the male, who has also an indistinct crimson moustache; in the female these parts are brown; tail above, dark-brown, barred with yellow, the tip ridged, and golden-orange; below brown, glossed with golden-yellow, the yellow bars also showing. Length, 10"; wing, 5" 3"; tail, 3" 9"'.

This singular bird presents a remarkable instance of the adaptation of creatures to the localities wherein their lot is cast. Though belonging to the woodpecker tribes, it never pecks wood, but bores its way into the banks of rivers, sides of hills, or the walls of mud-buildings, in search of its prey, and for a home for its young. It also seeks for food on the ground, in the same manner as the *Golden-winged Woodpecker* of North America; its flight likewise struck me as very similar.

It excavates a hole, sometimes several feet in depth, in which to deposit its eggs, which are pure white, and from three to five in number: axis, 13"'; diam., 11"'.

Families seem to keep in company until the breeding στοργη separates them. They feed together, and roost together in some deserted hole; and their loud, harsh cries, as they call to each other, may be heard to a considerable distance. It is common throughout the whole of the colony.

The Sub-Family, YUNCINÆ, or Wrynecks,

have the bill short, straight, with the tip acute; the wings moderate and pointed; the tail moderate, rounded, and composed of soft and flexible feathers; the tarsi short; toes moderate.

Genus YUNX, Linn.

Bill short, straight, with the culmen slightly sloping to the tip, which is rather acute, and the gonys of the lower mandible moderate and advancing upwards; the nostrils basal, lateral, partly closed by a membrane, and hidden by the frontal plumes; wings moderate, pointed, with the first quill rather shorter than the second, which is the longest; tail moderate, rounded, and composed of soft and flexible feathers; tarsi short, partly clothed with feathers, but mostly covered by broad scales; toes moderate, the two anterior ones united at the base by a membrane.

476. Yunx Pectoralis, Vigors; Pro. Z. S. 1831, 93; *Y. Ruficollis*, Licht., Mag. Zool. 1835, Ois, t. 33, Gr. et Mitch., Gen. B., Pl. 112.

ABOVE, brown, profusely variegated with fine markings of black, and different shades of grey; outside webs of the wing-feathers barred with rufous; tail barred with black; chin and throat, deep chesnut; belly and vent dirty-white; each feather with a dark-brown line down the shaft; under tail-coverts rufous. Length, 6″ 6‴; wing, 3″ 6‴; tail, 2″ 3‴.

This is evidently a rare bird in the neighbourhood of the colony, but one specimen having reached my hands from any quarter. This was sent by Mr. Henry Bowker from Butterworth, Transkei. Mr. Ayres, however, seems to have found it not uncommon in Natal.

The Fourth Family, CUCULIDÆ, or Cuckows,

have the bill of various length, generally slender, and more or less compressed on the sides; the culmen curved to the tip, which is more or less emarginated; the nostrils lateral, and placed in a membranous groove; the wings long, and generally pointed; the tail lengthened and rounded; the tarsi more or less long, and covered with broad transverse scales; the toes long and unequal; the claws moderate, and more or less curved.

The Sub-Family, INDICATORINÆ, or Honey-Guides,

have the bill short, broad at the base, with the culmen curved, and the sides compressed to the tip, which is entire; the nostrils lateral, and placed near the culmen, in a membranous groove; the wings long and pointed; the tail moderate and emarginated; the tarsi very short, and covered with transverse, broad scales; the toes unequal, and the outer anterior toe the longest; the claws moderate and strong.

Genus INDICATOR, Vieillot.

Bill more or less short, and broad at the base, with the culmen curved, and the sides compressed to the tip, which is entire; the lateral margins nearly straight, the gonys moderate, and ascending; the nostrils lateral, and placed in a membranous groove, with the opening linear, and near the culmen; wings long and pointed, with the first quill nearly as long as the third, fourth, and fifth, which are equal and longest; tail moderate, emarginated in the middle, and rounded on the sides; tarsi shorter than the outer anterior toe, and covered with broad scales; toes unequal, the outer anterior toe the longest, the anterior pair united at their base; the claws moderate, compressed, and curved.

477. Indicator Major, Steph. et Vieil.; *Indicator Levaillantii*, Leadb.; *I. Maculatus*, G. R. Gray; *Indicator Flavicollis*, Swain., W. Af., 2, p. 198; *Le Grande Indicateur*, Le Vail., No. 241.

GENERAL colour above, brown, with a tinge of yellow on the forehead; ear-covers whitish, and beneath them a dark-brown line; chin, throat, and neck, yellow, fading into white on the sides, belly, and vent; thighs immaculate. The tail, as in *I. Variegatus*, consists of twelve feathers, thus marked: first centre pair all brown; second pair following, brown, with *outer half* of the *inner* web white; the three outer pair, with the tip and outer lower portion of the outer web, brown; the outer pair very short. Size similar to that of *I. Variegatus*.

Le Vaillant found this species and *I. Variegatus*, which he mistook for the female, all along the South-Eastern Coast, as far as Kaffraria. I have received it from Messrs. Atmore and Cairncross, killed at Swellendam and George. Le Vaillant states they build in holes of trees, and lay four white eggs.

In November, 1865, while at Tygerhoek, on the River Zonder End, I shot a specimen clinging to the upright branch of a tree like a woodpecker. I subsequently saw a single bird at the entrance of Cogman's Kloof in December. I fancy it is pretty generally distributed throughout the colony.

478. Indicator Variegatus, Less.; *I. Levail-lantii*, Bp.; *I. Maculicollis*, Sunde.; *I. Sparrmanni*, Leadb.

GENERAL colour above, dull olive-green, changing to yellow on the wings; below dirty yellowish-white, immaculate on the centre of the belly and vent, but much variegated on the throat, breast, flanks, and thighs, with brown; forehead speckled with white. The tail consists of twelve feathers, thus marked: two innermost pair all brown; one pair following white, with portion of inner web brown; three outer pair white, with brown tips; the outermost pair very short. Length, 7″ 6‴; wing, 4″ 4‴; tail, 3″ 4‴.

I shot a solitary specimen of this *Honey-Guide* at the Knysna. It was flitting about some bushes in the forest. The bill in this species is stronger and more curved than in *I. Major*, and the edge of the upper mandible more scalloped.

With respect to the oft-repeated story of the *Honey-Guide* leading persons to the nest of the honey-bee, it is as well to mention that the bird will perform the same antics, and utter the same cries, to lead any one to a leopard, wild cat, or snake, or will even follow a dog with the same vociferations.

479. Indicator Albirostris, Temm.; Pl. Col. 867; Cuv., Vol. 2, p. 462; *I. Leucotis*, Swain. W. Af., Vol. 2, p. 193; *I. Flaviscapulatis*, Rüpp; *Cuculus Indicator*, Linn.

GENERAL colour, brownish-grey, lightest on the breast, and fading to pure white on the belly, vent, and thighs, these latter marked with broad streaks of brown; rump white, streaked in the same way; outer margins of all the wing-feathers dirty yellowish-white, those of the secondaries being the most marked; ear-tufts white; chin and upper part of throat black; the two middle tail-feathers entirely brown; the next pair have the inner webs, with the exception of the tip, white, and the outer three are all white, except at the tips; the outermost feather is in this, and *I. Major*, considerably the shortest. Length, 7″ 6‴; wing, 4″ 4‴; tail, 3″ 3‴.

A specimen was obtained near Swellendam by Mr. Cairncross; and the Messrs. Atmore sent me a second from the same locality. It is, however, a very scarce species.

480. Indicator Minor, Vieil.; *I. Minimus,*

Temm.; *I. Buphagoides,* Leadb.; *I. Diadematus,*
Rüpp., Neue Wirb., p. 61.; *Cuculus Minor,* Cuv. Vol.
2, p. 462; *Le Petit Indicateur,* Le Vail!, No. 242.

GENERAL colour above, yellowish-brown; the yellow brightest
on the wing-feathers; head ashy-brown; cheeks, chin, throat,
breast, and belly, cinereous; vent white; thighs faintly
maculated with brown; moustache blackish; tail composed
of twelve feathers, thus marked: two inner pair all brown,
four outer pair white, with base and tips brown. In this
species the outer pair are but very slightly shorter than the
next. Length, 5" 9"'; wing, 3" 7"'; tail, 2" 6"'.

The lesser Honey-Guide is found at the Knysna; and LeVaillant gives
the Swartkop and Sunday's Rivers as other localities. It probably
extends all along the South-East Coast. Mr. Atmore procured it at
Blanco, and writes thus: "May 26, 1864. I have had another good
opportunity of watching the habits of *I. Minor.* I used to wonder
where they got all the bees-wax that is usually in their gizzards, and
the other day I found out. There was a ♂ at a bee-hive as busy as
possible catching bees. After watching him for some time, Tom shot
him, and his gizzard was full of bees' legs, with the wax on them. He
is held in no repute here as an indicator; but *I. Major* is, and he is
scarce." Mr. Atmore has mistaken the pollen of the bee for wax. The
bird's habit of capturing bees like a fly-catcher is interesting; but his
most singular statement follows. I had asked him for information
upon certain points, and he writes: "I can't answer your queries yet,
but I can tell you that *I. Minor* kills and eats small birds as savagely
as *Lanius Collaris!* The very first I shot was in the act of eating a
sparrow *that I saw him kill in flight:* I suspect the others of similar
propensities." In another letter he writes: "Eggs white, in nests of
Picus Capensis and *Laimodon Unidentatus.*"

The Sub-Family, COCCYZINÆ, or Ground Cuckows,

have the bill generally elevated at its base, with the culmen
arched, and the sides much compressed to the tip, which is
entire; the gonys usually long and straight; the nostrils
basal, with the opening generally linear, and partly closed by
a scale; the wings moderate, and more or less rounded; the
tail long and graduated; the tarsi lengthened, and covered
with broad scales; the toes unequal, and armed with various-
sized claws.

Le Vaillant, in his "Oiseaux d'Afrique," gives as South African,
several birds belonging to this sub-family which are now well known
to inhabit Madagascar. I feel sure that Le Vaillant must also have

been cognisant of their habitat at the time he described them, as he has actually in one instance given the Malagash name. They are as follows :—

Genus COUA, Cuvier.

Bill moderate, with the culmen more or less curved, and the sides compressed to the tip, which is entire ; the lateral margins curved, and the gonys long and ascending ; nostrils basal, sunk in a broad short groove, with the opening anterior, linear, and oblique ; wings moderate, and much rounded, with the fifth, sixth, and seventh quills equal and longest ; tail long, broad, and much rounded on the sides ; tarsi longer than the middle toe, and covered in front with broad scales ; toes unequal, and armed with moderate, compressed, curved, acute claws.

481. Coua Cristata. (Linn.) Pl. Enl. 589 ; *Cuculus Cristatus*, Linn.; *Le Coua*, Le Vail., No. 217 ; *Serisomus Cristatus*, Sw. ; *Cuculus Madagascariensis Cristatus*, Bris. ; Shaw, Vol. IX., Pt. 1, p. 118.

HEAD and upper parts of body, ash-colour, inclining to green ; throat and fore part of neck cinereous ; lower part of neck and breast greenish ; belly and sides whitish, with a rufous tinge ; under tail-coverts white ; thighs white, marked with a band of light ash-colour; quills pale-green, with a blue and violet gloss ; beneath cinereous ; tail cinereous ; lateral tail-feathers tipped with white, the two middle tail-feathers longest. Length, 14".

Le Vaillant affirms that he found this species in South Africa ; but does not indicate the locality. Sundevall doubts the fact, the bird being an inhabitant of Madagascar.

482. Coua Cœrulea. (Linn.) Cuv., Vol. 2, 458, Pl. Enl. 295, f. 2 ; *Polophilus Cœruleus*, Shaw, Vol. IX., Pt. 1, p. 56 ; *Le Coua-tait-sou*, Le Vail., No. 218 ; *Cuculus Cœrulœus*, Linn.

GENERAL colour, beautiful blue-green, changing in certain lights to violet ; colours of female less vivid. Length, 14".

"Forests of Kaffraria " (Le Vaillant) ; "Tantummodo incola Madagascariæ" (Sundevall).

Genus LEPTOSOMUS, Vieil.

Bill about the length of the head, robust ; upper mandible curved, notch near the tip ; gonys straight ; nostrils oblong, oblique ; the margins elevated, naked, and placed towards the middle of the upper mandible; feet short; toes in pairs; wings lengthened, pointed, the first and second quills longest; tail moderate, even.

483. Leptosomus Viridis, Vieil.; *Cuculus Afer*, Gmel.; *Le Vouroug-driou,* Le Vail., Nos. 226, 227; Cuv., Vol. 2, p. 461; *Bucco Africanus,* Shaw, Vol. IX., Pt. 1, p. 25; *Cuc. Madagascariensis Major,* Bris.

HEAD, throat, and neck, ash-colour; crown of head blackish, with a green and coppery gloss; from the bill to the eye a white line; back, rump, scapulars, and upper wing and tail coverts, green, glossed with copper; breast, belly, sides, thighs, under wing and tail coverts, bright grey; greater quills, blackish; lesser quills, dull-green, with a greenish copper gloss; tail composed of twelve feathers of equal length: above copper and green-gold, beneath black. Length, 15″.

Le Vaillant affirms he found this bird in the forests of Kaffraria; Sundevall states it is exclusively a Madagascar species. From this island it is constantly received; and the very name of "Vouroug-driou" adopted by Le Vaillant shows whence he obtained his specimens, that being the Malagash designation of the species in question.

Genus CENTROPUS, Illiger.

Bill short, elevated at the base, with the culmen much curved, and the sides much compressed to the tip, which is entire; the lateral margins much curved, and the gonys long and straight; the nostrils sunk in a deep broad groove, with the opening anterior, linear, or oblique, and partly closed by a membranous scale; wings rather long, and much rounded, with the fourth, fifth, and sixth quills equal and longest; tail long, broad, and rounded on the sides; tarsi as long as the middle toe, strong, and covered in front with very broad scales; toes unequal, the anterior one the longest, the inner posterior toe armed with a long straight claw.

484. Centropus Nigrorufus, Cuv., Vol. 2, p. 460; *Le Coucal Noirou,* Le Vail., p. 220; *Polophilus Tolu?* Shaw., Vol. 1, P. 1, p. 52; *Cuc. Madagascariensis,* Bris.; *Corydonyx Bicolor,* Vieil., Encl. 1354.

GENERAL colour throughout, black, with a blue or green gloss; wings chesnut. Shaw (loc. cit.) describes the "lower part of

*E

the breast dirty-white;" but Le Vaillant does not allude to this, either in his description or plate. Length, 14" 3'".

Found, according to Le Vaillant, near the Zwart River ; procured also by Whalberg, but locality not given. I have not yet received it from any of my contributors ; but I procured a specimen answering the description in every particular in Boyana Bay, on the North-West Coast of Madagascar.

485. Centropus Rufinus, Cuv. ; Regn. An., t. p. 416, and Vol. 2, p. 461 ; *Le Coucal Rufin*, Le Vail., No. 221; *Polophilus Rufus*, Shaw., Vol. 1, Pt. 1, p. 44.

GENERAL colour, rufous, all the feathers of the superior parts of the body having a dash or streak of light-red, or white; wings reddish, the last feathers barred with fuscous, as are those of the upper part of the tail ; tail reddish, the two inter-mediate feathers with transverse lines of brown. Length, 11".

Le Vaillant states he found this species on the " Great Fish River ;" but Sundevall observes that the bird is not now known in Africa ; and suspects that Le Vaillant described his specimen from the young of *C. Affinis*, of Java, which his plate does not badly represent. Bona-parte, in his Conspectus, gives *C. Burchellii*, Sw., as a synonym of this, but with a mark of doubt (?).

486. Centropus Superciliosus; Ehrenburg et Rüppell, Neue Wirbelthier, p. 56, and Faun, t. 21, f. 1.

HEAD and back of neck, deep black-brown, the feathers on the latter stiff ; the shaft and part of the plumlets white, margined by an indistinct black line ; a white eyebrow, springing from the nostrils, mingles with these feathers at the back of the head ; back reddish-brown, some of the feathers with white shafts ; rump black, transversely barred with narrow light-brown stripes ; wings bright rufous ; tail glossy brown-green, faintly barred at the base like the rump, and narrowly tipped with white ; under parts fulvous, with white shafts ; flanks and thighs barred with brown ; irides red. Length, 12" 3'" ; wing, 6" 9'" ; tail, 8" 6'".

A fine specimen of this bird was sent to me in the flesh, during the winter months, from Swellendam, by the Hon'ble R. Southey. Its stomach contained insects. It has also been found in Natal by Mr. Ayres.

487. Centropus Burchellii, Swains. ; An. in Menag., p. 321.

ABOVE, cinereous ; wings red, beneath whitish ; crown, nape, and ears, black ; feathers of the neck and interscapulars with a central white stripe, margined with black ; rump and upper tail-coverts transversely lineated with greyish-white and black ; tail black, with greenish reflections, tipped with white, the feathers lineated at their basis. Length, 16″ 6‴ ; wing, 7″ 9‴ ; tail, 9″ 7‴.

Le Vaillant has evidently, in his description of "*Le Coucal Hou-hou*," confounded the Egyptian species *(C. Ægyptius)* with this one ; his figure, indeed, better accords with our bird than with the Egyptian.

I have received but two specimens of "Burchell's lark-heeled cuckoo." One was shot in "Grootevadersbosch," near Caledon, by the Hon'ble T. H. Vigne ; the other near Swellendam, by Mr. Cairn-cross.

Le Vaillant states that he first met with this species near the Gam-toos River, and that it makes its nest in holes of trees, laying four eggs, of a reddish-white colour.

488. Centropus Æthiops, Cuv. ; Vol. 2, p. 461,

and Regn. An. 1, p. 426 ; *Le Coucal Nègre*, Le Vail., No. 222 ; *Corydonix Nigerrimus*, Vieil. ; *Polophilus Maurus*, Stephens ; Shaw, Vol. IX., Pt. 1, p. 57.

ALL black ; the female being sooty-black on the belly. Length, 11″.

Le Vaillant states he discovered this species in Kaffraria, and describes its nest in a hollow tree, and its egg as white. Sundevall, in his critique on Le Vaillant's work, denies its existence, and stated that it is a manufactured species.

The Sub-Family, CROTOPHAGINÆ, or Anis,

have the bill more or less lengthened, with the culmen arched, and the sides much compressed ; the nostrils basal, lateral, and pierced in the substance of the bill ; the wings short and rounded ; the tail lengthened, broad, and gradu-ated ; the tarsi long, and covered with broad transverse scales ; the toes long, placed two and two, and the two outer ones the longest ; the claws short and curved.

Genus ZANCLOSTOMUS, Swainson.

Bill much compressed its whole length ; gonys curved downwards ; culmen and upper mandible greatly curved ; basal margin considerably dilated ; tarsus and middle toe equal ; lateral fore toes nearly equal ; claws short.

489. Zanclostomus Aereus, Vieil. ; Hartl. W.

Af., 187 ; *Z. Flavirostris*, Swain., Nat. Lib., Vol. 11, p. 183, Pl. 19 ; *Le Coucou Gris Bronzé*, Le Vail., No. 215.

BODY above, wings, and tail, glossy violet-purple ; head, neck, and body, beneath cinereous ; tail, beneath, with lilac

reflections ; bill yellow, with blackish spot in front. Length,
13″ ; wing, 4″ 6‴ ; tail, 9″.

Kaffraria and Port Natal—Whalberg (Hartl. loc. cit.)   Natal—
Gurney (Ibis, 1859, p. 248). Rare—frequents dense bush, and creeps
about like the colies ; feeds on locusts and grasshoppers.—(Ayres,
" Ibis," loc. cit.)

The Sub-Family, CUCULINÆ, or Cuckows,

have the bill broad, and rather depressed at the base ; the
culmen curved, and the sides compressed to the tip, which is
entire, or slightly emarginated ; the nostrils basal, and membranous, with the opening exposed ; the wings long, and
generally pointed ; the tail long, and usually graduated ; the
tarsi short, partly clothed with feathers, and partly covered
with broad scales.

Genus CUCULUS, Linn.

Bill broad, and rather depressed at the base, with the culmen curved, and the sides gradually compressed towards the
tip, which is entire and acute ; the gonys long and arched ;
the nostrils basal, lateral, and placed in a short broad membranous groove, with the opening round and exposed ; wings
lengthened and pointed, with the third quill the longest ; tail
long, graduated, or even, and the outer feather on each side
shorter than the others ; tarsi very short, feathered below the
knee, and the exposed part covered with broad scales ; toes
unequal, the outer anterior toe the longest, and united to the
inner one at the base.

490. Cuculus Solitarius, Cuv. ; Vol. 2, p. 454 ;

Le Coucou Solitaire, Le Vail., No. 206 ; Shaw, Vol.
IX., Pt. 1, p. 84 ; *C. Capensis*, Lath. ; *C. Rubiculus*,
Swain. Nat Lib., Vol. 12, p. 181.

GENERAL colour above, greyish-black, lightest on the forehead ; chin cinereous, passing into rufous on the breast ; body
beneath, fulvous-white, barred with black ; under tail-coverts
immaculate ; the tail-feathers have four white spots along the
shafts, and are otherwise mottled with white ; legs bright-
yellow. Length, 12″ 6‴ ; wing, 7″ 6‴ ; tail, 6″ 6‴.

This cuckoo, known among the colonists by the name of " *Pietmijn-
vrouw*," from its call resembling these words, is a periodical visitant
over the whole colony, extending even as far as the Cape peninsula.
They generally make their appearance from November till Christmas

time, and feed on caterpillars, several species of hymenoptera, and on beetles.

Le Vaillant states that they deposit three eggs, which are of a pinkish colour, dotted with clear brown spots, in the nests of the " Capocvogel," "Jan Fredrik" (*Pet. Superciliosa*), and other small birds. My friend Mr L. du Toit confirms this statement as regards the "Jan Fredrik."

Found also at Natal by Mr. Ayres.

491. Cuculus Gularis, Steph.; Gml. Zool. 9, p. 83; *Cuculus Capensis*, Shaw, Vol. IX, Pt. 1, p. 85; *Le Coucou Vulgaire D'Afrique*, Le Vail., No. 200, ♂, 201 ♀.

GENERAL colour above, bluish-grey; tail dark, and variegated with white spots near the shafts and edges of the feathers; inner webs of wing-feathers barred with white; under parts all white, barred with grey, and having a rufous tinge; chin and throat, light blue-grey, and immaculate. Length, 12″; wing, 9″; tail, 7″ 2‴.

Very rare near the sea-coast; but becoming more common towards the Interior, appearing frequently in Chapman's collection. Le Vaillant describes the egg as olive-grey, dotted with red.

492. Cuculus Clamosus, Cuv.; Vol. 2, p. 455; *Cuculus Nigricans*, Swain. Nat. Lib., Vol. 12, p. 180; *Le Coucou Criard*, Le Vail., No. 204, 205.

GENERAL colour, black, glossed with blue; feathers of wings and tail, irregularly barred and spotted with white; tips of the latter white; inside of wing albescent, with dark bars. Length, 12″; wing, 7″; tail, 5″ 9‴.

Le Vaillant states that the "noisy cuckoo" is common about the Sunday and Swartkop Rivers, and throughout Camdeboo, depositing its eggs in the nest of the "Capocier" (*Drymoica Capensis*). I have never seen a colonial specimen; but Mr. Andersson procured it in some abundance at Elephant Vley, in Damaraland. Mr. Ayres procured it in Natal, and remarks that it feeds on caterpillars, disappearing from the country during the winter months.

Sub-Genus CHALCITES, Lesson.

Plumage shining metallic-green; bill and general structure of *Cuculus*; tarsus very short, almost entirely plumed; rump and upper tail-coverts soft.

493. Chalcites Auratus. (Gmel.) Shaw, Vol. IX.,

Pt.1, p. 127 ; *Le Didric*, Le Vail., Nos. 210 and 211 ;
Lampromorpha Chalcopepla, Vigors ; *Chrysococcyx Auratus*, Bp. ; *Cuculus Cupreus*, Bodd. (nec Lath.)

ABOVE, shining coppery-green, mottled with white; a white stripe passes down the centre of the head, and over each eye ; under parts white, barred with coppery-green on the flanks and sides. Length, 7″ ; wing, 4″ 8‴ ; tail, 3″ 4‴.

This beautiful little cuckoo, known by the name of "*Didric*," from its oft-repeated mournful cry of "*Di-di-di-didric*," is extremely abundant throughout the Karroo. At Nel's Poort, near Beaufort, I have frequently seen a dozen or more in a morning, while their loud notes were incessantly ringing in my ears ; they are, however, so shy that I only procured three specimens in as many months. When calling, they perch on the summit of some dead branch, ready to do battle with any male, or engage in an amorous chase after any female that comes within their ken. They pursue each other with great ardour, turning, twisting, and dashing about with much rapidity. The stomachs of those examined contained nothing but small insects, chiefly swallowed whole.

I also obtained specimens at the Knysna, and from still farther to the Eastward. It appears in Mr. Chapman's collection, and has been sent from Colesberg.

Le Vaillant states that it lays a white egg, and gives an account of the manner in which it is carried in the mouth, to be placed in the nests of those birds which are selected as foster-mothers for its neglected offspring.

494. Chalcites Klaasii, Less. ; *Cuculus Klaasii*,

Vieill. ; Cuv., Vol. 2, p. 456 ; *Le Coucou de Klaas* Le Vail., No. 212 ; *Mietje* of Colonists.

ABOVE, all shining-green, with a very slight coppery tinge ; a small white streak over the eye ; under parts pure white, faintly barred on the flanks ; a green patch on each side of the chest, and one of the same colour on the thighs ; two centre tail-feathers all green, the rest white, with narrow bars of green sparsely placed up the inner webs, and a broad patch of the same colour near the tip, which is pure white. The female is, on the upper parts, green, profusely barred with brown ; on the under side white, similarly barred with coppery-green ; tail-feathers much as in the male. Length, 6″ ; wing, 4″ 2‴ ; tail, 3″.

Klaas's cuckoo is not uncommon in most wooded parts of the colony, extending even as far as the Cape peninsula. It is, however, not nearly so abundant as the preceding, from which it is easily distinguished by its cry.

495. Chalcites Smaragdineus. (Swain.) *Cuc.*
Cupreus, Lath. ; Cuv., Vol. 2, p. 456 ; *Cupreous Cuckow*, Shaw., Vol. IX, Pt. 1, p. 129.

GENERAL colour above, with the neck and breast, shining emerald-green, if held between the spectator and the light, if held from the light, a beautiful golden and copper gloss pervades the green ; the belly and vent vary in some specimens from white to bright-yellow. This colour also fades in death ; but I have had on the same stick, after a morning's shooting, specimens exhibiting all the shades from white to yellow ; outer tail-feathers broadly barred with white ; tail-coverts white, barred with green. Length, 8″ ; wing, 4″ 7‴ ; tail, 3″ 6‴.

The "*Golden Cuckoo*" is abundant at the Knysna, and throughout all the forest regions to the eastward along the sea-coast. The males are killed in a proportion of at least ten to one female, from their habit of perching on the topmost branches of trees, and from thence uttering their loud call of love or defiance. This is easily imitated by a whistler, and the bird will answer the whistle, and remain on the look-out for its supposed foe, until the shooter creeps up within range.

All the cuckoos of South Africa appear to be migratory, the majority of them making their appearance in the colony about October or November ; some a little later.

Genus OXYLOPHUS, Swainson.

Head crested ; bill slender, considerably and suddenly compressed from the nostrils, which are ovate ; upper mandible entire ; wings moderate, rather pointed, fourth quill longest.

496. Oxylophus Glandarius; *Cuculus Glandarius*, Linn., Pl. Col. 414, Ed. Birds, p. 57 ; *C. Andalusiæ*, Briss. ; *C. Macrurus*, Brehm ; *C. Melissophanus*, Vieil. ; *C. Pisanus*, Gmel.

GENERAL colour above, ashy, spotted with white, beneath white, tinged with yellow, darkest on the throat ; tail long, graduated, and tipped with white ; head crested. Length, 14″ ; wing, 8″ ; tail, 8″ 6‴.

This bird is very rare within the colony. All that I have seen, three in number, came from Kaffraria. One was sent by Capt. Bulger from Windvogelberg. It appears in Chapman's collection not unfrequently.

497. Oxylophus Coromandus, Linn.; Pl. Enl.
274, f. 2 ; Cuv., Vol. 2, p. 455 ; *Le C. à Collier
Blanc*, Le Vail, No. 213 ; Shaw, Vol. IX., Pt. 1, p.
119 ; *Cuculus Collaris*, Vieil.

HEAD crested and blackish, as is the upper part of the body ;
upper part of neck surrounded by a white collar ; throat and
thighs blackish ; fore part of neck, breast, belly, and under
tail-coverts, white; scapulars and wing-coverts, blackish in
the middle, with rufous margins, but the great wing-coverts
farthest from the body are rufous ; tail blackish, and wedge-
shaped. Length, 12" 6'".

Le Vaillant affirms that he found this bird on the Swartkop and
Sunday's Rivers ; but this is doubtful, it being an Indian species.
I found it in Ceylon : and had the bird appeared in any collection
formed here since my residence in the colony, I must have noticed it.

498. Oxylophus Edolius, Swain.; *Cuculus Edo-
lius*, Cuv., Vol. 2, p. 455 ; *Cuc. Serratus*, Sparm. ;
Oxylophus Serratus, Gray ; *Le Coucou Edolio*, male,
Le Vail., No. 207 ; *Nieuwejaarsvogel* of Colonists.

GENERAL colour throughout, black, with greenish reflections ;
a broad white bar extends across the middle of the wings ;
head crested. Length, 13" ; wing, 6" ; tail, 7".

Abundant in mimosa bushes throughout the Karroo ; extends into
the Cape peninsula, and has been received from all parts of the
colony to the Eastward ; plentiful in Chapman's collection. This bird
visits the Cape about the New Year, whence the name that it has
acquired among the colonists. It evidently lays at that season, as I
took a mature egg from the body of one that was killed at Rondebosch. The egg was white, glossy, and rounded at each end : axis,
13'" ; diam., 11'". The stomach contained caterpillars, beetles, mag-
gots, and flies, but the chief mass consisted of termites. I found
them in considerable abundance at Nel's Poort, usually in pairs,
♂ and ♀, frequenting the trees along the river banks. Mr. Atmore
writes that the "eggs are white, and usually deposited in the nest
of the 'Geelgat' *(Pycnonotus Capensis)*." This is singular, as the
eggs of parasitic birds usually resemble those of the birds upon which
they are intruded.

499. Oxylophus Melanoleucus. (Lath.) *Le
Coucou Edolio*, female, Le Vail., Pl. 208.

GENERAL colour above, black, with green reflections ; below
white; tail-feathers tipped with white ; a bar of the same
colour extends across the centre of the wing ; head crested.
Length, 12" 6'" ; wing, 6"; tail, 7".

This species inhabits the same country as the preceding, and was mistaken by Le Vaillant for the female of that bird. It is, however, not so common. Its food is the same, and its call-note not very unlike.

500. Oxylophus Afer, Gray; *Cuc. Afer*, Leach, Zool. Miscel., Pl. 31; Shaw, Vol. IX., Pt. 1, p. 115; *Le Coucou, Edolio* var., Le Vail., p. 209; *Oxylophus Vaillantii*, Swain.; Nat. Lib., Vol. 11, p. 183; Zool. Ill., pl. 18; *C. Levaillantii*, Lesson.

GENERAL colour above, black, glossed with green; wing-feathers brownish; under parts dirty-white; head crested; throat and neck faintly striped with black; tail graduated, each feather, with the exception of those in the centre, tipped with white. Length, 12″; wing, 6″ 3‴; tail, 8″.

Whalberg killed this species on the Limpopo. Le Vaillant did not observe it in South Africa, his specimen having been procured near the Line; but two specimens have been forwarded to me by Mr. W. Cairncross, who obtained them near Swellendam.

Genus EUDYNAMYS, Vig. and Horsf.

Bill long, broad, with the culmen curved, and the sides compressed to the tip, which is slightly emarginated; the gonys short and angulated; the nostrils basal, lateral, and placed in a short membranous groove, with the opening large and exposed; wings moderate, with the fourth and fifth quills equal and longest; tail lengthened and rounded; tarsi rather short, robust, covered in front with broad scales; toes unequal, the outer anterior toe the longest.

501. Eudynamys Niger; *Le Coucou à Grosbec,* Le Vail. Ois. d'Af., No. 214; *Eudynamys Niger*, Bp., p. 101; *Cuculus Crassirostris*, Shaw, Vol. IX., Pt. 1, p. 16; *Cuculus Honoratus*, Linn.; *Le Tachirou*, Le Vail., No. 216.

♂—All black throughout. This is "Le Coucou à Grosbec" of Le Vaillant. ♀—Bronze-brown throughout, speckled with white: "Le Tachirou" of Le Vaillant. Length about 12″.

Le Vaillant has here committed a great blunder. He has described the male and female of one bird as two different species! affirming that he found the sexes of both. Adhering to a resolution of describing all the species *said* to be South African, we describe both the sexes, though we fully agree with Sundevall, that the bird is not South African, and regret that in this and many other instances Le Vaillant should have shown himself so false and untrustworthy.

*F

Order IV. COLUMBÆ.

THE Fourth Order, COLUMBÆ, is composed of those birds only which are well known under the denomination of Pigeons; and as it embraces but one family, the characters will be given under it.

The Family COLUMBIDÆ, or Pigeons,

have the bill short, straight, compressed, with the apical half of the mandibles more or less vaulted and strong, and the base more or less weak, and covered by a soft fleshy membrane, in which are placed the nostrils; the wings moderate; the tarsi more or less long, and robust; the toes lengthened, divided, and padded beneath.

The Sub-Family, TRERONINÆ, or Tree-Pigeons,

have the bill short, with the ends of both mandibles vaulted, and of nearly equal thickness; the tarsi very short, and more or less feathered; the toes divided at the base, the inner much shorter than the outer; and the claws short and curved.

Genus TRERON, Vieillot.

Bill short, robust, with the tips of both mandibles much vaulted, especially that of the upper: the basal portion depressed and straight; the nostrils lateral, and longitudinal; wings moderate and pointed, with the second and third quills nearly equal, and longest; the third quill with the inner web notched near the middle; tail moderate, rounded, sometimes lengthened and wedge-shaped; tarsi very short, thick, and feathered below the knee; toes free at the base, the inner shorter than the outer, the hind toe long and broad; and the claws moderate, compressed, and arched.

502. Treron Delalandi ; Bp. Consp., II., 6.

FOREHEAD, top of head, and back, dull green; back of head cinereous; chin, throat, chest, collar round the neck, and thighs, bright yellow; lower portion of breast, and belly, light cinereous; the same colour extends in a ring over the back, just under the broad yellow collar; vent white, with dark cinereous blotches; under tail-coverts deep red, with white blotches; tail above, basal two-thirds greenish-yellow; the rest dark cinereous; below, basal two-thirds very dark; and dirty whitish; shoulders slightly vinaceous, the usual yellow edging to the lesser quills and some of the wing-covers. Length, 11″ 6‴; wing, 7″; tail, 4″.

A single specimen of this bird was received from Kaffraria.

503. Treron Australis. (Linn.) Steph. Jard. and Selby, Illus. Orn., t. 81.; *Vinago Nudirostris ?* Swain. Nat. Lib., Vol. 11., p. 205.

HEAD, neck, throat, breast, and belly, light yellowish-green; indistinct collar at back of neck, flanks, and tail-feathers, cinereous, the latter with tips palest; back and wings rather darker, with a distinct greenish tinge; shoulders vinaceous; the edging of the lesser quills and wing-coverts very pale-yellow; thighs bright-yellow; vent white, blotched with cinereous; tail-coverts partly white, partly deep-red; basal two-thirds of tail very dark; the rest almost white. Length, 11″; wing, 6½″; tail, 4½″.

This accords in many ways with Swainson's *T. Nudirostris*, the type specimen of which I have seen in the Cambridge Museum; but is larger. It was procured in Damaraland by Mr. D. Kisch, who has contributed largely to the specimens in the Museum. I am ignorant of the sex.

504. Treron Abyssinica. (Lath.) Tem. Pigeons, t. 8; *Le Colombar*, Le Vail., Nos. 276, 277; Cuv., Vol. 3, p. 93; *Vinago Abyssinica*, Swain. Nat. Lib., Vol. 11, p. 202.

HEAD, neck, and breast, light olivaceous grey; body beneath yellow; wing-covers vinaceous; under tail-covers cinnamon; back green; a bright yellow line, formed by the edges of some of the covers and lesser quills, extends down the wing; tail above, light cinereous; beneath with the basal half black, the other half cinereous white. Length, 11″; wing, 7″ 3‴; tail, 4″ 2‴.

Le Vaillant states that he discovered his "*Colombar*" in Namaqualand; Sundevall shows that he has taken his figure and description from *Treron Abyssinica*, which is not found in South Africa, but is replaced by the first species, *T. Delalandi*, which is found in Kaffraria. I have before me a specimen of the true *Abyssinica*, one of *Delalandi* from Kaffraria, and another from Damaraland, *T. Australis*. This latter may be the species seen by Le Valliant, and which he has confounded with *Abyssinica*.

The Sub-Family, COLUMBINÆ, or Pigeons,

have the bill moderate, slender, with the basal portion covered with a soft humid skin, and the apical part hard, slightly vaulted, and acute at the tip; the nostrils forming a longitudinal slit in the fore part of the soft basal portion of the bill; the wings moderate and pointed; the tail of various lengths, and generally rounded; the tarsi generally short; the toes lengthened; the lateral ones mostly equal, and the hind toe about the length of the tarsus.

Genus COLUMBA, Linn.

Bill moderate, straight, the basal half of the culmen covered with a soft cartilaginous substance, the apical half hard, arched and hooked to the tip, which is somewhat compressed; the nostrils placed towards the middle of the bill, above which the skin is swollen; wings moderate and pointed, with the first quill shorter than the second, which is the longest; tail rather short, even, or rather rounded at its end; tarsi very short, and moderate; toes moderate, and free at their base, with the lateral ones equal, and the claws short, strong, and curved.

505. Columba Guineæ. (Lin.) Temm. Pig., t. 16;

Le Ramier Roussard, Le Vaill., No. 265; Cuv., Vol. 3, p. 79; *Columba Trigonigera*, Wagler. Nat. Lib., Vol. 8, p. 212; *Bosch-Duif* or *Wilde-Duif* of Colonists.

HEAD, body beneath, rump, wings, and basal two-thirds of tail, on upper side, cinereous, or plumbeus; back, shoulders, and wing-coverts, vinaceous, the latter with numerous triangular white spots, which also appear on some of the cinereous feathers of the wings; feathers of neck and breast, ruddy vinaceous; each feather being bifid at the tip, stiff, and inclined to cinereous, with a glossy green tint; apical third of tail, black on the upper side ; legs and cere round the eye, in life, crimson ; bill dark cinereous. Length, 12″ ; wing, 8″ 9‴ ; tail, 4½″.

Common throughout the colony, nesting in rocky places, on inaccessible ledges and holes ; never in trees. They fly in flocks when the crops are on the ground, and do considerable damage to the agriculturist.

In the sea-face of the mountains, of which Cape Point forms the extreme south, there are numerous caverns tenanted by these birds. Some years ago I entered one of them in a boat, and for the first time had the pleasure of seeing this fine pigeon breeding in considerable numbers : every ledge of the cavern side was tenanted by as many nests as could be conveniently stowed away, while the parent birds were continually arriving or departing on their busy task of feeding their young. The cave was unapproachable except by water, and one would have thought that the birds would have been careless in consequence in the choice of their eyrie ; but not a nest was accessible. I have, however, obtained eggs of this species from other sources ; they are always two in number, glazy white, nearly similar at each end : axis, 17‴; diameter, 12‴.

506. Columba Delagorgui, Verreaux; *C. Johannæ*, Verr. ; *C. Lunigera*, Gr.

GENERAL colour, dark slaty grey, shading into purple brown, except the tail and wing feathers, which are blackish ; lower part of neck, towards the top of head, blotched with white ; and all the neighbouring parts, in front and back, from cheeks to breast, are iridescent, with emerald, amethyst, and purple tints, each colour prevailing according to the position of the bird ; the beak is black from base to centre, the rest yellow ; feet also yellow. The female differs from the male by having the head reddish-brown, dashed with amethyst, and by the total absence of the white blotches on the base of the neck. Length, 30 centimeters.

Inhabits the forests in the neighbourhood of Port Natal, but is very rare (Delagorgue's Voyages).

507. Columba Arquatrix, Temm. Pig., t. 5 ;
Le Rameron, Le Vail., No. 264 ; Cuv., Vol. 3, p. 78 ; *Olive-Dove* and *Bush-Dove* of Colonists.

GENERAL colour above, dark ashy, inclining to reddish on the back and shoulders ; light on the edge of the wing ; and very dark on the tail, over which there is a decided greenish tinge ; wings, at the shoulders, spotted with white ; forehead vinaceous, passing into light ash-colour on the back of the head ; chin, neck, and breast, obscure vinaceous, mottled with black ; belly and flanks, a mixture of vinaceous and cinereous, with a profusion of crescent-shaped markings ; bare place round the eye ; bill and legs bright yellow. Length, 13″ ; wing, 8″ 9‴ ; tail, 5½″.

In the months of November and December these pigeons congregate at the Knysna in vast flocks, to feed on the berries that are then ripe ; they also extend to the Cape peninsula, migrating thither at the season when the wild olive is in fruit, on the berries of which it greedily feeds, whence its name. It breeds upon trees in mountain ravines, and I have been informed lays four eggs, only two of which come to perfection. Two eggs which I have had given me as the eggs of this species, are of the usual white colour and shape : axis, 1″ 6‴; diameter, 14‴. The nest is a loose structure of sticks.

Mr. Atmore writes, under date 13th May, 1864, from near Blanco : "The large yellow-billed and yellow-legged bush-dove is breeding here now, high on the mountain. They make their nests on the tops of the tree-ferns. I am trying to get some young ones for you."

Genus ÆNA, Selby.

Bill very slender, and moderate ; wings lengthened, with the first three quills nearly equal, and longer than the others ; tail very long, and much cuneated, with the two middle feathers narrowed ; tarsi nearly the length of the middle toe, and slender ; toes moderate, with the lateral ones nearly equal.

508. Æna Capensis, Linn.; Selby, Pl. Enl., 140 ; *Columba Capensis*, Lath. ; *La Tourtelette*, Le Vail., Nos. 273, 274 ; Cuv., Vol. 3, p. 92 ; *Col. Atrogularis*, Wagl. Nat. Lib., Vol. 8, p. 214.

ABOVE, ash-coloured ; bluish on the secondaries, with a rich purple spot ; wing-feathers deep red, edged with dark-brown ; forehead, cheeks, chin, throat, and chest, glossy black ; as are also the under sides of the tail-feathers, the outermost of which are blotched with white ; a white bar almost hidden by the closed wing extends across the rump, succeeded after a little interval by a narrow black one—the ends of the tail-coverts, black, form a third bar on the tail ; bill and feet, in life, coral-red and orange. The ♀ wants the black on the head, throat, &c., and is altogether more dull in colour ; tail much elongated and graduated. Length, 10½″ ; wing, 4″ ; tail, 5″ 9‴.

These lovely little pigeons extend over the whole of the colony, and appear in all the collections made in the Interior. In the Karroo they abound, breeding in the mimosa bushes. Their nests are rarely placed above a man's reach, and are simply a loose platform of sticks, covered with fibrous roots, and on this are deposited their two delicate cream-coloured eggs : axis, 10″; diameter, 8″, similar at each end.

The young the first year are mottled. They generally fly in pairs, but several are often seen in company. The males are very salacious, cooing to any female they chance to fall in with. I kept one in confinement for many months : it became very tame, and would salute my first coming into the room in the morning with a deep, plaintive *"coo-coo-oo."* It was killed at last by some little love-birds *(Euphemia Elegans),* that tore it to pieces. They feed entirely on grass seeds and grain, and walk with such rapidity when feeding, that I have often mistaken them for rats when about the farmsteads where they congregate. It is probably migratory.

Genus TURTUR, Selby.

Bill slender and straight, with the top slightly arched and acute ; wings rather lengthened, with the first two quills rather graduated, and the second and third quills the longest ; tail moderate, even or rounded ; tarsi rather shorter than the middle toe, and scutellated in front ; toes very long and slender, with the outer one shorter than the inner ; the claws moderate and slightly curved.

509. Turtur Vinaceus. (Gmel.) *Turtur Erythrophrys,* Swain., *Columba Le Vaillantii,* Smith.

GENERAL colour of the back, interscapulars, wings, and tail, grey-brown, having a strong cinereous tinge on the back, and a slighter one on the rump and outermost wing-covers ; the crown is of a most delicate cinereous, graduating to white in front, and blending behind into the vinaceous red of the nape and upper neck ; the back of the neck has a broad, black semi-collar, margined by a narrow cinereous line ; the whole of the under plumage, as far as the belly, is vinaceous, becoming paler on the ears, and almost white on the chin ; the belly, vent, flanks, and under tail-covers clear cinereous ; tail broad, and rounded ; the outer half cinereous (almost white beneath), and the basal half black ; bill black ; feet dark ; orbits naked, and rich red. Total length, 11″; bill from gape, 1″; wing, 7″; tail beyond, 1½″; ditto from base, 5″; tarsus, $\frac{7}{10}$″; middle toe and claw, $1\frac{5}{10}$″; hinder ditto, $\frac{7}{10}$″.

As so much similarity exists between the two turtle-doves found in this country, I give full descriptions of both from the careful pen of that most accurate and minute of all describers, Swainson, whose specimen I have seen. I find, however, much difference exists in the measurements of that described by Mr. Swainson, or even in two different examples now before me, one of which came from Swellendam; the other shot at the Knysna by Mr. Butler, the taxidermist of the Museum. In this latter the length is 12½″; bill, 1″ 1‴; wing, 7″ 5‴; tail, 5½″; tarsus, 1″; middle toe and claw, 1″ 5‴; hinder ditto, 9‴.

The collector may at once distinguish the species even on the wing, by the white outside tail-feathers of *T. Semitorqua-tus* : these are particularly visible when the bird alights, as it then spreads its tail, probably to break its fall.

510. Turtur Semitorquatus, Swain., Nat. Lib.
11, p.208; *T. Albiventris*, Gray; *T. Risorioides*, Temm.

GENERAL colour of the upper plumage, drab-brown ;. nearly of the same tint as the last; but the top of the head and nape, instead of being delicate French grey, is of the same vinaceous as the under parts, the front, as is usual in this family, being much paler ; behind the neck is a semi-circular black collar, as in the last; the wing-covers are very light cinereous, with whitish edges ; but this tint blends on the scapulars into the drab-brown of the back ; the under plumage is vinaceous, palest on the chin, and changing into a cream-coloured white on the belly, vent, thighs, and under tail-covers, these parts being of a full cinereous in the last species ; the basal half of the four outer pair of tail-feathers is deep black; the other portion is more or less white, according to their distance from the two central pair, which are drab-brown, tinged with cinereous ; on their under surface the terminal half of these feathers is much whiter, and the base much blacker ; bill black ; feet pale-red; the orbits are more feathered than in the last, and there is no red tinge on the naked skin ; above all, the *inner* toe is one-twentieth of an inch *longer* than the outer. In *Erythrophrys* this proportion is almost reversed, or at least the inner toe is not even equal to the outer. Total length, 10″; bill gape, $\frac{8}{10}$″; wing, 5½″ ; tail, base, 4″ ; tarsus, hardly $\frac{7}{10}$″; middle toe and claw, $\frac{9}{10}$″.

In specimens before us there are differences of admeasurements equally as great as those quoted in the description of the last species ; indeed, they are so apparent in some specimens as to lead one almost to believe there are several races, if not species. The same differences are also observable among our *Francolins*, not attributable to locality or age, or any ascertainable cause, as both large and small may be killed from the same covey.

T. *Semitorquata* is very abundant all over the colony ; *T. Erythro-phrys* rather less so. Both frequent the same places, breeding even in Cape Town. The nests are simply rude platforms of sticks, covered with a thin layer of fine roots, through which the eggs (two in number, oval, and pure white: axis, 13'''; diameter, 11''') are, in most instances, plainly visible from below.

511. Turtur Senegalensis. (Linn.) Temm., Pig., t.
45 ; *C. Maculicollis*, Wagler ; *La Tourterelle Maillée*, Le Vail., No. 270 ; *C. Cambayensis*, Gm. ; Cuv., Vol. 3, p. 84 ; *C. Ægyptiaca*, Lath. ; *Laughing-Dove* of Colonists.

HEAD, neck, and breast vinaceous; chin almost white; breast marked with black ;* wings, flanks, rump, and upper side of tail cinereous, darkest on the tail, the two centre feathers of which are tinged with ashy, while the three outer pair have the apical half white ; shoulders deep rufous ; belly whitish ; vent quite white. Length, 10'' ; wing, 5'' 3''' ; tail, 5''.

This elegant dove is found throughout the colony, and is well known from the peculiarity of its note, which somewhat resembles a human laugh. It frequents the same localities as do the two preceding species, and is often mistaken for them when on the wing. A hen-bird, attracted by the call of the male of the Australian *Geotrupes Tranquilla* in my aviary, entered the room and was captured. I kept her in confinement a few days, and finally let her loose. To my astonishment she appeared in the evening on the window-sill to pick up the seed spilt there accidentally. She remained about the house and garden, sometimes venturing into the room, and in the breeding season sought a mate and bred in one of the oak-trees close to the window. Her first nest was destroyed by a mischievous boy ; but she made another, and brought off the young one, which together with herself and mate are daily to be found morning and evening, feeding in the garden, unmindful of our presence. They have also associated with them *Turtur Erythrophrys* and a Java sparrow. Subsequently they attracted many of their own species, and I have counted as many as sixteen in the garden at once. Eggs taken from the nest of this bird measure—axis, 13''' ; diameter, 10'''. They are the usual pure glossy white.

The Sub-Family, GOURINÆ, or Ground-Pigeons,

have the bill moderate, more or less slender, straight, with the apical portion generally strong and vaulted ; the wings moderate, and pointed ; the tail moderate, and rounded on the sides ; the tarsi as long as, or longer than, the middle

* The peculiar marking of the breast represented by the French "Maillée" is caused by the black of the basal half of the feathers appearing through the bifurcated tips of those that overlap them, *tile-wise.*

toe, and robust; the toes long and margined on their sides with a membrane, and the hind toe usually long and slender; the claws short and curved.

Genus PERISTERA, Swainson.

Bill lengthened, slender; wings pointed; the second quill longest; tail moderate; tarsi rather short; lateral toes equal; hinder toe very short, not more than half the length of the middle toe.

512. Peristera Tympanistria. (Temm.) Pig., t. 36; *Columba Tympanistria*, Temm.; *Tympanistria Bicolor*, Bp.; *Le Turterelle Tambourette*, Le Vail., No. 272; Cuv., Vol. 3, p. 83.

GENERAL colour above, on the vent and tail-feathers fuscous-brown, with a few dark-green, iridescent spots about the centre of the wing; forehead, stripe over the eye, and the under parts, pure white; inner webs of the wing-feathers and flanks, clear rufous. Length, $8\frac{1}{2}''$; wing, $4\frac{1}{2}''$; tail, $2''\ 9'''$.

Confined to the forest districts apparently. I found it common at the Knysna; but though continually heard, it was difficult to procure, as it possesses the power of throwing its voice to a distance like a ventriloquist, thus deceiving the hunter, who often creeps from the tree on which the bird is actually perched.

513. Peristera Afra. (Linn.) Pl. Enl. 160; Temm. Pig., t. 38, 39; *Le Tourterelle Emeraudine*, Le Vail., No. 271; Cuv., Vol. 3, p. 83; *Turtur Chalcospilos*, Wagler (Swain.), Nat. Lib., Vol. 11, p. 210.

GENERAL colour above, cinereous-brown; forehead, light cinereous; top of head, lead-coloured; across the lower part of the back is a white stripe, on each side of which is a broad black one; three narrow bands of the same colour extend across the tail-coverts, the last one being on the tips of the feathers; on the centre of the wings are two or three large spots of brilliant purple and green, while the inner webs of the wing-feathers are bright rufous; under plumage, vinaceous, lightest on the vent; under surface of the tail, and tail-coverts, black; the basal half of the outer web of the former, and part of the three outer pair of the latter, white. Length, $7\frac{1}{2}''$; wing, $4''\ 5'''$; tail, $3\frac{1}{2}''$.

Le Vaillant appears to have found this species abundant about the Gamtoos, the Lourie, and Van Staden's Rivers, and throughout Kafirland. Only one specimen has, however, reached my hands, and for this the Museum is indebted to Mr. J. D. Hugo, of Worcester, near which place he shot it on the 3rd of September, 1862. Mr. Atmore recently forwarded another specimen from the Long Kloof (Sept., 1865).

514. Peristera Larvata. (Temm.) Pigeons, t. 31 ;

La Tourterelle à Masque Blanc, Le Vail., 269 ; *Columba Erythrothorax*, Temm. Pig., t. 7 ? *Cinnamon-Dove* of Colonists.

GENERAL colour above, dark-brown ; back of the neck and head vinaceous, iridescent, with a beautiful coppery-green ; forehead, top of head, and throat, white ; under parts vinaceous, inclining to rufous on the belly and vent, and iridescent on the breast, with a coppery gloss ; tail-feathers above, with the exception of the two central pair, cinereous-brown ; cinereous on the tips, below the same, but the brown is darker, the cinereous lighter ; thighs brown. Length, $10\frac{1}{2}''$; wing, $6\frac{1}{2}''$; tail, $3''$ $9'''$.

Not uncommon about Rondebosch, near Cape Town. Found also plentifully at the Knysna. Keeps to wooded places, and feeds on berries. I have not yet obtained its nest.

Besides the species here enumerated, four other pigeons are spoken of by Le Vaillant as being found in South Africa ; of these, two are natives respectively of Mauritius and Madagascar, and the remaining two of Le Vaillant's imagination. It is very improbable that the two former ever reached the Cape ; certain it is that if they did so in Le Vaillant's time, when the colony was in the charming state of nature and innocence personified by the lovely Narina, they have deserted it in disgust in these present days of refinement and crinoline. I include them in this catalogue, however, for the reasons already given.

515. Columba Franciæ, Lath. ; Shaw, Vol. 11, p.

3 ; *Le Ramier Herissé*, Le Vail., No. 267 ; *Columba Jubata*, Wagler ; *Alectroena Nitidissima*, Bp.

GENERAL colour, indigo-blue ; head, neck, and throat covered with stiff white feathers ; tail red.

Le Vaillant states that he procured five of these birds in Great Namaqualand ; but Sundevall doubts this, and says that the species has not been found except in Mauritius. I have never heard of anything like the bird being seen in South Africa.

516. Carpophaga (?) Madagascariensis,

Gmel., Temm., Pig. t., 17 ; *Columba Madagascariensis*, Shaw, Vol. 11, p. 9 ; *C. Phœnicura*, Wagler ; *Le Ramier Founingo*, Le Vail., No. 266 ; Cuv., Vol. 3, p. 80.

GENERAL colour throughout (with the exception of the tail, which is red), dark indigo-blue, with a violet lustre.

Le Vaillant states that this species arrives in Kaffraria in the month of February, from the Island of Madagascar ; but that, being very shy, and frequenting the tops of the highest trees, they are very difficult to obtain. Sundevall doubts their ever reaching South Africa; and I have never heard of any being seen here of late years, at all events.

517. Columba Carunculata, Temm., Pig. t., 2;
Le Colombi-galline, Le Vail., No. 278; Cuv., Vol. 3, p. 72; *Goura Carunculata*, Shaw, Vol. 11, p. 123.

GENERAL colour, slaty-blue, with a white rump; large wattles pendant from the bill, and extending along the chin, crimson.

Stated by Le Vaillant, loc. cit., to inhabit Namaqualand; but Sundevall shows that two specimens from Le Vaillant's collection now exist in the Leyden Museum, which are nothing more than the common European *C. Livia*, with false wattles attached.

518. Columba Hottentota, Temm., Pig. t., 15;
Columba Gularis, Wagler; *Le Colombi-Caille*, Le Vail., No. 283; Cuv., Vol. 3, p. 75; Shaw, Vol. 11, p. 131.

GENERAL colour above, deep-red; forehead, chin, and throat, white; under parts, grey-brown, with scale-like markings of black, bordered with white. Length about five inches.

This species, stated by Le Vaillant to inhabit the mountains of Great Namaqualand, is affirmed by Sundevall to be fictitious.

Order V. GALLINÆ. Linnæus.

THE Fifth Order, GALLINÆ, contains the Game Birds, which have the tarsi lengthened, robust, and sometimes armed with a spur, or spurs; the toes more or less lengthened, but always connected at their base by a membrane; the hind toe more or less developed, and when present more or less elevated from the ground.

The Third Family, PHASIANIDÆ, or Pheasants,

have the bill moderate, with the culmen arched to the tip, which overhangs that of the lower mandible, and the sides compressed, the wings moderate, and much rounded; the tail more or less lengthened and broad; the tarsi moderate, usually armed with a spur or spurs; the toes moderate, the anterior ones united at the base by a membrane; the hind toe short and elevated.

The Sub-Family, MELEAGRINÆ, or Turkeys,

have the tail more or less short and pendant. The head and neck denuded of plumes, or only covered with scattered hairs, or carunculated; the base of the lower mandible sometimes wattled.

Genus NUMIDA, Linnæus.

Bill moderate; the sides compressed, and the culmen arched to the tip, which hangs over that of the lower mandible, and the lateral margins smooth and curved; the nostrils large, oval, and partly covered by a membrane; wings moderate, with the first four quills graduated, and shorter than the fifth, which is the longest; tail short and pendant; tarsi longer than the middle toe, covered in front with broad, divided scales; toes moderate: the anterior ones united by a membrane at their base; the inner toe shorter than the outer; and the hind toe rather short and elevated; the claws short, and very slightly curved.

519. Numida Mitrata; Pallas. Spic. Zool. IV., t. 3, f. 1.

GENERAL colour, dark-grey, profusely spotted with round white spots; on the outer edges of the wing-feathers these spots are replaced by dashes; and on the neck they are very small, run into each other, and become indistinct bars; head bare, with a casque in some specimens 1″ 6‴ in length; tip horny, yellow; base, with top and back of head, cere, and tip of wattles, pendant from base of upper mandibles, bright crimson; neck bare, sparsely sprinkled with hairs, and together with bare space round eye, brilliant sky-blue. Length, 16″ to 17″; wing, 11½″; tail, 6″.

This species extends over the whole of the frontier districts, into Ovampoland on the West, and to the Moçambie on the East. It is still abundant in some places within the colony, where the mimosa bush affords it sufficient shelter. It feeds on grain and insects, and lays from seven to ten eggs, rather sharply pointed at the small and rounded at the obtuse end, of a dark cream colour, minutely dotted all over with pin-point spots of brown: axis, 2″ 1‴; diameter, 1″ 7‴.

In reply to some queries of mine, as to whether my friend Mrs. Barber had ever seen a wild bird with white quill-feathers, she writes :—" With regard to guinea-fowl. They are still very numerous in several partially-wooded districts of the Eastern Province, but more especially in the valley of the Great Fish River, which is their stronghold, and I have been accustomed to see them occasionally all my life; but I have *never seen a wild one* with white feathers in its wings. It is contrary to the laws of natural selection (in which I am a believer) for nature to produce any form that is *useless* or *hurtful* to her, such as white wing-feathers would be to wild guinea-fowls, for they would at once point out to wild-cats, owls, hawks, and sportsmen, the direction in which the bird had flown or ran (for in crossing all rough places they open their wings while running); and I am inclined to think that if they do occur in any part of the colony with white wings, it is when they have accidentally been crossed with the tame, white-breasted guinea-fowl, that is so common (especially amongst the Dutch colonists) nearly all over the colony. Our wild ones are easily tamed, and you frequently see *both sorts together* upon farmers' homesteads on the frontier. Our South African species is altogether a much handsomer and larger bird than the tame, white-breasted sort, and is of a much deeper colour, with the white spots larger and more conspicuous. They are good layers, and rear their young much in the same way as our pheasants do. If the female bird is startled she flies off and leaves her young ones, who at once disperse in every direction, and hide so cunningly amongst the grass and bushes that they are seldom found; and the dear little creatures will remain

for any length of time in their hiding-places until they are called
together again by the shrill note of the parent bird. In the Fish
River valley they roost upon the willow branches that project over the
large holes of water, out of the reach of wild-cats. Sir Walter Currie
has upwards of an hundred of these beautiful birds upon his property
(Oatlands) at Graham's Town: they are thoroughbred South African
ones, with dark wings. I will try and get you a skin of one somewhere. As I have friends in the Fish River, where they are numerous,
I will ask some of them to get me a skin.—M. E. B."

520. Numida Cristata, Pall.; Shaw, Nat. Misc.
Pl. 757; Lath., Gen. His., VIII., p. 148; N. Ægyptiaca, Lath.; Hart. O. W. Af., p. 200.

GENERAL colour, black, profusely spotted with small blue
spots running into bars on some of the wing-feathers; head
and neck bare, blue, with the exception of the chin and
throat, which are red; top of head ornamented with a tuft of
black, stiffish feathers. Length 18″; wing, 11″.

An inhabitant of Natal, according to M. J. Verreaux; quoted by
Hartlaub, loc. cit.; but I have never heard of any species from that
locality except the common N. Mitrata. I obtained the specimen
from which my description is taken at Zanzibar.

The Fourth Family, TETRAONIDÆ, or Grouse,

have the bill more or less long, broad at the base, and the
sides compressed, with the culmen arched to the tip, which
is obtuse; the nostrils basal, lateral, sometimes covered with
feathers, or protected with a naked hard scale; the wings
short and rounded; the tail more or less lengthened and
rounded; the tarsi strong, sometimes clothed with plumes, or
naked and scutellated; the hind toe moderate and elevated.

The Sub-Family, PERDICINÆ, or Partridges,

have the margins of the bill entire, and the nostrils protected
by a naked, hard scale; the tarsi long, naked, covered in
front with divided scales, and sometimes armed with spurs,
or blunt tubercles.

Genus FRANCOLINUS, Stephens.

Bill more or less long, with the culmen at the base dividing
the frontal plumes, and the apical half arched to the tip,
which is obtuse, and sometimes advancing much over that of
the lower mandible; the sides compressed; the nostrils
lateral, basal, the opening placed in a nasal groove, and

covered by a hard rounded scale; wings moderate and rounded, with the third, fourth, and fifth quills longest; tail short, sometimes nearly concealed by the coverts; tarsi strong, as long as, or shorter than, the middle toe, armed in the males with a spur, or tubercle, on each leg; toes more or less long, the fore toes united at their base by a membrane, with the lateral ones nearly equal, the outer rather the longest; the hind toe short; the claws moderate, and slightly curved.

521. Francolinus Clamator; *Perdix Clamator*, Temm.; Cuv., Vol. 3, p. 49; *Pheasant* of Colonists.

BLACKISH-BROWN, each feather longitudinally marked with narrow whitish lines, converging to the shaft; top of head, dark-brown; chin and part of throat, white; feathers of neck brown, with white edges, giving this portion a mailed appearance; feathers of belly and flanks, dark-brown, sparsely mottled with white, and with a broad white line down the centre. Length, 14″ to 16″; wing, 8″; tail, 4″.

The "pheasant" is found throughout the whole of the maritime districts of the colony, delighting in bushy kloofs and water-courses, from which it is driven with difficulty, owing to its habit of perching on branches, just out of the reach of dogs. It at all times prefers to escape by running, instead of flying; and on Robben Island, where it abounds, having been placed there some years ago, whole flocks may be chased for a mile or more in full view without once taking wing. They usually, on these occasions, make for the rocks on the beach, and will run out to the farthest extremity, regardless of the surf breaking over them. Like the other *Francolins* of the colony, they feed during the morning and evenings on bulbs, grain, and insects; form their nests under a bush, and lay from eight to fourteen eggs, of a greenish-brown colour: axis, 1″ 10‴; diameter, 1″ 6‴. They are noisy clamorous birds, their harsh calls being heard to a great distance. At Zoetendal's Vley they consort with the domestic fowls, coming close up to the house to feed.

I have kept them in confinement, and observe that they always seek the highest lateral branch as a roost at night. A strange disease affects them when in a cage—a kind of opthalmia—their eyes swell to a great size, and they become perfectly blind, if the complaint does not end in death. They have been introduced into New Zealand and Australia, and will, I am convinced, soon become naturalized.

522. Francolinus Nudicollis, Lath.; *Tetrao Nudicollis*, Cuv., Vol. 3, p. 49; *Perdix Capensis*, Lath; *Red-necked Pheasant* of Colonists.

GENERAL colour, brown, the feathers of the back having a black stripe down the centre; those of chest cinereous, with black centre stripe; those of neck white, with black stripe, very broad; feathers of breast, belly, and flanks, dark brownish-black, with a white stripe running up each side of the shaft; chin, fore part of throat, and space round the eye, bare, and bright crimson. Length, 13″ to 15″; wing, 7″; tail, 4½″.

The "red-necked pheasant" is only found in wooded districts, such as the forests of George and the Knysna, where it replaces *F. Clamator*, and which it resembles in all its habits,—the call is, however, different. Mr. Atmore has it on his farm at Kykoe, in the Long Kloof; and I am informed that it is very common on the frontier. About Swellendam, likewise, it is plentiful; and at Mr. Moodie's farm, Grootvadersbosch, it is equally abundant as the common pheasant. On this farm I killed all four *Francolins* usually found in the colony, viz., *F. Clamator*, *F. Nudicollis*, *F. Afer*, and *F. Levaillantii*.

523. Francolinus Adspersus, Waterhouse;
Alex. Exp. of Discov., Vol. 2, p. 267.

THROUGHOUT of a mottled, brownish-grey, brownest on the back and wings, where the marking is finest; below, the white and dark-brown are pretty evenly divided into narrow, irregular transverse lines; on the neck they assume the appearance of scales; from the base of the bill over the nostrils is a small black patch; a black line also extends from the angle of the bill under the eye; bill and legs red; and there appears a bare space round the eye, probably red. Length, 12″; wing, 7″; tail, 3¼″

Found about the Orange River, and northward through the country to Ovampoland. It appears from all accounts to be very common, and to replace the " pheasant," which it somewhat resembles.

524. Francolinus Swainsonii, Sm., Z. S. A.,
p. 13.

GENERAL colour, brown; each feather of the back with a dark stripe down the centre, and finely mottled; under parts rusty, or yellowish-grey; the breast and lower part of neck darkest, each feather marked in the course of the shaft by a very dark umber stripe; chin, throat, and space round the eyes bare, and of a colour between lake and vermilion-red; top of head brocoli-brown. Length, 14″; wing, 8″ 3″; tail, 3″ 6‴.

The expedition under Dr. A. Smith first found this *Francolin* in a valley immediately south of Kurichane, perching on the branches of decayed trees near the margin of a small rivulet. To these places they appear to resort at night, feeding by day on the banks of rivers. In

*H

this, and in their habit of effecting their escape by fleetness of foot, rather than by taking wing, they resemble the colonial *Clamator* and *Nudicollis*, to which they are closely allied. Mr. Andersson procured it in Damaraland.

525. Francolinus Afer, Tem.; Shaw, Vol. 11, p. 323 ; *F. Afra*, Lath. ; *Perdix Capensis*, Lath., apnd. J. W. Grill, Victorin's list ; *Greywing-Partridge* of Colonists.

GENERAL colour, light cinereous, approaches to ashy on the breast and belly, variegated on the back with very dark blotches, rufous transverse bars, and a light streak down the shaft of each feather ; breast and flanks much blotched with deep rufous ; belly and vent much mottled with transverse, narrow bars of blackish-brown, assuming an arrow-headed form on the shafts of many of the feathers ; chin, throat, and stripe extending from the back of the eye almost to the shoulder, white, mottled with black, disposed more or less in the shape of a circle ; a broad rufous band, more or less mottled with black, extends from the forehead over the head down the neck ; a similar band less in breadth, and scarcely mottled, extends from the bill under the eye, and joins the rufous marking of the chest ; tail dark-brown, barred with rufous. Length, 12″ ; wing, 6″ ; tail, 3‴.

Inhabits the maritime districts in considerable number, feeding on bulbs and insects, which it digs up from the ground with its powerful hooked bill. About Beaufort, it is only found on the mountain tops. It nests among bushes, forming a loose structure of grass and roots, in a depression in the soil. Its eggs, six to eight in number, vary in colour from a greenish-brown to a dark-brown, or a light-greenish, almost white. All are, however, minutely spotted with brown pin-points : axis, 1″ 6‴ ; diameter, 1″ 2‴.

It feeds during the early morning and evening during hot weather, sheltering itself from the rays of the sun during the hotter portions of the day under thick bushes, from which it rises readily on being pursued. It can, however, run with great swiftness, *and a winged bird*, unless followed by a good dog, is sure to escape.

This species, and *F. Clamator*, live tolerably well in confinement, and may be acclimatised, I think, in Mauritius, St. Helena, Australia, New Zealand, and elsewhere. *F. Afer* might perhaps be introduced into England ; but it is a dry, tasteless bird compared with the European partridge.

526. Francolinus Levaillantii, Tem.; Pl. Col., Pl. 477 ; Smith, Z. S. A., Pl. 85 ; *Perdix Levaillantii*, Cuv., Vol. 3, p. 49 ; *Red-wing-Partridge* of Colonists.

MUCH resembling the preceding species, but easily distin-guishable from it by its larger size, darker and brighter markings, the uninterrupted broad, rufous bands of the neck, and black, mottled collar.

General colour grey, mottled on the back with very dark markings on each feather; down the centre of which, and on the shaft, is a rather broad light buff stripe, inclining often-times to white; top of head mottled-brown; a narrow white band, much mottled with black, extends from the nostrils over each eye, unites at the back of the head, and extending down to the nuchal collar behind; a similar band extends from the same place under the eye across the ear, and sweeps round to the chest, where it extends into a broad crescent-shaped gorget; between these two bands is one of deep orange rufous, enclosing the eye, widening as it passes over the ear, and spreading backwards to the nuchal collar and forward to the gorget; the fore part of the throat is of the same colour; the chin albescent; below the gorget, the chest, belly, and flanks are beautifully mottled with deep rufous, and brown; wing-feathers, inside, deep rufous. Length, 13″; wing, 6″ 6‴; tail, 3″ 3‴.

This very handsome bird is in the Western districts extremely local, frequenting isolated spots, generally secluded vallies between high mountain ranges, through which flows a mountain stream. In the palmiet which crowds the morasses formed by the unrestrained waters, and crouching amid the tufts of coarse grass and reeds, these birds may be found; and I have often killed one with my first barrel, and a snipe with the second.

They lie very close, and on several occasions I have actually parted the grass under the pointer's nose to allow the birds to rise. If flushed a second time and well marked down, they may often be caught with the hand, as they will hardly rise again. In dry weather they keep so close to the dense palmiet, that it is impossible to get them out. I am informed that in the Eastern districts it affects the hill sides, and does not frequent morasses. This I found to be the case at Grootvadersbosch, except that the places in which they lived were always covered *with long grass and rushes*, indicating that they were always more damp than the surrounding country, and in the rainy season were probably swamps.

Mr. Atmore makes the following observations on these birds:— "Here's a query for you about partridges. The drought appeared to have nearly destroyed them, particularly the red-wings, and now they are swarming. They could not have increased so much in the time, and I can't make out where they come from We had a little work near Heidelberg, and in one kloof, say one mile long, we saw more than one hundred birds. From that place we went to Riversdale: killed five out of the cart; between Riversdale and Gouritz River, killed seven and one korhaan; between Gouritz River and Blanco fifteen, and shot only what presented themselves. On this journey we killed more than I saw altogether in two years of the drought. Again, two years ago, in this Outeniqualand, Red Kestrils were numberless, preying upon locusts. Last year rain had fallen, and very few *Kestrils* were seen, but *Harriers* were numerous; at present neither Kestril nor Harrier is seen but rarely. South African birds appear to be very migratory, and irregularly so—probably food is the cause."

527. Francolinus Gariepensis, Smith, Zool. S. Af., Pl. 83, 84.

FOREHEAD, upper surface of head, and back of neck, dull reddish-brown; all the feathers edged with pale rusty orange; sides of neck, with two arched liver-brown bars, mottled with white, the upper one originating at the outer feathers of the eye; the other at the base of the lower jaw, and separated from each other by a rusty orange stripe, mottled with umber-brown; the superior of these arches terminates at the base of the neck, the other joins its fellow on the opposite side, and forms a dark semicircle in front of the neck; chin and upper portion of throat, white; lower portion, breast, belly, and vent, reddish-orange, blotched on the throat, breast, and flanks with deep rusty, and on the belly with very dark-brown; upper surface of back, ashy-brown, mottled with rusty, transversely barred with brown, and longitudinally streaked down the shafts of the feathers with brownish-white. Length, 14"; wing, 6" 9'''; tail, 3" 9'''.

Found by Dr. Smith towards the sources of the Caledon and Vaal Rivers, from whence Mr. Arnot has likewise forwarded it to the South African Museum. It bears, as Dr. Smith remarks, a strong resemblance to *F. Levaillantii*, but is not so dark. Found about grassy slopes.

528. Francolinus Pileatus, Sm., Z. S. Af., p. 14; *Perdix Sephœna*, Smith, Rept. S. Af. Exp.

UPPER surface of head, rusty-grey, clouded with brown; stripe of the eye, chin, and throat white; sides of neck and nuchal collar, white, spotted with red; back, rufous-grey, the shoulders most rufous, each feather having a broad whitish line down the shaft; central tail-feathers light-brown, delicately mottled with darker brown; outer feathers dark-brown; breast and belly, cream-yellow, the former marked with triangular reddish-brown spots at the point, and a similar coloured line in the centre of each feather; the latter minutely barred with fine brown transverse lines. Length, 13" 6'''; wing, 6" 6'''; tail, 4".

Found by the expedition under Dr. A. Smith on the banks of the Marikwa River, north of Kurrichane, frequenting open localities, and feeding in grassy places, on bulbs, seeds, and insects.

529. Francolinus Natalensis, Smith, S. A. J., Vol. 2, p. 48 ; Ill. Zool, Pl. 13 ; *Francolinus Lechoho*, Smith.

Top of head, back, scapulars, shoulders, and tail, light-brown, finely mottled with brownish-black and tawny white, the former in the shape of longitudinal blotches upon the scapulars; eyebrows, sides of head, neck, breast, and anterior part of belly, variegated black and white; posterior part of belly, vent, and under tail-coverts, a mixture of black, tawny white, and light-brown ; bill reddish-brown , legs, toes, and claws, reddish-yellow ; eyes brown. Length; $12\frac{1}{2}''$; wing, $7''$; tail, $3''$.

Inhabits brushwood thickets in the vicinity of Natal.—Dr. A. Smith loc. cit., and Mr. Ayres, Ibis., Vol., 1860, p. 215. This species appears to represent *F. Clamator* at Natal, and bears a considerable resemblance to it. Its food, according to Mr. Ayres, consists of insects and seeds ; and, like its congener, it runs with great rapidity, and takes refuge in trees when roosting or disturbed.

530. Francolinus Subtorquatus, Smith, Zool. S. Af., Pl. 15, ♀; *Perdix Coqui*, Smith, Report S. Af. Expedition.

Upper part of head, deep rufous-brown, dashed with a tinge of purple ; eyebrow, ear-coverts, back, and sides of the neck, pale ochry-red ; sides of head crossed by two fine black lines, one above and one below the eye, the former terminates on the side of the neck, behind the ear-coverts, the latter, which springs from the base of the bill, descends towards the throat, and with the corresponding one of the opposite side, forms a narrow lunated collar across the throat ; chest and narrow nuchal collar, vinaceous, with white shafts to the feathers ; belly, flanks, and vent, creamy-white, barred with broadish-black or dark-brown markings ; back rusty-grey, with rufous and dark-brown transverse markings, and white centre to the feathers, resembling in the distribution of the colours *F. Afra.* Length, $10''$; wing, $5''\ 3'''$; tail, $2''\ 6'''$.

Found by Dr. Smith's expedition near the Tropic of Capricorn, upon the slopes of stony hills covered with brushwood. Mr. Arnot has forwarded specimens obtained in the Free State ; and Mr. Ayres has procured it at Natal

Dr. Smith's description is taken from an old female. The male differs considerably: its colours are much brighter and darker on the back; the throat and chin are bright rufous; there are no signs of the black collar and stripes on these parts; the chest has the same ground colour as the lower parts, and is closely and deeply marked with black bars.

Specimens of both sexes now before me are armed with a spur on each leg; that of the male being the most developed. This armature is common to all our *Francolins;* the ♀ assuming it as she gets old.

Genus COTURNIX, Maehring.

Bill short, more or less elevated at the base and arched to the tip, which is obtuse; the sides compressed; the nostrils basal, lateral, and covered by a hard scale; wings moderate, with the second, third, and fourth quills the longest; tail very short, mostly hidden by the coverts, and pendant; tarsi short, covered in front with divided scales, and unarmed; toes moderate, united at their base, with the inner toe shorter than the outer; the hind toe short; the claws short and slightly curved.

531. Coturnix Dactylisonans. (Temm.) *Coturnix Major,* Bris.; *C. Europœa,* Swain.; *C. Vulgaris,* Jard., Gould. B. of Eur., Pl. 263; *Tetrao Coturnix,* Lin., Pl., Enl. 170; *Coturnix Communis,* Bonn.; *The Quail.*

ABOVE brown, variegated with grey and black; the shafts of many of the feathers with a broad white stripe; head dark-brown, with a light buff stripe down the centre, and over each eye; throat and chest, deep rufous; the former in the male with a black patch down the centre, the latter with faint-whitish lines down the shafts of the feathers; flanks longitudinally richly variegated with dark-brown, black, and pale buff; belly light yellow-brown, immaculate. Length, 6″ 8‴; wing, 3″ 9‴; tail, 1″ 7‴.

The common quail arrives at this, the most southern limit of its migration, about the end of August (sometimes as early as the 15th), in great numbers. At first, if the corn crops are not sufficiently high to afford it the necessary cover, it frequents the grassy plains and stunted bushes It breeds in the standing crops, depositing its eggs in a mere depression of the soil, sometimes without even a few shreds of grass to protect the eggs from the ground. The eggs, from six to twelve in number, are of a yellowish ground, more or less spotted and blotched with dark-brown : axis, 1″ 3‴; diameter, 12‴.

The young birds run the instant they are excluded, and are attended by both parents, who will feign lameness and tumble about before the dog, or hunter, in order to draw him from their brood. The male generally begins this manœuvre, while the female leads off the little chicks; but should he not succeed, she will perform the same tricks with tenfold more boldness, and frequently falls a prey to her maternal solicitude. I have at this moment in my aviary a female thus captured, with two of her brood. I have observed from these birds that the migratory desire is evidently strongest at night. At this season, though perfectly quiescent during the day, my birds fly up and dash themselves against the wires at all hours of the night, particularly during moonlight. This could not have been from any terror, as they were quite tame at the time, feeding from my hand, and scratching on my palm to obtain some desired seed that their little quick eyes discriminated in the mass thus offered to them. Some quails remain with us all the year round. On Robben Island, for instance, ten or twelve brace may be shot any day in the year. Why they should choose this barren spot, eight or ten miles from land, in the mouth of Table Bay, I cannot conceive. The farmers declare that every seven years the numbers of quail exceed those that visit us during the intermediate six. During one of these "years of plenty," I bagged forty brace in one day, and lost many more.

532. Coturnix Histrionica, Hartl., Rev. et Mag. de Zool., 1849, p. 495; *C. Delegorquei,* Deleg.; *C. Crucigera,* Heugl. Uebers, p. 51.

ABOVE, fuscus cinereous, with black and white transverse markings; feathers of the back and the wing-coverts marked with longitudinal white patches, bordered and centered with black; top of head and back of neck brown; eyebrows, and a little mark on the top of head, white; a short band between the nostrils and the eye, and others beneath the eyes, black; throat and fore part of neck white; the centre spots black, and anchor-shaped; the greater part of chest black; belly intense rufous, the larger spots black; under tail-covers white; under the wings white; bill black; legs yellow; irides yellow. Length, 7″ 2‴; wing, 3″ 8‴.

Inhabits Kaffraria, teste Delegorque et Wahlberg.—Hartl., loc. cit.

I think that a specimen of this quail was purchased by the Count de Castelnau in a collection of birds made near Swellendam. I could not obtain a description of it; but, as far as my memory serves me, the foregoing is not unlike the bird.

The Sub-Family, TURNICINÆ, or Bush-Quails,

have the bill moderate, straight, and the sides compressed to the tip, which slightly overhangs that of the lower mandible; the nostrils lateral, and placed in a nasal groove, that reaches beyond half the length of the bill, with the opening linear, and protected by a long scale; the wings rather short, and rounded; the tail short, and almost concealed by the dorsal feathers; the tarsi moderate and strong; the toes usually three in number, long, and free at their base, the outer toe longer than the inner.

Genus TURNIX, Bonnaterre.

Bill moderate and straight, with the culmen more or less elevated at the base, but always curved, and the sides compressed to the tip, which hangs over that of the lower mandible ; the gonys moderate and ascending ; the nostrils lateral, basal, and placed in a long groove, with the opening linear, and closed by a scale ; wings rather short, with the first, second, and third quills equal and longest ; the tertials rather shorter than the primaries ; tail short and graduated ; tarsi longer than the middle toe, strong, and covered in front with transverse scales ; toes long, and rather slender, with the outer toe longer than the inner, the hind toe wanting ; the claws short, slender, and slightly curved.

533. Turnix Hottentotus, Vieil. ; Pig. et Gal., Vol. 3, p. 557; *Hemipodius Hottentotus*, Temm. ; Cuv. Vol. 3, p. 64 ; Shaw, Vol. 11, p. 321 ; *Riet-Quartel*, and *Sand-Quail* of Colonists.

ABOVE, variegated black, brown, and white : the colours so disposed on each feather as to make the bird appear scaley ; sides of the head, chin, throat, and breast, clear rufous, the sides of the latter variegated with a few black and white bars ; belly and flanks albescent ; vent, pale-rufous. Length, 6″ 4‴ ; wing, 3″ 4‴ ; tail, 1¼″.

This handsome little bird is found sparingly throughout the colony, evidently in the most southern part of the continent, taking the place of the next species, *H. Lepurana.* I never saw more than two together, rarely even that. It frequents grassy places, and reeds in dry vleys, whence its name. When flushed, it flies a short distance, drops into cover, and instantly decamps, running with great rapidity. It feeds on seeds and insects, and is usually very fat. It remains with us all the year round, and I am told breeds in vleys, laying from seven to ten eggs, like those of the common quail, but smaller, and only minutely spotted.

534. Turnix Lepurana; *Hemipodius Lepurana,* Smith, Zool. S. A., Pl. 16.

MALE, above, ground colour, intermediate between pale rufous and light chesnut ; upper surface of head barred with brown ; on the neck, back, and shoulders, numerous slender, black-brown bars, or irregular crescents ; eyebrows, sides of head, and stripe between base of bill and nape of neck, rusty-white ; chin and throat dull-white ; middle of breast, pale Dutch-orange, with a few minute brown dots ; sides of breast and belly, white, with a yellowish tinge; each feather with an arrow-shaped brown spot near the point.

Discovered to the north of Latakoo, in grassy valleys, feeding on seeds and small insects. Like the colonial species, *T. Hottentotus*, they seem difficult to flush a second time, and are never to be found in any numbers. Mr. Ayres has procured it in Natal.

The Sub-Family, PTEROCLINÆ, or Sand-Grouse,

have the bill short, with the culmen curved to the tip, and the sides compressed; the wings and tail lengthened and pointed; the tarsi longer than the middle toe, robust, and more or less covered with feathers; the toes moderate and robust; the lateral toes more or less united to the middle one, and the hind toe rudimental.

Genus PTEROCLES, Temminck.

Bill small, the culmen rounded and curved to the tip, the sides compressed, and the lateral margins arched and entire; the nostrils basal, lateral, with the opening partly closed by a membrane, which is nearly hidden by the frontal plumes; wings very long and pointed, with the first and second quills the longest; tail moderate, and wedge-shaped; tarsi robust, longer than the middle toe, with the front and inner sides clothed with feathers; toes short, thick, much united at their bases by a prominent membrane, which extends along the sides of each toe, the upper surface covered with transverse scales, and the under surface with granulated scales; the hind toe rudimental; the claws short, curved, and robust.

535. Pterocles Tachypetes, Tem.; Gal. des Ois., p. 715; *Namaqua Grouse*, Shaw., Vol. 11, p. 313; *Tetrao Namaqua*, Gmel., Cuv., Vol. 3, p. 42; *Namaqua Patrys* of Colonists.

♂: ashy-brown; throat and chin, yellowish; head, neck, and chest, ashy; feathers of back, dark-brown at the shafts, succeeded by a band of bright nankin, then by a light-purple spot, and finally tipped with dark-brown, giving the back a spotted appearance; wing-feathers dark-brown; lower part of chest, vinaceous; between it and the belly a double band, first of white, then of dark brownish-red; belly, purple-ash; vent and under tail-coverts, ochrecus-

*I

yellow; tail-feathers pointed, tipped with the same, but lighter. Length, 10½"; wing, 7"; tail, 3".

♀: colour throughout, rich nankin, confusedly barred with deep brown; paler on the under side; under tail-coverts immaculate; wing-feathers as in male.

The "*Namaqua Patrys*" of the colonists is very abundant on the arid karroo plains throughout the colony and Namaqualand. It congregates in small parties of one or two families, as may easily be seen from the age of the various individuals, which is marked by their plumage. It runs with considerable swiftness, and its flight is extremely rapid. While on the wing, it utters a shrill "tweet," which may be heard at a vast distance, and resembles the piping of a plover, for which, till well accustomed to the sound I have often mistaken it—indeed, the whole appearance of the bird, whether flying or running, reminds one of the plover. It occasionally, during great droughts, extends its migrations as far as the Cape Flats; but of late years their visitations have been few and far between. It lays from seven to ten eggs, of a light cream-coloured ground, spotted with brown and purple, oblong, rounded, or semi-circular at both ends: axis, 17‴; diameter, 12‴. The crops of those I have examined contained masses of small grass seeds.

536. Pterocles Bicinctus, Temm.; Vieill., Gal. de Ois., t. 220; Cuv., Vol. 3, p. 43; Shaw, Vol. 11, p. 307.

♂: above, ashy-brown, with triangular white spots and reddish-brown bars; top of head, reddish-brown, with longitudinal dark stripes; forehead black, with a small. white patch at the base of the bill, and a -larger over each eye; chin, sides of head, and throat, dirty nankin colour; chest ashy; between it and the belly, two bands, the first white, the second black; belly dirty-white, minutely barred with very dark-brown; under tail-coverts, bright rufous nankin, sparingly barred with arrow-headed black markings.

♀: not unlike that of *Pterocles Tachypetes*. Length, 9½"; wing, 7"; tail, 3½".

Specimens of this sand grouse have been received from Kuruman, through the late Mr. R. Moffat, and from Mr. Arnot, at Colesberg. It does not appear south of the Orange River. Mr. Arnot's specimens came from the northern bank.

537. Pterocles Gutturalis, Smith, Zoology of S. Af.. Pl. 3 and 31.

♂: upper surface of head, dull-green, faintly freckled with black; sides of head and chin, straw-yellow; eyebrows, yellowish-white; space between the eye and the bill, black;

the neck, the breast, and a portion of each shoulder, inter-mediate between oil-green and sulphur-yellow; chest crossed by a narrow crescent of deep brownish-black; back and upper tail-coverts, pearly-grey, strongly tinged with brown; belly and under tail-coverts, deep reddish-brown.

♀ : above, nankin-yellow, profusely variegated with black; chin and throat immaculate; belly and vent, deep-chesnut, transversely barred with black; under tail-coverts, immaculate. Length, 11½″; wing, 8″ 3‴; tail, 4″.

The expedition under Dr. Smith discovered this species about 80 miles east of Latakoo, repairing in large flocks to water at fixed periods, and feeding on grass seeds. I have never seen this species in any collection that I have inspected. A solitary female in the Museum, probably one of those collected by the expedition, was presented by the South African Institute.

538. Pterocles Variegatus, Burchell, Travels
S. Af., Vol. 2, p. 345 ; *Pt. Maculosus*, A. Smith, Zool. S. Af., Pl. 10.

♂ : general colour above, greenish-yellow, spotted with white; top of head dark-brown, mottled with ochry-yellow; eyebrows, chin, and sides of head, silver-grey; bare space round eye, light-yellow; under surface, rufous-brown; the breast spotted with white; the belly and vent clouded with ochreous; wing-feathers, dark ashy-brown, tipped with white.

♀ : chin, sides of head, and stripe over the eye, ochry-yellow, instead of silver-grey; belly, pale yellowish-brown, barred with dusky white; the rest much the same as in the male. Length, 9½″; wing, 6″ 3‴; tail, 3″ 3‴.

This species was also found by the expedition at Tsining, 30 miles to the west of Latakoo. In habits it resembles the preceding. A solitary male is in the South African Museum, obtained from the South African Institute, and is the only specimen I have seen.

Dr. Smith (Illus. Zool. S. Af., Pl. 31), alludes to another species, *Pt. Simplex*, Roux, as inhabiting South Africa; but I cannot find a description of it in any author.

Order VI. STRUTHIONES. Latham.

The Fourth Order, STRUTHIONES, embraces a series of Birds which are mostly of large size, with the wings much abbreviated, and generally incapable of affording the means of flight.

The First-Family, STRUTHIONIDÆ, or Ostriches,

have the bill of various forms; the wings very short or imperfect, and the toes short, broad, and unequal, varying from two to four in number.

The Sub-Family, STRUTHIONINÆ, or Ostriches,

have the bill broad, depressed, and rounded in front; the toes two or three in number, and all placed anteriorly.

Genus STRUTHIO, Linnæus.

Bill broad and depressed, with the culmen flattened, and the tip strong, rounded, and overlapping that of the under mandible; the nostrils placed in a broad membranous groove, near the middle of the bill, and the opening oval; wings short, imperfect, and furnished with long, bending, soft plumes; tail moderate, and composed of curved, pendant feathers; tarsi very long, robust, and covered with hexagonal scales, except in front, near the toes, where the scales are transverse; toes short and robust; the outer short and much padded; the inner one not apparent; the claws short, broad, and flattened.

539. Struthio Camelus, Auct.; *The Ostrich*, or *Struysvogel* of Colonists.

GENERAL colour, black in the male; cinereous in the female and young male; feathers of the spurious wings and tail, pure white, or sometimes mottled with black; neck covered with brown hair-like feathers; stands about eight feet high.

The Ostrich is still found in most of the Karroo country, within the borders of the colony. It usually runs in small families of from three to six individuals—except, of course, in the breeding season, when a hen and her whole clutch of chickens may be seen together.

The hen lays a large number of eggs in her nest (thirty-six have been found in one), while many more are dropped in the neighbourhood. The farmers affirm that these latter are broken when the young are excluded, and serve for their first meals. They also affirm that more than one female lays in a nest, which is always on a sandy patch, and that they incubate by turns. The eggs measure : axis, 6''; diameter, 4'' 9''' or 5.''

The South African ostrich is now recognised as distinct from the North African bird, to which probably the name of "*Camelus*" was originally given. The egg of the former may at once be distinguished by its finer texture.

Order VII. GRALLÆ. Linnæus.

THIS Order comprehends a large series of Birds, that have the lower portion of their tibiæ, or thighs, naked, and the tarsi lengthened, rounded, and slender.

The First Family, OTIDIDÆ, Selys.,

have the bill rather short, stout, broad at the base, somewhat compressed towards the tip; upper mandible convex, and slightly curved; nostrils in a large membranous groove; wings ample, more or less pointed; legs long, rather stout; tarsi reticulated; toes, three before, which are united at their base by a membrane; hind toe always absent; claws short, and blunt.

The Sub-Family, OTIDINÆ, or Bustards,

have the bill more or less lengthened or compressed on the sides, with the culmen straight above the nasal groove, and then vaulted to the tip, which is strongly emarginated; the nostrils basal, lateral, and placed in a large membranous groove, with the opening large and suboval; the wings moderate, and rather pointed; the tail moderate, broad, and rounded; the tarsi long, and covered with small scales; the toes short, and covered with small narrow scales above; the claws short, blunt, and broad.

Genus EUPODOTIS, Lesson.

Bill more or less lengthened, and rather slender, broad at the base, and the sides gradually compressed to the tip, which is emarginated; the culmen rather depressed, and straight above the nasal groove, and then slightly curved to the tip; gonys generally long and straight; the nostrils basal, lateral, placed in a large nasal groove, partly closed by a membrane, leaving the opening large and of a lengthened oval ; wings long, with the second, third, and fourth quills usually equal and longest, and the tertials as long as the quills ; tail moderate, broad, and rounded ; tarsi much longer than the middle toe, and covered with small scales ; toes short, broad, with the inner toe shorter than the outer, and all the fore ones covered with narrow transverse scales ; the claws short, very broad, and blunt.

540. Eupodotis Cristata, Scop.; Sonn. Voy. N.
Guin., t. 49 ; *Otis Kori*, Burch. Trav. S. Af., Vol. 1, pp. 394, 402 ; *Otis Luçoniensis*, Vieil. ; *Gom-Paauw* of Colonists ; *Kori-Bustard*.

♂ : general colour above, ashy-grey, with a tinge of rufous in some examples ; paler on the wings (white in some), all profusely mottled ; the wings with large black blotches ; wing-feathers black ; head and neck grey, transversely barred with black ; top of head black, much crested ; breast and belly, white ; a half-collar of black between the grey of the neck and the white of the chest ; legs yellow. Female much smaller and duller coloured than the male, which weighs from 30 to 35 lbs. Male, length, 56″ ; wing, 31″ ; tail. 16.″ Female : length, 44″ ; wing, 23″ ; tail, 14″.

The "Gom-Paauw" is pretty generally distributed in favourable localities ; open plains dotted with mimosa jungle in the Northern and Eastern parts of the colony. It is a noble bird, and when seen stalking about in its proper haunts, affords a sight to a hunter's eyes never to be forgotten. It is migratory, as are all our *Otidæ*, and is more plentiful in some years than others ; their numbers depending on the drought of the Interior, whence they come. In habits and food it resembles the rest ; but it is never found far from the mimosa jungle that skirts the rivers. It can swallow a lizard or snake of considerable size ; and a female shot by my friend, Mr. A. V. Jackson, and myself disgorged the largest chameleon we had ever seen ; besides this, its crop contained a mass of locusts, small snakes, &c. ! !

541. Eupodotis Caffra. (Licht.) Less. Cat. Berl.
Mus., 1793 ; Cuvier, Vol. 3, p. 305 ; *Otis Stanleyii*, Grey ; *Otis Ruficollis*, Cuv.

ABOVE, ashy, black waved and streaked ; eyebrows, chin, nape, band on sides of neck and belly, white ; crown, quills, and tail, white, with three black bands. Male, throat slate-coloured ; female, band on crown and throat black, waved. Length, 38″ ; tarsi, 6″.

This bird is common in the Northern portions of the colony, particularly about Beaufort and that neighbourhood. It rarely comes to the sea-coast, where its place is supplied by *E. Ludwigii*. The food of the two species is similar—seeds, insects, and small reptilia; nor is there any difference in their habits, both frequenting open plains. The males when "playing" before the females, expand the feathers of the throat and strut about, uttering a loud booming noise, which can be heard at a great distance. Like *E. Scolopacea*, they will squat and lie close to the ground, to avoid detection, and may thus be approached and killed with a charge of No. 7 shot.

It lays two large eggs, of a reddish-olive ground, spotted with brown and indistinct purple blotches: axis, 3″; diameter, 2″ 2‴.

Capt. Bulger, of H. M.'s 10th Regt., thus writes of two young birds in his possession : — "Feb. 9, 1864. Batho had two young *Paauws* give him to-day. I have no idea what their age is; but they are as large as a chicken, and completely feathered. Their irides are a light-yellowish brown. They make a most extraordinary low, plaintive noise, like one of the high notes on a glass flute, or still more like the sound produced by moving your finger rapidly round the rim of a tumbler : their note is a semitone so soft and sweet, that it is almost impossible to imitate it. They also make a sort of clucking or barking noise, apparently when alarmed. As yet they have voluntarily eaten nothing. Batho took them out in the garden, and one of them immediately squatted like a partridge; the other wandered about."

542. Eupodotes Ludwigii, Rüppell, Mus. Senck.

1837, t. 14; *Otis Colei*, A. Smith.

♂, in full plumage : general colour of back, ochraceous, profusely variegated with minute dark-brown wavy lines; top of head black; eyebrows and chin white; front of neck and chest white, tinged more or less with slate-colour; ruff at back and side of neck rufous, under parts white; wings, when closed, appear mottled black and white; tail white, with four broad black bars; legs and bill yellow. Length, 3′ 6″; wing, 23″; tail, 13″.

♀ much resembles the male, except that she is smaller, has only a coronal of black round her head, and is altogether less highly coloured.

This Bustard comes to us in its migrations, apparently from the Westward; at least most of them that I have seen have been procured on the north-western side of Cape Town. In its habits it resembles the preceding species—its eggs, two in number, are similar in size and appearance to those of *E. Caffra*, but are usually of a lighter and greener ground.

543. Eupodotis Scolopacea. (Tem.) Pl. Col.

576; *O. Torquata*, Cuv.; *Otis Vigorsii*, Smith, Less.
Proc. Z., 5, 1830, p. 11; *Vaal-Knorhaan* of Colonists.

GENERAL colour, cinereous, here and there passing into rufous, minutely mottled with dark-brown and black; quill-feathers black, with the inner webs more or less isabella-coloured;

chin, and top of throat, jet-black, surrounded by a pale-yellow edging; a black halfmoon-shaped mark at the back of the head, which is slightly crested. The whole of the body plumage in the live bird is glossed with a beautiful pink lustre, which fades after death; each plume on being withdrawn shows a delicate ferruginous and very lax web throughout two-thirds of its basal portion. Length, 20″; wing, 14″; tail, 8″.

The *Vaal Knorhaan* is common on the Karroo, about Beaufort West and Zoetendals Vley. It is usually found in pairs, and prefers running among the scanty herbage, and trusting to its dusky plumage to effect its escape, to taking flight. If it fancies itself unobserved, it will suddenly squat, and unless the spot is correctly marked, so great is its similarity to the soil and stones among which it is found, that it is next to impossible to detect it. It is so well aware of this, that it will remain immoveable till the sportsman walks direct towards it, on which it instantly takes flight; but if it is approached in a series of concentric circles, it remains until the sportsman is within a few paces.

It feeds on seeds, insects, and small reptiles, constructs no nest, but deposits its two eggs in a depression of the soil in the open veldt. The eggs vary much in colour—some are olive, some light-brown, and others rich nankin. All are, however, more or less spotted and blotched with brown and indistinct purple. Axis, 2″ 8‴; diameter, 1″ 10‴.

544. Eupodotis Cærulescens. (Vieil.) Pl. Col.
532; *Otis Verrauxii*, Smith; *O. Cana*, Licht; *Blue-Knorhaan* of Colonists.

UPPER parts, ferruginous, minutely mottled with dark-brown and black; less so on the wings, the quills of which are black, in some places edged with blue; the ends of the tail-feathers are also black; top of head, back, and lower portion of front of neck, breast, belly, and vent, blue; forehead, eyebrow, spot under the eye, and upper part of throat, black; a white patch arising on the chin extends backward to the hinder part of the head; a similar but smaller patch extends from the forehead, and joins it over the ears; legs yellow. Length, 20″; wing, 13″; tail, 7″.

This beautiful bird is locally distributed. I have received it from Riversdale, and from the neighbourhood of Hanover, and have been told that it is plentiful on the Eastern frontier in certain favoured spots.

545. Eupodotis Senegalensis. (Vieil.) Less. Enc.
Meth., p. 333; *Otis Rhaad*, Rüpp.; *Otis Barrowii*, Gray; *Blue-necked Bustard*, Lath.; Cuv.,Vol. 3, p. 304.

GENERAL colour above, bright rufous, variegated with black; tail barred with four cross bars, that near the point being the

broadest ; top of head rufous, minutely mottled, and changing
into light ash-colour at the back ; cheeks and chin, white ;
throat, and crescent-shaped mark at back of head, jet-black ;
lower part of throat and breast rufous, with here and there a
bluish tinge ; large wing-feathers black, the rest bright
rufous ; under parts white. Length, 16″ or 17″ ; wing, 10½″ ;
tail, 5″.

Several pairs of this handsome bustard were procured in the Free
State by Mr. Arnot ; but I know nothing of its habits.

546. Eupodotis Ruficristâ. (Smith.) Z. S. A., Pl. 4.

TOP of head, neck, and upper part of chest, blueish-ash,
darkest on the head ; ground colour of back deep rufous, mottled with deep brown and black ; the prevailing marking on
this part is of a rufous colour, V shaped, with a black centre ;
chin and sides of head dirty-white ; head crested ; crest
deep ferruginous ; under parts all black ; legs light-yellow.
Length, 17″ ; wing, 11″ ; tail, 7″.

The female wants the crest on the head, and is otherwise less
brightly coloured.

Does not occur within the colony ; but is found beyond the Orange
River. Mr. Chapman procured it as far as the Great Lake.

547. Eupodotis Melanogaster ;. *Otis Melano-gaster*, Rüpp. ; Faun. Abyss., t. 7.

ABOVE, fuscous yellow, transversely streaked, and with large
longitudinal black markings down the centre of many of the
feathers ; cheeks dirty-white ; chin, throat, a long line
extending down the neck, breast, belly, and quill-feathers of
wings and tail, black ; a band of the same colour springs
from above each eye, and joins at the occiput ; shoulder and
large portion of the wing, white. Length, 24″ ; wing, 13″
4‴ ; tail, 6″.

This bustard has not yet, that I am aware of, been found within the
colony. Hartlaub, on the authority of M. J. Verreaux, gives South
Africa as a habitat ; and it is included by Mr. J. H. Gurney as among
the birds received from Natal. Mr. Chapman procured it towards the
Zambezi.

548. Eupodotis Afra. (Gmel.) Lath., Syn., Pl. 69 ; *Otis Atra*, Linn. ; *Black Koran*, Shaw, Vol. 11, p. 449 ; *Knorhaan* of Cape Colonists.

TOP of head, back, part of wings, and tail, brown-black,
varying in intensity, and crossed with irregular streaks of
rufous, paling into white ; stripe over the eye, reaching to

the back of the head, ear-coverts, half-collar on back of neck, and part of wings, pure white; the rest of the plumage deep black; legs bright yellow. Length, 19″; wing, 12′; tail, 5″. The female differs from the male in having the whole of the head, neck, and breast of the same colour and markings as the back.

The Knorhaan (lit. *Scolding-Cock*) is abundant throughout the whole colony, frequenting the open country. It feeds on insects, small reptiles, and seeds. The female deposits her two eggs in a depression of the soil, in which she places a few bents of grass. They are of an olive-green or brownish-ground, spotted and blotched with brown and indistinct purple : axis, 2″ 3‴; diam., 1″ 8‴.

It usually goes in pairs, and when followed by the sportsman always endeavours to evade pursuit by swiftness of foot rather than by flight. In this way I have seen it run before a pointer for several hundred yards, squatting at intervals, but all the time keeping a careful watch on its approaching enemy.

549. Eupodotis Afroides, Smith, Zool., S. Af., Pl. 19.

RESEMBLES the preceding in every respect, except the one great distinguishing mark, viz. : on opening the wings the quill-feathers of this species will be found with a large blaze of *white* on them, while those of *E. Afra* are entirely *black*.

Frequents the country to the north of the Orange River, and is rarely seen south of it. "Its call differs considerably from that of *E. Afra*, but in most of its habits they closely resemble each other."—Dr. Smith, loc. cit.

The Second Family, CHARADRIADÆ, or Plovers,

have the bill short, with the basal portion of the culmen rather depressed and weak, and the apical portion strong and swollen; the nostrils placed in a deep longitudinal groove of various lengths; the tarsi lengthened; the hind toe totally wanting, or small or elevated.

The Sub-Family, ŒDICNEMINÆ, or Thick-knees,

have the bill as long as, or longer than, the head, with the culmen slightly depressed at the base and swollen at the tip, and the gonys more or less angulated; the tarsi lengthened, with three rather short toes in front.

Genus ŒDICNEMUS, Temminck.

Bill rather longer than the head, the culmen straight, with the apical half arched and curved to the tip, the sides compressed, and the gonys nearly half the length of the bill, angulated, and advancing upwards to the tip; the nostrils in a subtriangular membranous groove, with the aperture longitudinal and anterior; wings of moderate length, pointed, with the first quill shorter than the second, which is longest, and the tertials the length of the quills; tail moderate and wedge-shaped; tarsi lengthened, three or four times the length of the middle toe, and covered with hexagonal scales; toes short, the inner shorter than the outer, and both united to the middle one by a membrane at their base, especially the outer; the claws short and slightly curved.

550. Œdicnemus Maculosus,[*] Temm., Less., Vol. 2, p. 337; Cuv., Vol. 3, p. 306; *Œdic. Capensis* Licht.; *Dikkop* of Colonists.

GROUND colour, pale-rufous, fading on the belly and part of the wings into white, everywhere (with the exception of the chin, and upper part of throat, which are pure white), mottled with dark-brown blotches, which become longitudinal stripes on the neck, chest, and flanks; under tail-feathers, rufous and immaculate; wing-quills, deep brown, approaching to black, the first three crossed near the tips with a white bar; a few of the others tipped with white; base of bill and legs green; eye yellowish-green, and very large. Length, 16" to 17"; wing, 9" 6'''; tail, 5".

Kaffraria and Cape Colony generally, and Natal, frequenting open, unwooded country. It feeds on seeds, insects, and small reptiles, and lays, in a mere depression of the soil, unprotected by stone or bush, two eggs of a light fawn-yellow ground, profusely spotted and blotched throughout with brown of various shades: axis, 2" 1'''; diam., 1" 6'''.

The Sub-Family, CURSORINÆ, or Coursers,

have the bill moderate, slender, with basal portion of the culmen weak, and somewhat cultrated above the nostrils, beyond which it is slightly arched to the tip; the nostrils lateral, placed in a short, subtriangular membranous groove, with the opening longitudinal and exposed; the wings lengthened and pointed; the tail short; the legs lengthened, scutellated before and behind, and with only three slender toes in front.

[*] J. W. Grill includes *Œ. Senegalensis*, Sw., in the list of birds collected by Victorine in South Africa. I conclude it is the present species, which he does not mention, but which I have seen in abundance in the localities frequented by M. Victorine.

Genus CURSORIUS, Lath.

Bill moderate, broader than high at the base, laterally compressed to the tip, the culmen straight at the base, and then gradually ·arched to the tip; the nostrils placed in a membranous groove, longitudinal and exposed; wings lengthened, with the first two quills the longest; tail short and nearly even; legs lengthened, with the apical portion of the thigh naked and scutellated for the length of the ·middle toe; tarsi not less than twice the length of the mid·dle toe, and covered with broad transverse scales both in front and behind; toes three in front, the outer longer than the inner; the claw short, and slightly curved.

551. Cursorius Burchellii; *Tachydromus Burchellii*, Swain., An. in Menag., p. 340; *Cursorius Capensis* (Swain.), Schlegel., Museum des Pays-bas. p. 130.

SIDES of neck and fore part of head, rufous; under part and nape, cinereous; crown bordered (behind the eye) with a pointed white collar, margined by black; upper and under plumage, fawn-coloured; rump cinereous; tail the same, the outermost feather white; two next with a black bar and white tip, the rest tipped with black only; chin, vent, and thighs, white; legs clothed with white scales; in the centre of the belly, a blackish patch; larger wing-feathers, black, the rest broadly tipped with white, which appears most when the bird is on the wing. Length, 7" 4'''; wing, 4" 7'''; tail, 1" 9'''.

This and the next species, *C. Bicinctus*, are not uncommon in the Karroo, in the neighbourhood of Nel's Poort and Beaufort. They are usually seen in small flocks, varying in number from five to twenty or more, running along with great rapidity between the tufts of stunted herbage which cover these extensive plains. They have a curious habit of swaying themselves to and fro on their long thin legs, when halting after a run, as if they had overshot themselves, and were trying to recover their balance. They feed on seeds and small insects.

Specimens have been received from Colesberg; and the Messrs. Chapman brought them from the Interior. Found also at Natal. Two eggs, said to belong to this species, were brought to me by a shepherd lad, who pointed out the bird. They are of a delicate pale cream-coloured ground, indistinctly blotched throughout with darker cream-colour, rounded at each end, nearly oval: axis, 14'''; diam., 10'''. After seeing eggs of allied species in the collection of Messrs. Newton, Tristram, Walters, and others, I am satisfied of the correctness of this identification.

552. Cursorius Bicinctus, Temm., Man. d'Orn., 11; Cuv., Vol. 3, p. 325; Jard. and Selby's Ill. Orn., t. 48; *Tachydromus Collaris*, Vieil.; *Cursorius Grallator*, Leach.

UPPER parts, variegated rufous, black, dirty-white, and brown; each feather being the latter colour at the base, more or less tinged with rufous; then follows an indistinct black mark, extending in a point down the shaft to the tip, which is dirty-white, tinged more or less with rufous; throat dirty-white; chest and belly rufescent, separated by two black crescent-shaped bars; the feathers of the throat and chest have each a dark-brown line down their centres; on the belly only the shafts of some of the feathers are of this colour; vent and rump white; the first three pair of outer tail-feathers the colour of the belly; the rest brown; legs long, covered with hard white scales. Length, 9″ 6‴; wing, 6″ 3‴; tail, 3″ 3‴.

I procured what I took to be, from the anxiety of the birds hanging about the place, two eggs of this species. They are of a light nankin-yellow ground, densely covered with thin compressed streaks of dark-brown, so closely distributed as almost to hide the ground-colour: axis, 14‴; diam., 12‴.

Mr. Atmore forwards an egg identical in all respects, save size, which he says belongs to this bird. I therefore see no reason to doubt my first supposition. Since this was written, I have seen eggs of allied species, and am satisfied that my specimens are the eggs of *C. Bicinctus*.

553. Cursorius Senegalensis, Lichtenstein; *Cursorius Temminckii*, Swain. Zool. Ill., Pl. 106; *Cursorius Asiaticus*, Temm. nec. Lath.

MUCH resembling *C. Burchellii*, but rather larger; on the occiput and chest red-brown, with a large black patch in the centre of the belly. Wing, 4″ 7‴; tail, 1″ 6‴.

Inhabits South Africa.—Teste Rüppell, Monogr. der gattung Otis. Mus. Senck., 1837, p. 211.

554. Cursorius Chalcopterus, Temm., Gray's Gen. of B., p. 536; *Cursorius Chalcopterus*, Sw., Nat. Lib., Vol. 12, p. 233.

LIGHT-BROWN above; front, throat, rump, and body beneath, white; quills black, tipped with shining violet; breast pale-brown, with a black band. Length, 10″; wing, 7″ 9‴.

Obtained at Natal by Mr. Ayres; but has not yet, that I am aware of, been procured within the limits of the Cape Colony. Mr. Andersson found it in Damaraland, and presented a specimen to the Museum, from whence this description is taken.

The Sub-Family, GLAREOLINÆ, or Pratincoles,

have a short bill, which is broad at the base, and laterally compressed at the tip; the wings very long, with the first quill the longest; the legs moderate, with the tip of the tibia naked; the toes three in front, and one posteriorly, which is elevated.

Genus GLAREOLA, Brisson.

Bill short, broad at the base, much compressed to the tip, with the culmen depressed at the base, elevated and arched to the tip, the lateral margins curved; the nostrils basal, lateral, and oblique; wings lengthened, pointed, extending beyond the end of the tail, with the first quill longest; tail moderate, more or less forked; legs moderate and slender, with the tarsi scutellated, and the middle toe and claw lengthened; the outer toe longer than the inner, and united at the base to the middle one; the hind toe very short, elevated, but touching the ground; and the claws rather long, nearly straight, that of the middle toe slightly pectinated on one side.

555. Glareola Nordmanni, Fisch.; Hartl., Or. W. Af., p. 211; *Gl. Pratincola*, Pall., Zoogr. Ross. As. II., p. 150; *Gl. Melanoptera*, Nordm., Bullet. Mos., 1842, p. 314, Pl. 2; *G. Nordmanni*, Fisch., ib.; *G. Pallasii*, Schleg. Krit. Uebers; *Small Locust-bird* of Colonists.

GENERAL colour above, cinereous-brown, glossed with green; this shade extending over the wing-feathers, which are black; throat tawny-white, with faint brown streaks, and separated from the breast, which is cinereous, by a crescent-shaped collar; belly, vent, and rump, white; tail long and forked, the two outer and longest pair of feathers white, tipped with black; the rest white at the base, brown at the ends; legs black in a dried skin, but reddish in a fresh one; eyelids and cere at base of bill also red. Length, 10"; wings, 7" 3'''; tail, 4" 3'''.

This *Pratincole* differs from the European bird, and may at once be distinguished from it by the under side of the wings, which in this species is *black*, in the other *white*. It appears in great numbers on the Eastern frontier and Natal during the visitations of the locusts, upon which it feeds. I am informed that it hawks about the clouds of the destroying-insect, now and then darting into the mass, and never failing to secure a victim. It is said to run and fly with great swiftness.

556. Glareola Ocularis, A. Smith; *G. Geoffroyi*, Pucher.

HEAD and shoulders, green-black; under each eye a fine transverse white stripe; upper parts of neck and back green-grey; chin white; under parts of neck and the breast, brownish-grey; anterior part of belly pale-chesnut; hinder parts, and under and upper tail-coverts, white; base of tail partly black and partly white, the tips of a few of the feathers white, the other parts pure black; bill black, with the edges of the mandibles towards the angles of the mouth, red. Length about eight inches and a half.

Inhabits South Africa.—Dr. A. Smith, S. A. Journal of Science. Sed non vidi.

The Sub-Family, CHARADRINÆ, or Plovers,

have the bill more or less long and slender, the culmen depressed at the base, but vaulted at the tip, the sides compressed and grooved; the nostrils basal, linear, and placed in the groove of the upper mandible; the wings long and pointed; the tail moderate, broad, and generally even; the tarsi usually long, and rather slender; the toes sometimes four in number; the outer toe longer than the inner, and more or less united at the base; the claws small, compressed, and curved.

Genus CHETTUSIA, Pr. Bonaparte.

Bill moderate, and more or less strong, with the culmen depressed at the base, and vaulted at the tip; the sides compressed and grooved; the nostrils lateral, basal, and placed in the groove of the upper mandible, which extends for two-thirds of its length, with the opening linear; wings long and pointed, with the first, second, and third quills nearly equal and longest; tail moderate, broad, and even; tarsi much longer than the middle toe, slender, and covered in front with divided broad scales; toes four; the three anterior toes long and rather slender; the outer toe longer than the inner, and united at the base; the hind toe short and elevated; the front of the head sometimes lobed, and the wings sometimes armed with a spine.

557. Chettusia Lateralis; *Vanellus Lateralis*, Smith, Zool. S. A., Pl. 23.

FRONT of head white, with a brown bar extending over the vertex; back of the neck, back, and rump, brown, tinged and glossed with green; wing-feathers black; tail white, with a broad black bar across the apical half; chin white; throat black; neck streaked with black-brown; breast and belly, dove-colour; flanks and thighs, dark grey-brown; vent white; a yellow wattle tinged with orange-red extends upwards and downwards from before the eye; bill greenish-yellow, tipped with black; legs greenish-yellow; wing; armed with a spur. Length, 13″ 6‴; wing, 9″; tail, 4″.

According to Dr. Smith, this Plover is an inhabitant of the South-East Coast, and does not extend its range west of Natal, or to the southern extremity of the continent. It has never appeared in any collection examined by me, and must be very rare.

Genus HOPLOPTERUS, Pr. Bonaparte.

Bill moderate and slender, with the basal portion of the culmen depressed, and the tip vaulted, the sides compressed and grooved; the nostrils lateral, basal, and placed in the upper lateral groove, which extends for two-thirds the length of the bill, with the opening linear; wings long and pointed, with the first quill nearly as long as the second and third, which are equal and longest; tail moderate, broad, and even; tarsi much longer than the middle toe, and covered in front with larger scales, which are more or less divided in the middle; toes three, more or less long and slender, with the outer toe longer than the inner, the hind toe wanting; the claws small, compressed, and slightly curved. The front of the head sometimes more or less lobed; and the wings sometimes armed with an acute spine or blunt tubercle.

558. Hoplopterus Armatus, Jardine and Selby;

Hoplopterus Speciosus, Wagler; *Spur-winged Plover.*

GENERAL colour, black; top of head, and forehead, patch at back of neck, lower front of belly, vent, and basal two-thirds of tail, white; lesser wing-feathers, grey; legs and bill black; wing armed with a sharp spur. Length, 12″; wing, 8″ 7‴; tail, 4″ 6‴.

Forwarded in some abundance from Colesberg by Mr. David Arnot, who describes their habits as similar to those of *H. Coronatus.* It occurred constantly in all the collections formed in the Interior by my several contributors.

559. Hoplopterus Albiceps. (Temm.) Pl. Col.

526; *Lobivanellus Albiceps,* Gould; *Sarciophorus Albiceps,* Fras. Zool. Typ., Pl. 64; Strickland's Proceed. Zool. Soc., 1841, p. 33; Hartlb. Orn. W. Af., p. 214.

HEAD, throat, middle of the wing, rump, belly, and vent, white; sides of head and neck, cinereous-purple; back brownish; scapulars, three external feathers, and apical half

*K

of the tail, black; bill greenish-yellow; tip black; wattles yellow, attached at right angles to the beak, pendulus, narrow, pointed, 1″ 6‴ long; wing-spur about an inch long. Length, 13″; wing, 8″; tail, 4″.

Found near the Buffel's River, teste Alexis Verreaux.—Hartb. loc. cit. Sed non vidi.

Dr. Smith says it does not extend into South Africa. Can M. Verreaux have mistaken his *H. Lateralis* for it?

560. Hoplopterus Coronatus; *Charadrius Coronatus*, Temm., Vol. 2, p. 328; Shaw, Vol. 11, p. .489; Cuvier, Vol. 3, p. 316; *C. Alexandrinus*, Lath.; *C. Atricapillus*, Pen. Less., Vol. 2, p. 328; *Kiewit* of the Dutch Colonists.

GENERAL colour above, grey-brown, glossed with purple; paler on the throat, and darkening on the breast, to a bold black bar, which is succeeded by the pure white of the belly and vent; top of the head black, divided into two portions by a pure white crown streak, which springs from the forehead, and passes round to the back of the head; a white line extends down the closed wings, the quills of which have their inner webs more or less white; tail-feathers white, broadly barred at the ends with black; extreme tips white. Length, 13″; wing, 8″; tail, 4″ 6‴.

Scattered throughout the country in small parties, and frequenting grassy places, where it subsists on worms and small insects. Its nest, usually a mere depression in the soil, contains two, and sometimes three eggs, of a deep greenish-brown ground, blotched with rather coarse-brown and indistinct purple, chiefly in a ring at the obtuse end: axis, 1″ 9‴; diam., 15‴.

During the day this bird does not stir about much; but as soon as the cool shades of evening fall on the earth, the "Kiewit" makes known its presence by its loud plaintive call. This, with the pipe of the "Dikkop," (*Æ. Maculosus*) and the wail of the jackal, are nearly the only sounds that break the silence of the night about the solitary dwelling of the Cape farmer.

561. Hoplopterus Melanopterus, Rüpp.; Atlas, T. 31.

GENERAL colour of back, brown, with a purple gloss; forehead, chin, belly, rump, thighs, vent, and stripe along the wing, white; wing-feathers and band across the chest, black; back of head, neck, and chest, grey; basal half and tip of tail-feathers, white. Length, 10″ 6‴; wing, 8″ 9‴; tail, 3″ 8‴.

This Plover has been found at Natal, and is abundant in certain circumscribed localities in the neighbourhood of Mossel Bay, where it was procured by the Messrs. Atmore, who forwarded the only specimens which have fallen under my notice.

Genus SQUATAROLA, Cuvier.

Bill nearly as long as the head, more or less strong and straight, with the basal portion of the culmen depressed, the apical part strong, vaulted, and curved, the sides compressed and grooved on both mandibles ; the gonys short and ascending ; the nostrils basal, lateral, linear, and placed in a groove, which extends beyond half the length of the bill; wings long and pointed, with the first quill the longest ; tail long, broad, and rounded ; tarsi longer than the middle toe, slender, and covered with reticulated scales in front ; toes four ; the outer toe longer than the inner, and united at the base by a membrane ; the hind toe very small, and not touching the ground ; the claws small, compressed, and slightly curved.

562. Squatarola Helvetica; Pl. Enl. 853, 854, 923 ; *Tringa Squatarola*, Gmel. Sys. Nat. 1, 252 ; *Tringa Helvetica*, Linn. Sys. Nat. 1, 250 ; *Grey-Plover ; Vanellus Melanogaster*, Bechst.

♂ in summer plumage : fore part of head, neck, chest, and belly, deep brown-black, bordered by a white line, which blends into the ground of the back and top of the head, both of which are more or less mottled with dark-brown and black ; vent and thighs pure-white ; tail white, transversely barred with brown-black ; the plumage in winter is dark-grey, variegated with light markings, the black plumage disappearing entirely. The ♀ is similar to the male in winter dress. Length, 12" ; wing, 7" 10''' ; tail, 3" 6'''.

I have never seen the Grey Plover in the summer or breeding plumage in this country ; but many specimens in winter dress have occurred to me both here and on the East Coast of Africa, where, as far as 1½° South, I shot it in considerable numbers.

Genus CHARADRIUS, Linn.

Bill more or less short, robust, and straight ; the culmen, for two-thirds its length, usually depressed, and the tip vaulted and curved ; the sides compressed, and furnished in both mandibles with a groove, which extends on the upper mandible for two-thirds of its length ; the nostrils basal, linear, and placed in a groove ; wings long and pointed, with the first quill the longest ; tail moderate, broad, and rounded ; tarsi longer than the middle toe, more or less slender, and covered in front with small reticulated scales ; toes three, moderate, the outer toe longer than the inner, and more or less united at the base by a membrane, the inner one usually free, the hind toe wanting ; the claws small, compressed, and slightly curved.

563. Charadrius Pluvialis, Linn., Pl. Enl. 904 ; Char. Auratus, Suckow ; Char. Virginicus, Bechst. ; Gould. B. of Eur., Pl. 294 ; Golden-Plover.

In winter, upper parts brownish-black, spotted with yellow ; cheeks, neck, and fore part of breast greyish, variegated with brown ; throat and abdomen white. In summer, upper parts black, spotted with bright-yellow ; fore part of neck and breast black ; forehead, line over the eye, band bordering the black of the lower parts, and lower tail-coverts, white.

This well-known bird is given by Dr. Hartlaub (Orn. W. Af., p. 215) as a native of South Africa. I have not met with it further south than Lamoo, on the East Coast, nor is it included in Mr. Ayres' lists from Natal.

564. Charadrius Cantianus, Lath. ; MacGil. Brit. Birds, Vol. 2, p. 125 ; Bewick's Brit. Birds, Vol. 2, p. 18 ; The Kentish Ring-Plover ; C. Albifrons, Meyer ; Ch. Littoralis, Bechst. ; Ch. Alexandrinus, Hassel. ; Gould's B. of Eur., Pl. 298.

ADULT, with the upper parts light brownish-grey ; the forehead with two bands, a white and a black ; the hind part of the head light brownish-red ; the loral space and a band behind the eye, black ; the throat, and a band crossing the hind neck, white ; a large patch of black on each side of the lower part of the neck ; bill black ; the lower mandible flesh-coloured at the base ; feet dusky ; claws black. Young without black on the head ; the bands on the sides, and the patches on the neck, brown ; feathers of the upper parts margined with whitish. Length, 6″ 9‴ ; wing, 4″ 6‴ ; tail, 2″.

The only specimen of the " Kentish-Plover " which has fallen under my notice occurred in a collection of birds made by the late Mr. Villet at the Knysna. I purchased the specimen, which is now in the South African Museum.

565. Charadrius Tricollaris, Vieil. ; N. Dict. D'His. Nat., p. 147 ; C. Indicus, Lath. ; Lesson, Vol. 2, Pl. 320 ; Charadrius Bitorquatus, Wagl. Ency. ; Meth. Ois., t. 233, f. 4.

GENERAL colour above, brown, glossed with green ; forehead white ; a white line extends from thence over the eye, and unites behind the head, at the base of the occiput ; throat grey ; this is succeeded by a black collar, next by a white one, and that again by a broad black one ; from this to the end of the tail it is all pure-white ; wing-feathers black and dark-brown ; the latter with the secondaries tipped with white ; tail-feathers brown, glossed with green, and tipped with white, the outer pair most, the middle pair least so ; bill black ; base yellow ; cere round the eye and bill, red ; iris hazel. Length, 6″ 3‴ ; wing, 4″ 7‴ ; tail, 3″.

This very pretty little Plover is common throughout the colony, frequenting equally the sea-shore, the natural vley, artificial dam, or river. I found it abundant in the water holes of the Karroo rivers, and equally so on the sea-shore of the coast. It is a solitary species, never appearing more than a pair, ♂ and ♀, at once ; and though five or six pairs may often be found within a few hundred yards from each other, they keep to their respective puddles, and always return to them if driven up by a passing foot.

They run with considerable rapidity, often squat behind a stone, or amid tufts of grass. to evade detection; but when on the wing, fly strong and well. They breed on the sea-shore or along the banks of streams and vleys, never far from water. Their nest is a mere depression in the soil, unprotected by stone or bush ; and their eggs, enormous for the size of the bird, are two in number, of a dirty-white ground, profusely and minutely covered with hair streaks, having a tendency to run in zones, there being always one at the obtuse end : axis, 15‴ ; diam., 11‴.

Their food consists of minute crustaceans, shells, and insects.

566. Charadrius Kittlitzi, Reichenbach.

GENERAL colour above, ashy-brown, variegated with light ash ; forehead, semi-circular band round the back of the head, chin, throat, and vent, white ; a black band extends across the head, just behind the white of the forehead, through the eyes and behind the white collar, a black patch extends from the bill to the eye ; wings black ; chest and belly, isabella-yellow ; four lateral tail-feathers on each side, isabella-yellow, tinged with grey, the rest dark ashy-brown ; bill and legs black. Length, 6″ 6‴; wing, 4″ 5‴ ; tail, 1″ 9‴.

This little Plover is common about the chain of lagoons formed by the "Salt River," and along the sea-shore near Cape Town. I have also seen it in the month of September on the rocks at Green-Point and on Robben Island, and it is abundant near Zoetendals Vley.

Hitherto our Cape bird has been confounded with *Ch. Pecuarius,* Temminck ;* but the acquisition of the true *Pecuarius* by myself in St. Helena has cleared up the error. They are, however, very closely allied ; but the St. Helena bird inhabits stoney places, instead of the banks of vleys and the sea-shore.

---

* Which that author states was procured at the Cape.

567. Charadrius Hiaticula, Linn., Pl. Enl., 921 ;

Hiaticula Torquata, Leach ; *The Common Ring-Plover ;* MacGill. Brit. Birds, Vol. 3, p. 116 ; Bewick's Brit. Birds, Vol. 2, p. 16.

ADULT, with the upper parts greyish-brown ; the forehead with two bands, a white and a black ; a dark-brown band under the eye ; a ring of white including the throat, succeeded by a broader ring of brownish-black ; bill orange at the base, black at the tip ; feet orange ; claws black. Young, without the black band on the forehead, the other bands lighter ; the feathers of the upper parts margined with pale-brownish ; the bill entirely black ; the feet dull greenish-yellow ; the claws black. Length, 8″ 6‴ ; wing, 5″ 4‴ ; tail, 2″ 8‴.*

M. Victorin procured this species at the Knysna, and Mr. Ayres at Natal. My son shot a fine specimen in full plumage at Salt River Mouth, near Cape Town, in April, 1865.

568. Charadrius Marginatus, Vieil. ; N. Dic-

d'Hist Nat., XXVII, p. 138 ; *Ch. Leucopolius,* Wagler; *Hiaticula Heywoodii,* Gray.

ABOVE, cinoreous-brown ; the edges of the feathers rufescent ; forehead white ; nuchal band faint-rufous ; lores black ; below all white ; wing-coverts brown, with white shafts ; webs of inferior wing-feathers margined and tipped with white ; four middle tail-feathers brown ; those at the side white. A specimen shot April 29th, 1863, measured : length, 7″ ; wing, 4″ 9‴ ; tail, 2½″. Irides brown ; bill black ; the base livid ; legs and feet livid-blue.

This little Sand-Piper is common along all parts of our coast, and on any extensive inland pieces of water. It breeds with us in the month of November, generally laying two eggs, of a fawn or nankin colour, profusely spotted with black, sometimes streaked, and sometimes the spots form a circle at the obtuse end : axis, 16‴ ; diam., 11‴.

The young run immediately on their exclusion from the egg. They are covered with white down, and on being pursued, cower down with such success among the rounded white pebbles, which they closely resemble, that if once the eye is taken off them, it is a mere chance if they are again detected.

---

* I copy entire the descriptions of *C. Hiaticula* and *O. Cantianus* from MacGillivray's work on the Birds of Britain, in order that the two species may readily be distinguished from each other. I also use his excellent descriptions for such of our birds as belong to this class, as none can exceed them for clearness and accuracy.

569. Charadrius Leschenaultii, Lesson; *C. Fuscus*, Cuvier; *C. Rufinus*, Blyth, Journal As. Soc., Beng. XII, p. 180; *C. Subrufinus*, Hodg.

GENERAL colour above, greyish-brown, each feather slightly margined with pale-rufous; a white line springs from the nostril and passes over the eye, but does not extend to the back of the head; throat, and all under parts, white; loral space, greyish-brown, as are also patches on each side of the lower part of the neck, which nearly meet across the chest, and form a collar; tail-feathers all brown, the outer pair almost white; under side of wings, white; upper side of quill feathers, very dark-brown; bill robust, black, thickened towards the point, flesh-coloured at the base; legs (apparently) flesh-coloured. Length, 8″; wing, 5″ 9‴; tail, 2″ 9‴; tarsus, 1″ 6‴; middle toe, 11‴.

I shot a single specimen (♀) of this plover on the Salt River, near Cape Town, in 1858, since which time no other specimen has occurred to me. On comparing it with Indian specimens received from my friend, Mr. Blyth, the Curator of the Asiatic Society's Museum in Calcutta, not the slightest differences can be observed. My bird was accidentally killed from a flock of sand-pipers (*Charadriadæ et Tringæ*) feeding in the marsh. On dissection it proved a ♀; its stomach contained minute crustaceans, worms, and the insects found in these brackish waters.

570. Charadrius Asiaticus *et Caspius*, Pallas; *C. Jugularis*, Wagler.; *C. Gigas*, Brehm (juv.); *C. Montanus*, Townsend; *C. Veredus*, Gould (juv.); *Morinellus Caspius*, Bp.

ABOVE, greenish-brown, the edges of the feathers buff; forehead, before the eye, chin, and throat, white; chest, deep buff-coloured, followed by a black band; all the rest of the under parts white; tail-feathers tipped with pale-buff. Length, 9″; wing, 5″ 9‴; tail, 2⅓″.

. Mr. Arnot, who forwarded the only three specimens that have reached me, writes: " Found together in flocks of 15 or 20, very far away from water. They are scarce, and I only see them after showers of rain, which bring out small coleoptera and animal life of that sort, on which these birds seem to feed, and get enormously fat." Procured near Colesberg.

The Sub-Family, HÆMATOPODINÆ, or Oyster-Catchers,

have the bill lengthened, strong, with the apical half much compressed to the tip, which is obtuse; the nostrils linear, and placed in a membranous, lateral groove; the wings long and pointed; the tail moderate; the tarsi robust, and covered with small scales; the toes three; the lateral toes united at the base by a membrane, especially the outer one; the claws strong, broad, and curved.

Genus HÆMATOPUS, Linnæus.

Bill longer than the head, strong, straight, with the culmen slightly depressed at the base, and the apical portion much compressed to the tip, which is obtuse; the nostrils placed in a membranous groove, which reaches nearly to the middle of the bill, with the opening linear; wings long, with the first quill the longest; tail moderate and even, or slightly rounded; tarsi strong, longer than the middle toe, and covered with small reticulated scales; toes moderate, strong; the lateral toes united to the middle toe by a basal membrane, especially the outer; the claws strong, broad, and slightly curved.

571. Hæmatopus Moquini, Bonap; *Hœmatopus Niger*, Cuv., Reg. An., Vol. 1, p. 469; *Ostralagus Atra*, Lesson, Vol. 2, p. 301; *H. Capensis*, Licht.; *The Oyster-Catcher* of Colonists.

ALL over a deep black; bill and cere round the eyes in life, coral red; legs deep-crimson; bill, 2″ 8‴ long, flattened throughout, and very much compressed at the point.

The "Oyster-Catcher" is not uncommon along the shores of South Africa, extending far towards the Line on both sides of the continent. It is generally found in pairs, and feeds on small fish, mollusca, &c., which it picks up along the margin of the retreating tide. The colours of all the naked parts change considerably after death. It breeds with us; and I have been favoured with eggs, collected by Mr. Hugo, of Simon's Town, who procured them along the shore towards Cape Point. The eggs are generally two in number, laid in a simple depression in the sand, in the debris accumulated just beyond high-water mark. They are of a greyish cream-coloured ground, generally, but rather sparsely covered with coarse, irregular wavey black and dark-brown broken lines: axis, 2″ 6‴; diam., 1″ 9‴. My son found it breeding on Robben Island about Christmas, 1865.

The Sub-Family, CINCLINÆ, or Turnstones,

have the bill rather short and straight, with the culmen straight, and sometimes vaulted at the apical portion, the sides compressed to the tip, which is obtuse or acute; the nostrils placed in a membranous groove, with the opening linear and longitudinal; the wings lengthened, with the first quill the longest; the tail rather short; the tarsi short, robust, and covered with small scales ; the toes long, the lateral ones unequal, and free at their bases, the hind toe rather long, slender, and elevated.

Genus CINCLUS, Mœhring.

Bill rather shorter than the head, straight, and slightly depressed at the base, with the culmen straight, and the sides much compressed to the tip, which is truncated ; the lateral margins of both mandibles curved upwards at the tip; the gonys moderate and ascending ; the nostrils lateral, and placed in a membranous groove, that extends half the length of the upper mandible, with the opening linear, and longitudinal ; wings very long and pointed, with the first quill the longest : tail moderate, and slightly rounded; tarsi as long as the middle toe, robust, and covered in front with broad scales ; toes long, the outer rather longer than the inner one, both free at the base, and the sides of all margined by a narrow membrane ; the hind toe elevated, with the tip resting on the ground.

572. Cinclus Interpres. (Linn.) Pl. Enl., 856 ;
Strepsilas Collaris, Temm. ; Charadrius Cinclus, Pall.

GENERAL colour of the upper parts, dark-brown, glossed with green and purple, the feathers edged with pale-brown ; fore part and sides of the head, brownish-white, mottled with black ; throat white ; a band on each side from the lower mandible, the side of the head behind the ear-coverts, and the fore part of the neck, black ; the feathers slightly edged with whitish ; the rest of the lower parts pure white, as on the hind part of the back and the upper tail-coverts, some of the rump-feathers, however, being black ; the tail is white at the base, brownish-black towards the end, edged externally with light-brown, the outer web of the outer, and the tips of all, excepting the two middle, white ; legs and feet orange ; claws black, as is also the bill ; the lower mandible tinged with red at the base. Length, 9" 9''' ; wing, 6" 4''' ; tail, 2" 9'''.

The *Turnstone* is a constant resident on these shores, extending upwards as far as the Equinoctial Line. It frequents rocky places in preference to sandy beaches, feeding on small crustaceans, shells, and the various insects which infest the decaying seaweed. They usually keep in small families of from four to eight ; do not appear to mingle with the sand-pipers ; and are wary and difficult to approach. I fancy it must breed on Robben Island, among the rocks at the northern end, having seen young birds in that neighbourhood.

*L

The Third Family, ARDEIDÆ, or Herons,

have the bill more or less long, and generally much compressed on the sides, with the culmen curved at the tip, which is sometimes emarginated and acute ; the wings moderate and rounded; the tail mostly short, and rounded on the sides; the tarsi lengthened and rather slender ; the toes more or less long, and rather slender, with the outer toe longer than the inner, and both united at their bases, the hind toe more or less short, and generally placed on the same level with the anterior toes.

The Sub-Family, GRUINÆ, or Cranes,

have the bill more or less lengthened, straight, and strong, with the tips of both mandibles pointed and equal in length ; the nostrils placed in a deep groove, which extends beyond the middle of the bill ; wings long, the tertials lengthened and pendant; the tail short and even ; the tarsi very long and slender ; and the toes rather short.

Genus GRUS, Linnæus.

Bill longer than the head, straight, both mandibles of equal length, with the tips pointed, the sides compressed, the basal part of the culmen flattened, and the apical part slightly curved to the tip ; the nostrils placed in a broad deep nasal groove, which reaches beyond the middle of the bill, with the opening placed anteriorly, large, and closed posteriorly by a membrane ; wings long, with the third and fourth quills the longest, and the tertials lengthened and pendant; tail rather short; tarsi very long, slender, and covered with transverse scales ; toes rather short, the outer united at the base to the middle one, the lateral ones equal, the hind toe very short and elevated ; the claws short and strong.

573. Grus Carunculata. (Gm.)   Cuv., Vol. 3, p. 332 ; *Ardea Carunculata*, Gmel.; Lath. Syn., V. 82, t. 78 ; *Wattled Crane; Grus Carunculata*, Shaw, Vol. 11, p. 533 ; *Ardea Palearis*, Forst. Desc. Mam., p. 47.

GENERAL colour above, grey-slate, darkest on the back, at the end of the wings, and top of the head ; neck pure white ; the rest of the plumage black.   In the male, the fore half of the head is bare, wrinkled, and of a red colour; in the female this is much reduced in size, and the dark mark on the top of the head is indicated by a few dark spots only ; below the chin depend two flaps of skin, covered with short dense feathers ; legs black ; bill red ; irides red. Length, 4′ 7″ ; wing to the end of drooping feather, 3′ ; tail, 13″.

This great Crane is only found in a few favoured localities, scattered over the country. A single pair take up their haunts and maintain it for years, breeding constantly in the same nest, which is repaired as occasion requires. I had the pleasure of watching a pair, through my binoculars, engaged in this proceeding : both birds contributed to the work, stopping now and then to do a little courting, like an ordinary sparrow or canary, but surely undignified in so grave a bird ! It suggested to me the possibility of an Archbishop or a Lord-Chancellor making love ! Still " something came of it," for my friend Hugo took two glorious eggs out of that selfsame nest, and presented them to me, and they now form part of the treasures of the South African Museum. They are of a dull olive-brown, irregularly blotched throughout with reddish-brown, closely resembling those of *Anthropoides Virgo* and *S. Stanleyanus.* Axis, 4″ 3‴ ; diam., 2″ 9‴.

Genus ANTHROPOIDES, Vieillot.

Bill the length of the head, straight, the sides compressed, the apical part of the culmen slightly curved to the tips, which are equal in length ; the gonys long and slightly advancing upwards ; the nostrils placed in a basal groove, which reaches beyond the middle of the bill, with the opening in a longitudinal slit ; wings long, with the third and fourth quills the longest, and the tertials lengthened and pendant ; tail rather short ; tarsi long and slender, and covered with transverse scales ; toes moderate, slender, and covered above with transverse scales, the lateral toes equal, the outer united at the base to the middle toe, and the hind toe very short and elevated. ·

574. Anthropoides Stanleyanus, Vigors ;

Zool. Journ. II, 234, Pl. 8 ; Cuv., Vol. 3, p. 330 ; *Grus Paradisea*, Licht. ; *Tetrapteryx Capensis*, Thunb. ; *Blue Crane* of Colonists.

ENTIRELY of a leaden-blue, with the exception of the upper portion of the head, which is white, and the ends of the long drooping plumes of the wings, which are black. Length, 4′ 4″ ; wing, 3′ 10″ (including the drooping plumes) ; tail, 1′ 2″.

The " Stanley " or " Blue " Crane is not abundant in any locality, but seems very generally distributed. I fancy that certain pairs frequent the same district for the whole term of their lives, and may always be found within a certain radius. I saw it thus continually at Nel's Poort, and at the Knysna. It is wary, and difficult to approach within gunshot; but may always be got with a rifle. It goes in pairs, ♂ and ♀, the latter being considerably the smallest. It becomes very tame when in confinement, and feeds readily from the hand, eating bread, fruit, vegetables, &c. Its harsh, rattling, guttural cry may be heard at an immense distance. In a state of nature it feeds on fish, reptiles, locusts, and small mammalia, and is often found permanently inhabiting the Karroo country at a great distance from water.

Mr. Arnot has forwarded eggs from Colesberg, and informs me it breeds in that neighbourhood in some abundance. The eggs resemble those of *G. Carunculata* in every particular.

Genus BALEARICA, Brisson.

Bill shorter than the head, strong, thick, the basal half of the culmen flattened, and the apical half slightly arched and gibbose, the sides compressed, the gonys of the upper mandible short, and advancing upwards; the nostrils placed in a broad basal groove, which reaches to the middle of the bill, with opening anterior, large and oval; wings lengthened, with the third quill the longest, and the tertials broad, reaching to the end of the tail; tail short and even; tarsi lengthened, slender, and covered in front with transverse scales, and posteriorly with small subquadrate scales; toes moderate, slender, and covered above with transverse scales; the lateral toes unequal, and the hind toe very short and elevated; the cheeks naked; the base of the bill and the throat beneath wattled.

575. Balearica Regulorum. (Licht.) Cat. Dupt.
Berl. Mus., 1793; *Grus Balearica*, Pl. Enl. 265; *Crowned-Crane*; and *Kafir-Crane* of Colonists.

UPPER parts, leaden-grey; lightest on the hackle-shaped feathers of the throat; top of the head, velvety black; chin black; back of head ornamented with a curious tuft of wiry bristles, radiating from an elongated centre, each bristle about 3½″ long, flat, and twisted throughout its entire length, one side being white, the other yellow, thus causing the bristle to appear ringed with white and yellow; tip black; beneath and behind the eye, and below the black of the chin, a bare red spot; outer feathers of wing white, the inner red; and covering them, when the wing is closed, are some elongated, decomposed yellow plumes; tail black; under parts leaden-grey. Length, 3′ 8″; wing, 1′ 9″; tail, 1′.

The "*Crowned*" or "*Kafir-Crane*" is more common in the eastern frontier of the colony, than about the western extremity, perhaps owing to the latter being more settled and cultivated, as it is found in considerable numbers to the north and north-west. It has never fallen under my notice in its wild state, nor can I find any one who has seen its nidification.

Mr. Chapman obtained it as far north as Lake N'Gami.

The Sub-Family, ARDEINÆ, or Herons,

have the bill more or less lengthened, and generally acute, with the gape extending beneath each eye; the nostrils placed in the lateral groove, with the opening usually longitudinal, and partly closed by a membrane; the wings usually long, and the first quills more or less graduated; the tail short and even; the tarsi long, slender, and covered with transverse scutellations or reticulated scales; the toes long and more or less slender; the outer toe usually united at its base; sometimes both toes are united at the base for some distance.

Genus ARDEA, Linnæus.

Bill lengthened, and more or less slender, with the culmen nearly straight to the tip, which is acute and emarginated, the sides compressed, and the lateral margins straight and sometimes serrated; the gonys moderate and ascending; the nostrils lateral, basal, and placed in a groove, which extends for more than half the length of the bill, with the opening linear and closed by a membranous scale; wings long, with the first quill nearly as long as the second and third, which are equal and longest; tail rather short, and even; tarsi longer than or as long as the middle toe, rather slender, and covered in front with transverse scales; those near the toes large, and of a hexagonal form; toes long, and rather slender; the outer toe longer than the inner, and united at the base; the hind toe long; the claws moderate, slight, curved, and acute.

576. Ardea Goliath, Temm., Pl. Col. 474; *A. Gigantodes*, Licht.; *A. Nobilis*, Blyth, Ann. Nat. His., 1844, p. 176; Rupp. Faun. Abyss., t. 26.

UPPER part of back, and drooping plumes of chest, leaden-grey; head, neck, and under parts of body, deep-rufous; chin white; down the front of the neck extends a line of black, mottled with white, coalescing and mingling with the drooping plumes of the chest. Length, 4' 6"; wing, 1' 11"; tail, 11".

This gigantic Heron has only once fallen under my notice, a fine male bird in full plumage having been received from Mr. Arnot, of Colesberg. Mr. Chapman is familiar with it towards the Zambezi; and Mr. Andersson found it in Damaraland.

577. Ardea Cinerea, Linn., S. N., Vol. XII., p. 236, Pl. Enl., 755; A. Major, Gmel.; Ardea Leucophœa, Gould; Bonap. Consp., Vol. 2, p. 111; The Common Heron, Bewick's Brit. Birds, p. 36.

UPPER parts, blueish-grey; head, neck, and under parts, white; a broadish black stripe commences just before and extends over the eye to the back of the head, terminating in a lengthened pendant crest; a double row of black spots commences about one-third of the way down the front of the throat, and terminates among the pendant feathers of the chest, on each side of which is a broad black patch, extending down each side to the thighs; quill-feathers of wings black. Length, 3' 3"; wing, 1' 6"; tail, 8".

The Heron is very abundant throughout South Africa, in all suitable localities, feeding upon fish and frogs, and an occasional snake. I am told they breed on the Cape Flats, placing their nests in company on tufts of grass and rushes, surrounded with water. This seems a strange habit, and very different from their European practice; but I see no reason to doubt my informant, who has volunteered to show me the breeding place, and whose children make a yearly jaunt thither to feast on the eggs.

578. Ardea Atricollis, Wagler; Smith, Zool. S. Af., Pl. 86; A. Melanocephala, Child. and Vig.

UPPER parts, blue-grey, darkest in centre of back; upper portion of head and back of neck, black; chin and upper portion of throat, pure white; centre portion profusely mottled with black and white; lower part, blue-grey; under portion of body, light-blue grey; tail black. Length, 3' 3"; wing, 1' 4"; tail, 8".

I suspect that the black-necked Heron is more common than is generally supposed; but, owing to its resemblance to the common species, is not often brought to notice. I once shot one without being at all aware that I was obtaining such a prize. It was on the marshy ground close to the Observatory, near Cape Town, and at the moment when I fired it was engaged in a fight with another. I took them both to be A. Cinerea till I picked up my bird. I have received it from Colesberg in the young state; and Mr. Chapman procured it along the Zambezi route; Mr. Atmore in the George district. I also saw a specimen in the Worcester Valley. Three eggs, said to be of this species, were sent to me from Verloren Vley; they are blue-green, like those of the common herons: axis, 2" 2'''; diam., 1" 6'''.

579. Ardea Purpurea, Linn., Pl. Enl. 788; A. Caspica, Purpurata, Rubiginosa, (?) et Botaurus, aut Rufa, Gmel.; A. Variegata, Scop.; A. Monticola, Lafrey; The Purple Heron, Bewick's Brit. B., p. 40.

UPPER part of back, ashy-grey, variegated with a few red plumes on each side; top of head black; neck red, with a broad black stripe down the back, a narrow one on each side, and a blue one in front, formed of a succession of spots, and spreading out on the breast, which is bright-buff; centre of belly, very dark-rufous; sides lighter; thighs bright-rufous; as are also the shoulders. Length, 2' 10"; wing, 1' 2"; tail, 7".

The Purple Heron is abundant in all marshy places throughout the colony, and has been brought by Mr. Chapman from the Zambezi. It feeds on the same food as does *A. Cinerca*, and may be found in similar localities. I have never obtained its nest here.

580. Ardea Calceolata, Du Bus. Bull. Acad. Brull, IV., p. 40.

GENERAL colour, black; with chin and upper part of anterior portion of throat, white; two long narrow pendant plumes proceed from the occiput; the feathers of the breast and back are much decomposed and elongated; in the latter projecting slightly beyond the tail, and curved upwards at the tip; irides bright-yellow. Length, 17"; wing, 8" 3'''; tail, 4½".

Hartlaub, on the authority of Delalande, gives the Cape as a habitat of this curious heron. I have not met with it from hence; but shot a specimen in Boyana Bay, on the North-West Coast of Madagascar.

581. Ardea Bubulcus, Savig., Pl. Enl., 610; *A. Ibis*, Hasselq.; *A. Veranyi*, Roux; *Bubulcus Ibis* Bonap.; *A. Coromanda*, Bodd.

♀, in breeding plumage: general colour white; top of head, crest at the back, pendant feathers of breast, and decomposed feathers of back, rufous-isabella colour; bill and legs yellow, irides yellow. Length, 1' 7"; wing, 9"; tail, 4½".

Two specimens of this Egret, in non-breeding dress (which is all white, without the crest and pendant plumes,) have been received from Mr. Arnot.* Mr. Chapman's Zambezi collection contained several in both phases of plumage.

582. Ardea Flavimana; *Herodias Flavimana*, Sund. Ofvers. K. V. Ac. Forh., 1850, p. 111.

GENERAL colour, grey-black; toes yellow.

Of this bird I have not been able to obtain any other description than that given above; neither have I seen any specimen. It is given as a native of South Africa, loc. cit.

---

* A specimen also in non-breeding dress was brought to me by Mr. Shaw, of Wynberg, in May, 1867.

583. Ardea Egretta, Gmel. et Wagl.; *Ardea Flavirostris*, Temm.; *A. Magnifica*, Verr; *Ardea Alba*, Temm.; *Egretta Flavirostris*, Bonap. Consp. 11, p. 116; Cuv., Vol. 3, p. 339.

PLUMAGE, pure white; bill and bare space round the eye, and between it and the bill, bright-yellow; legs greenish-black. The male in breeding plumage has the feathers of the lower part of the neck elongated and decomposed, and those of the back also decomposed and prolonged to an enormous extent, projecting far beyond the tail, drooping gracefully over the end of the wings, and turned slightly upwards at the tips; irides bright-yellow. Length, 2′ 2″; wing, 10″; tail, 5″ 6‴.

A pair of these most lovely Egrets, ♂ and ♀, have been shot at different times in Zeekoe Vley, between Cape Town and Simon's Bay. I have myself seen white egrets wading about that piece of water, and also on the marsh nearer to Kalk Bay, but always too far off to determine the species. I have also seen white egrets on the marsh below the Observatory, and on the rocks at Robben Island. They are, however, generally so wary as to be unapproachable. Mr. Atmore tells me it is common at the Knysna, but very shy, and breeds on a small rock in the sea, east of the Heads; he was too late (December) for eggs.

584. Ardea Garzetta, Linn.; *Garzetta Egretta*, Bonap. Consp., II, p. 118; *A. Xanthodactyla* et *A. Nivea*, Gmel.; *A. Orientalis*, Gray; *A. Melanopus* Wagler; *A. Nigripes*, Temm.

ENTIRELY white, with two long plumes depending from the back of the head; feathers of the breast and back elongated and decomposed; the latter not extending beyond the tail, and curved considerably upwards; bill and legs in full breeding plumage, black: at other times the legs and feet are greenish-yellow; irides bright-yellow. Length, 2′ 4″; wing, 12″; tail, 5″.

Specimens of this beautiful Egret, both in winter and nuptial dress, have been procured in the neighbourhood of Cape Town; also at Colesberg and at the Knysna. The stomachs of this and the preceding, which I have examined, contained multitudes of small aquatic shells (*Physopsis Africana* and *Succinea Delalandi*).

585. Ardea Comata, Pall.; Pl. Enl., 348, 315; *A. Ralloides*, Scop.; *A. Senegalensis*, Gmel.; *A. Castanea* et *A. Pumila*, Lepech.; *The Squacco Heron.*

TOP and back of head, pale ochraceous-yellow; occipital feathers elongated, pointed, white; back rufous, the feathers elongated and decomposed; all the rest of the plumage white; bill blueish, with black tip; legs yellow; irides yellow. Length, 16″ to 17″; wing, 8½″.

Included by Mr. Gurney in his fourth Additional List of Birds received from Natal.—Ibis Vol., 1862, p. 157. Sed non vidi.

586. Ardea Leucoptera, Bodd., Pl. Enl. 911; *A. Grayi*, Sykes, Gray, Illus. Ind. Zool.; Pl. 48; *A. Malaccensis*, Gmel.; *A. Speciosa*, Horsf.

UPPER part of back, dirty-brown; wings and tail, pure white; head and neck white, closely streaked with brown; breast and under parts white, sparingly streaked with brown, and tinged with russet-brown on the flanks; back of head crested; bill greenish, and horn-coloured at the tip. Length, 16″; wing, 9″; tail, 3″ 9‴.

Two specimens of this little Heron were sent from Colesberg by Mr. Arnot. On comparing them with Indian specimens, I could not detect the slightest difference between them. I saw it abundantly up the East Coast and Mozambique Channel, frequenting the coral reefs which line the shore; but it has not been found, to my knowledge, along the coast of the colony.

587. Ardea Atricapilla, Afzel.; *Egretta Thalassina*, Swain. Ani. in Menag., p. 333; *A. Brevipes*, Hempr.; *Butorides Atricapilla*, Bp.

ABOVE, dark slate-coloured, tinged with green, which prevails on the wings; back, sides of neck, flanks, and under parts, cinereous; top of head and crest black, shot with green; chin and fore part of neck white, tinged with ferruginous; dorsal plumes long, narrow, and lanceolate. Length, 15″; wing, 7″ 2″; tail, 2″ 9‴.

I obtained specimens of this pretty little Egret at Mauritius, Madagascar, the Islands to the North-East, and along the whole East Coast of Africa. Two examples were procured by Mr. Arnot somewhere in Mahura's country; and Mr. Ayres has it from Natal. It frequents equally fresh and salt water, finding its food, which consists of small fish, frogs, water-insects, and shells, with equal facility in either. It is very partial to the mangrove thickets which line the tidal rivers on the coast, and roosts among them, perching on the branches.

588. Ardea Minuta, Linn., Pl. Enl., 323, S N., XII, p. 240; *Ardea Minuta*, Bp.; *Ardea Danubialis* et *A. Solonensis*, Gmel.; *The Little Bittern*, Bewick's Brit. B., p. 52.

TOP of head, back, wing, and tail, black; remainder of the plumage buff; chin white, and faint white longitudinal lines

*M

arc visible down the throat; a large patch on the wing is also buff, In the ♀, the top of the head and the wing and the tail feathers are black ; the back, deep rufous-brown ; the feathers edged with buff; neck deep-rufous ; the fore part lightest, distinctly marked longitudinally with white lines ; sides of chest beautifully variegated with a rufous-brown so deep as to appear black in some lights. Length, 13″ ; wing, 5″ 5‴ ; tail, 2″.

This pretty little Heron is not uncommon in all suitable localities in the colony ; and Mr. Chapman brought it in considerable abundance from the Zambezi. It conceals itself among reeds, and lies so close, that the shooter often treads within a few feet of it before it will rise. My son has shot several pair in the neighbourhood of Cape Town.

589. Ardea Pusilla, Vieil. ; *Ardeola Podiceps,*
Bonap. Consp., Vol. II, p. 135.

RESEMBLES *A. Minuta,* the preceding species, but is smaller ; the upper wing-coverts are reddish-grey, not whitish ; under, dirty-red, not white ; neck, obscure rufous, not grey ; abdomen ruddy, Length, 12″.

Inhabits Natal, from whence it has been forwarded by Mr. Ayres. I have not met with it in this colony.

590. Ardea Gutturalis, Smith, Zool. S. Af., Pl.
91 ; *A. Sturmii,* Wagl., Bp. Consp. 2, p. 131.

ABOVE, dark slate-colour ; beneath buff, variegated from the chin down to the belly with elongated blotches of dark slate-colour, here and there edged with rufous ; flanks tinged with rufous. Length, 12″ ; wing, 6″ ; tail, 2″.

I have not yet seen this species from within the colony ; but have received it from just beyond the borders. Mr. Andersson procured it in Damaraland, and Mr. Chapman brought several specimens from the Zambezi ; but the former tells me it is a very rare species.

Genus BOTAURUS, Stephens.

Bill long and straight, with the culmen straight, flattened at the base, rounded, and curved to the tip, which is strongly emarginated, and the sides compressed ; the gonys short and ascending ; the nostrils basal, and placed in a deep groove, that extends for two-thirds the length of the bill, with the opening linear ; wings long, with the first three quills equal and longest ; tail short and even ; tarsi as long as the middle toe, rather strong, and covered in front with broad transverse scales ; toes very long, and rather slender, the outer toe longer than the inner, and united at the base by a membrane, the hind toe long and rather slender ; the claws very long, slightly curved, and very acute.

591. Botaurus Stellaris. (Linn.) Pl. Enl., 789 ;
Ardea Stellaris, Linn. ; *The Common Bittern*, Bewick's Brit. Birds, Pl. 47 ; *Ardea Stellaris Capensis*, Sch., Mus. de Pays-bas ; "*Roerdomp*" of Colonists.

GENERAL colour, ochreous, profusely variegated with wavy lines of dark-brown ; top of head deep-brown ; chin white, or pale-ochreous ; the markings on the back assume a bolder character than those on the other parts of the body ; the neck is furnished with a ruff, which can be expanded at pleasure. Length, 2' 4" ; wing, 12" ; tail, 4".

The Bittern is occasionally found in a few favoured localities, even in the neighbourhood of Cape Town ; and I am told it breeds in Verloren Vley. At Zoetendals Vley I found it in great abundance ; and early in the morning, or during the evening twilight, their booming call resounded from every part of the lake. Mr. Hugo, of Fransch Hock, brought me a pair of eggs, unfortunately both broken, which he states are of this bird. The colour is a dark olive-green, smooth, and shining as a duck's. Having seen bitterns' eggs in various collections in England, I am satisfied that Mr. Hugo is right in his identification.

Genus NYCTICORAX, Stephens.

Bill rather longer than the head, strong, with the culmen gradually curved, and the sides compressed to the tip, which is emarginated ; the gonys long and ascending ; the nostrils lateral, and placed in a groove, with the opening linear, and closed by a membranous scale ; wings long, with the first quill shorter than the second and third, which are equal and longest ; tail short and even ; tarsi as long as the middle toe, rather strong, and covered with large irregular scales ; toes long, rather slender, the outer toe longer than the inner, both united at their base, especially the former ; the hind toe long, rather slender, and on the same plane with the others ; the claws moderate, curved, and acute.

592. Nycticorax Griseus. (Linn.) Pl. Enl. 758,
759 ; Bewick's Brit. B., p. 55, Bp. Consp. 2, p. 140 ; *Nycticorax Europœus*, Steph. ; *Ardea Nycticorax*, L.

GENERAL colour, light-slate ; paler on the side of the head and neck, with a vinaceous tinge ; chin, centre of throat, and under parts, white ; the back, top of the head, and a portion of the hinder part of neck, dark iridescent-green ; from the back of the head emanate two long, narrow, and pure white plumes, eight inches long ; legs yellow ; irides in adult bird the same ; in young birds light reddish-madder. The young bird is of a dirty-grey, mottled with white. Length, 24″; wing, 13″; tail, 5″.

The Night-Heron is generally distributed throughout the colony ; but appears to be nowhere common. It certainly breeds somewhere in the neighbourhood of Cape Town, as a young bird, still showing the down, was brought to me from Zeekoe Vley, on the Simon's Bay road.

Genus SCOPUS, Brisson.

Bill long, with the culmen elevated to the base, keeled and curved to the tip, which is hooked, the sides much compressed, and grooved near the culmen from the base to the tip ; the gonys long, and curved upwards ; the nostrils basal and lateral, with the opening linear, exposed, and partly closed by a membranous scale ; wings long, with the third and fourth quills equal and longest ; tail short and even ; tarsi longer than the middle toe, compressed, and covered with small reticulated scales ; toes moderate ; the outer longer than the inner ; both are united at the base by a membrane, which extends along the sides to the tip ; the hind toe moderate, and even with the others ; the claws rather short, and slightly curved.

593. Scopus Umbretta, Gmel. ; Pl. Enl. 796 ;
Ardea Fusca, Forst. ; *Cephus Scopus*, Wagl. ; *Hammerkop* of Colonists.

GENERAL colour, a warm brown ; glossed with purple ; tail with a broad bar at the tip, and numerous narrow ones ; head much crested ; bill black. Length, 20″; wing, 13½″; tail, 6″.

The "*Hammerkop*" (literally, Hammerhead) is found throughout the colony, and all the way to the Zambezi, frequenting ponds, marshes, rivers, and lakes. It is a strange, weird bird, flitting about with great activity in the dusk of the evening, and preying upon frogs, small fish, &c. At times, when two or three are feeding in the same small pool, they will execute a singular dance, skipping round one another, opening and closing their wings, and performing strange antics.

They breed on trees and on rocky ledges, forming a huge structure of sticks, some of them of considerable thickness. These nests are so solid, that they will bear the weight of a large, heavy man on the domed roof without collapsing. The entrance is a small hole, generally placed in the most inaccessible side. The eggs, three to five in number, are pure white : axis, 1″ 9‴; diam., 1″ 4‴.

On my late friend Jackson's farm, at Nel's Poort, there is a singular rocky glen between two hills. In this spot, a beautiful permanent spring, called "Jackalsfontein," takes its rise. Of course, in consequence, there are a few wild almond and other trees, and the place is a little oasis amid the barren mountains. It is a favourite resort of wild animals, hyænas, leopards, jackals, &c., and here Mr. Jackson has constructed one of his most successful hyæna-traps. On the ledges of the rocks in this secluded spot a colony of Hammerkops have built for years. Some of the nests are quite inaccessible, while others can be reached with a little trouble. I counted six or eight within fifty yards, all exhibiting the same form and structure, and some of them containing at least a large cart-load of sticks. Mr. Jackson told me they occupied the same nest year after year, and added to it or repaired it as required. About some that I visited, I found brass and bone buttons, bits of crockery, bleached bones, &c. Mr. Jackson said if a "Tottie" lost his knife or tinder-box on the farm, or within some miles of the place, he made a point of examining the Hammerkops' nests, and frequently with success; the birds, like the " Bowerbird " of Australia, embellishing their dwellings with any glittering or bright-coloured thing they can pick up.

In the Karroo, between Worcester and Robertson, I saw a nest placed on the ground on the side of a trifling rise: it was at least three yards in length, by one and a half across, with a small entrance hole at one end.

Genus PLATALEA, Linn.

Bill lengthened, straight, thin, much depressed, and broadly dilated at the tip, which is spoon-shaped, with a lateral groove commencing on the forehead, extending and somewhat parallel with the edge to the tip, which is slightly bent downwards ; the nostrils basal, and placed in the groove, with the opening oval, and partly closed by a membrane; wings long, the second quill the longest; tail short ; tarsi longer than the middle toe, rather slender, and covered by reticulated scales ; toes long, with the anterior toes much united at their base by a membrane, which extends along the sides of the toes to the tip, the hind toe long, rather elevated, and only partly resting on the ground ; the claws short, scarcely curved, and obtuse.

594. Platalea Tenuirostris, Temminck ; Man. d'Orni., Ed. 1816, P. CIII ; *P. Nudifrons*, Cuv. ; *P. Chlororhynchos*, Drap. ; *P. Nivea*, Burch. ; *C. Telfairii*, Vig., Pro. Zool. Soc., 1831, p. 41.

GENERAL colour, white ; the shafts of the larger wing-feathers, and the tips of the first four or five, being dark-brown ; fore part of head bare, and of a bright vermilion ; the bill very narrow, but expanding at the tip into a broad spoon. Length, $33''$; wing, $15\frac{1}{4}''$; tail, $4\frac{1}{2}''$.

Procured at Natal by Mr Ayres. The only specimens which have fallen under my notice were brought from the Great Lake by Mr. Chapman, where he states they are very common.

The Sub-Family, CICONINÆ, or Storks,

have the bill lengthened, usually straight, and conical, with the sides compressed to the tip, which is acute; the gonys long and ascending; the nostrils lateral, pierced in the substance of the bill, with the opening linear; the wings long and ample; the tail moderate and broad; the tarsi lengthened, and usually covered with reticulated scales; the toes moderate, with the anterior toes more or less united at the base; the hind toe long, elevated, and partly resting on the ground.

Genus CICONIA, Linn.

Bill lengthened, straight, and strong, with the culmen keeled and straight towards the tip, which is acute; the sides compressed, and the gonys long and gradually ascending; the nostrils lateral and pierced in the substance of the bill, with the opening linear and small; wings long and ample, with the third and fourth quills equal and longest; tail short and broad; tarsi twice the length of the middle toe, and covered with reticulated scales; toes moderate, with the anterior ones united at the base, especially the outer ones; the hind toe elevated, and partly resting on the ground.

595. Ciconia Alba, Bris.; *Ardea Ciconia*, L., Pl.

Enl., 866; *The White Stork*, Gould's B. of Eur., Pl. 283; *Great Locust-Bird* of Colonists.

ALL white, except the wings, which are black; bill and legs red.   Length, 3′ 9″; wing, 1′ 11″; tail, 1′.

This Stork is only a migratory visitant to the colony, following the flights of locusts, on which it feeds.   It is not confined to any locality, but pursues its prey to all parts of the country.

My valued correspondent, Mrs. Barber, however, informs me that it breeds in the Interior.   She writes as follows :—" My authority for saying that the White Locust-Crane build their nests in the Interior is good, or I should not have mentioned it.   When my brother Septimus was lion-hunting in the Free State, he saw the hills where great numbers of their old nests were.   Mr. William Stubbs (of the Queen's Town district) a great observer of Nature, and ' a good man and true,' is another of my authorities : he told me that he saw their nests upon the low rocky hills near St. John's River.   Like the small locust-birds, they always build their nests in the neighbourhood of large swarms of *young locusts*, so that they may have plenty of food for their young ones ; for young locusts remain long near the spot where they are hatched."

596. Ciconia Nigra, Bechst., Pl. Enl.,399 ; *Ardea Nigra*, Linn., Sys. Nat., 1, p. 235 ; *Melanopelargus Niger*, Bonap. Consp., Av. 11, p. 105 ; *Ciconia Fusca*, Briss. ; *The Violet-Stork* of Colonists.

GENERAL colour, very dark brownish-green, shot with green and coppery reflections; belly, thighs, and vent-feathers, pure white; bill, space round the eyes, chin, and legs, red. Length, 3′ 8″; wing, 1′ 10″; tail, 11″.

The Violet Stork is not common, but generally distributed. I have killed it at the Knysna, and Nel's Poort (Beaufort), and have received specimens from most of my correspondents, including Mr. Chapman. It feeds upon fish, frogs, and crabs, and will take a small mammal if it falls in with it. It is solitary in its habits, more than two being rarely, if ever, seen together. Its great height gives it a command of vision which puts all the shooters' powers of woodcraft to the tax, to enable him to circumvent his quarry ; and if the bird is only wounded, he proves no mean antagonist, striking fiercely with its sharp-pointed and powerful bill at the face and hands of his pursuers. On one occasion, having only broken the tip of a wing with a long shot, I had to dispatch my bird with half a charge of dust-shot, so great was his resistance ; he, moreover, as if aware of his greater chance, took to the water, and would not be driven from it.

597. Ciconia Abdimii. (Licht.) Rüpp. Atlas, t. 8 ; *Sphenorhynchus Abdimii*, Licht. p. 76 ; *Abdimia Sphenoryncha*, Bonap. Cons., 11, p. 106.

ABOVE, greenish-purple ; neck brown, with purplish-gloss ; all the under parts white ; chin, bare part of forehead, space round the eye, and tip of bill, orange-red ; the rest of the bill greenish ; bare space round the eye, blue. Length, 2′ 9″; wing, 16″.

I have not seen this bird from any locality, save the Zambezi, whence it was brought by Mr. Chapman, who informs me it is excellent eating, and is found in vast numbers about the borders of Lake N'Gami. Mr. Andersson procured it in Damaraland ; Hartlaub ranks it among the South African species ; and M. Van Horstock procured it somewhere in the colony (Schlegel's Mus. Pays-bas, *Ciconiæ*, p. 6).

598. Ciconia Leucocephala, Gmel., Pl. Enl., 906 ; *C. Episcopus*, Bodd.; *C. Umbellata*, Wagl., Syst. Av. Sp., 11 ; *C. Microscelis*, Gray, Gen. of Birds, Pl. 151, juv.; *Melanopelargus Leucocephalus*, Bonap. Consp., 11, p. 105.

VIOLET-BLACK, shining-green ; plumes of lower neck, blackish-green, long, wide, truncated, violet-purple at the tip; forehead, chin, and neck, white ; belly and rump, white ; top and back of head, black ; tail white ; the webs of the four outside feathers, black, shining coppery. Length, 2½' ; wing, 18½".

This bird appears rather frequently in Natal; but I have not met with it in this colony, nor in any collection made to the Westward, or in the Interior. It could not have passed me unnoticed, as I am well acquainted with it from seeing it in Ceylon.

Genus LEPTOPTILUS, Lesson.

Bill very large, straight, and the base higher than broad, with the culmen keeled, and straight to the tip, which is acute ; the sides compressed ; the gonys long and gradually ascending ; the nostrils small, pierced in the substance of the bill, with the opening linear ; wings long and ample ; tail moderate and broad ; tarsi robust, much longer than the middle toe, covered with reticulated scales; toes long, with the anterior ones united at the base, especially the outer toe ; the hind toe long, and partly resting on the ground. The head and neck denuded of feathers.

599. Leptoptilos Crumeniferus, Cuvier ;
Ciconia Argali, Temm., Pl. Col., 301 ; *C. Vetula*, Sundevall.

ABOVE, dark greenish-black, slightly glistening; below, dirty-white ; head and neck bare ; the back of the latter furnished with a stripe of scanty woolly hairs : a few of these are also scattered on the pouch which hangs in front, and on the bare spot of the breast beneath it. In life, these bare places are of a bright carmine. The bill is very strong, straight, and of a dirty yellowish-white ; the legs and feet jet-black. Length, 5' ; wing, 28"; tail, 10".

The Adjutant has not occurred to me, but is included in this list on the authority of Dr. Hartlaub, Sys. der Orn. W. Af., p. 228. I saw it abundantly on the East Coast ; and a specimen in the Museum was brought from the mouth of the Zambesi.

Genus MYCTERIA, Linn.

Bill long and strong, with the base higher than broad, the culmen straight towards the tip, which is turned upwards, as well as that of the lower mandible ; the sides compressed, and the gonys long and curved upwards ; the nostrils lateral, pierced in the substance of the bill, with the opening linear ; wings long and ample, with the second and third quills longest; tail moderate and broad; tarsi much longer than the middle toe, and covered with reticulated scales; toes long, and the anterior ones united at the bases, especially the outer toe; the hind toe moderate, with the tip resting on the ground.

600. Mycteria Senegalensis, Shaw; Lath.

Gen. His., IX., p. 19; *Ciconia Senegalensis*, Vieil. Galer., Pl. 255; *Ciconia Ephippioryncha*, Temm. Pl. Col. Rüpp. Atlas, t. 3.

WHITE; head and neck black, glossed with bronze-green; scapulars, and wing and tail feathers, raven-black; base of bill to nostril, bright crimson; as is also the skin round the eye and under the bill (the bill from the nostril for $3\frac{1}{5}$ inches is black; from thence to the tip bright crimson); frontal shield, bright yellow; shanks and tarsi black; knees and feet brick-dust red; the bare spot on the breast, crimson. Length, $4\frac{3}{4}''$; wing, 23''; tail, 10''; tarsus, $11\frac{1}{2}''$; bill, 12''.

This large Stork has not to my knowledge been found anywhere within my limits, except at Natal. Here it has been observed by Mr. Ayres, whose description of the bill and bare places, taken probably from a fresh-killed specimen, I copy from the "Ibis," Vol. 1862. Mr. Anderson procured it Damaraland.

Genus ANASTOMUS, Bonnaterre.

Bill long, higher than broad at the base, strong, with the culmen nearly straight, and the sides much compressed towards the tip; the lateral margins sometimes worn away, leaving an opening between the mandible, even when closed; the lower mandible very much compressed, and the gonys lengthened and curved upwards; the nostrils lateral, pierced in the substance of the bill, with the opening linear and narrow; wings long, with the third and fourth quills the longest; tail moderate and broad; tarsi longer than the middle toe, and covered with reticulated scales; toes long, rather slender, with the anterior toes united at their base, especially the outer; the hind toe long and slender, with the tip partly resting on the ground.

601. Anastomus Lamelligerus, Temm., Pl.

Col. 236; *Hians Capensis*, Less.; *Hiator Lamelligerus*, Bonap. Consp. 11, p. 109.

GENERAL colour black, shining-green and purple, with the feathers of the lower part of the neck, chest, belly, and thighs, lamellated at their tips, and along the shafts: those on

*N

the back of the neck and chest, deep-brown; those on the body black; bill very large and gaping. Length, 2″ 9‴; wing, 1′ 4″; tail, 9″.

This curious bird is not uncommon about the Zambezi; and the specimen from which the above description was taken was brought thence by Mr. Chapman. I have not seen it from this side of the Tropic; but I believe it has been found in Kaffraria. Hartlaub also gives it as an inhabitant of South Africa. Its enormous gaping bill, and the curious structure of its plumage, at once distinguish it from all our other storks.

The Sub-Family, TANTALINÆ, or Ibises,

have the bill lengthened, more or less slender, and curved throughout its length; the sides gradually compressed to the tip, which is obtuse; the nostrils lateral, and sometimes placed in a lengthened groove, with the opening always linear and exposed; the wings rather long; the tail moderate and even; the tarsi of various lengths, usually robust, and covered in front with transverse or hexagonal scales; the toes long, the inner toe shorter than the outer, both more or less united at the base; the hind toe long, and usually strong.

Genus TANTALUS, Linnæus.

Bill lengthened, broad, and elevated at the base, the basal half nearly straight, the apical portion curved, with the culmen gradually curved, and the sides compressed to the tip, which is emarginated; the gonys long and curved; the nostrils basal and lateral, with the opening pierced in the substance of the bill, longitudinal, and exposed; wings long and pointed, with the first quill rather shorter than the second and third, which are equal and longest; tail short and even; tarsi rather slender, much longer than the middle toe, and covered with hexagonal scales; toes long and rather slender, the inner toe shorter than the outer, and both united at the base; the hind toe long and slender; the claws short, broad, and obtuse; the head, and sometimes the neck, denuded of feathers.

602. Tantalus Ibis, Linn.; S. N. 241, Pl. Enl. 389;

Ibis Candida, Pen.; *T. Rhodinopterus*, Wagl. Syst. Av. Spec. 3.

WHITE; fore part of head bare of feathers, red; some of the feathers of the wing washed with rose-colour or purple; wing and tail black; bill yellow, and curved downwards at the tip;

thighs and legs red. Length, 3'; wing, 18''; tail, 7½''. The young bird is of a dull brownish-grey; wing and tail black.

A single specimen of this bird in the young stage of plumage was shot at Nel's Poort by Mr. Henry Jackson, and another appeared in Mr. Chapman's collection.

February, 1866.—A full-grown bird was received lately from Mr. H. Jackson, who states that it was killed by him on his dam. It was in company with two others, which would not, however, permit any approach after the death of their comrade.

Genus IBIS, Mœhring.

Bill lengthened, slender, and curved throughout its whole length; the sides compressed to the tip, which is obtuse; the nostrils lateral, basal, and placed in a slender groove, that extends from the base to the tip, with the opening linear and exposed; wings rather long and pointed, with the first and second quills equal and longest; tail rather short, and nearly even; tarsi slender, and longer than the middle toe, covered in front with broad transverse scales, some of which are divided irregularly; toes long and slender, the inner rather shorter than the outer, and both united at the base by a membrane; the hind toe long and slender; the claws long, and rather weak; the head is partly denuded of plumes.

603. Ibis Falcinellus. (Linn.) Sys. Nat., Vol. 1, p. 241; Pl. Enl. 819; *Tantalus Igneus* et *T. Viridis*, Gmel.; *Falcinellus Igneus*, Becht., Bonap. Consp. 2, p. 158.

GENERAL colour, very dark-brown, resplendent with iridescent-green, purple, and bronze reflections; shoulders deep ruddy-brown, as are also the neck and under parts; the cheeks and chin mottled with white. In some specimens the neck and under parts are not ruddy, and the former is mottled all the way down.

Hartlaub cites this as a South African species; but it has not come under my notice, dead or alive; and it is quite familiar to me in either state, having shot many of them in Ceylon. Mr. Ayers has obtained it in Natal.

Genus GERONTICUS, Wagler.

Bill more or less strong, elevated at the base, and curved throughout its length; the sides compressed to the tip; which is obtuse; the nostrils basal, lateral, and placed in a groove, which extends from the base to the tip, with the opening linear and exposed; wings long and pointed, with the first quill generally, and sometimes the second quill, shorter than the third and fourth, which are equal and longest; tail broad and even; tarsi as long as or rather shorter than the middle toe, strong, and covered in front with hexagonal scales; toes long and robust, with the inner toe shorter than the outer; the lateral ones united at the base, and all the fore toes margined at the side; the hind toe long and strong; the claws moderate, compressed, and curved; the head and neck more or less denuded of feathers; the scapulars of some species long, and composed of decomposed plumes.

604. Geronticus Æthiopicus. (Lath.) *Numenius Ibis*, Cuv., Vol. 3, p. 359; Bp. Consp. 2, p. 151; *Ibis Religiosa*, Savig.; *Sacred Ibis*.

GENERAL colour, pure-white; the tips of the wing-feathers being brilliant dark-green, and the supplementary plumes assumed by the male in his nuptial livery, deep rich-blue and white, and very lax; head, and the whole of the neck in the male, bare and black; in the female, clothed with short black and white feathers; chin and space round the eyes, bare, black; legs black. Length, 2' 6"; wing, 15½"; tail, 7".

A few specimens of the *Sacred Ibis* have come under my notice, killed in the colony; one, a female, in fine plumage, having been shot at Green-Point, within three miles of Cape Town, feeding about the rocks which line the coast in that direction. It appears in Mr. Chapman's collection; and he informs me they are very common towards the Lake. I saw a large flock of them at Zoetendals Vley in December, 1865; they walked rapidly about a mud bank in the river near the mouth, probing for worms; and I noticed that they turned about in the usual quick manner of the curlew. I was informed that they occasionally bred in that neighbourhood.

605. Geronticus Hagedash. (Vieil.) *Tantalus Hagedash*, Sparm.; *T. Caffrensis*, Licht.; *J. Chalcoptera*, Vieil., Gal. Des. Ois., t. 246; *Ibis Hagedash*, Wagl. Syst. Av.; *Hagedashia Capensis*, Bp. Consp. 2, p. 153.

GENERAL colour above, dark-brown, glossed with shining-green and purple, especially on the wings and tail; head, neck, and under parts, dull brownish-grey; ridge of the bill near the head, red, the rest black; legs red-brown. Length, 2' 5"; wing, 14½"; tail, 7".

The *Hadadah* (so called from its cry of "Ha-ha-hadadah!") is only found in wooded places, such as the Knysna, where it is not uncommon, associating in small flocks, and making the wooded streams resound with its loud and not unmusical call. It feeds on worms and small crustaceans, which it finds in the shallows and muddy creeks. Mr. Chapman procured it at the Zambezi. It is everywhere very shy. and difficult to procure, the best way being to watch their roosting place, to which they nightly repair, and where several may be obtained from an ambuscade by a successful discharge.

606. Geronticus Calvus. (Bodd.) Pl. Enl. 867 ; Cuv., Vol. 3, p. 361 ; *Tantalus Niger*, Gmel. ; *T. Capensis*, Forst. Descrip. An., p. 48 ; *Ibis Calva*, Vieil. ; Shaw, Vol. 12, Pl. 1, p. 7 ; Bp. Consp. 2, p. 153.

GENERAL colour throughout, shining-green, glossed with purple, with a patch of rich golden-copper on each shoulder ; head, and portion of the neck, bare, and of a deep red-colour ; as are also the legs. Length, $2'$ $6''$; wing, $16''$; tail, $8\frac{1}{2}''$.

The *Wilde Kalkoen* (wild-turkey) of Dutch colonists is not rare on the Eastern frontier ; but seldom appears to the westward. I am told it is a very foul feeder, frequenting the neighbourhood of native villages, and acting the part of scavenger. It thrives well in confinement, and becomes very tame. One in the possession of His Excellency Sir George Grey was at large in the Government Gardens ; he walked with great rapidity, and was always seeking for food, probing the ground with his long curved bill, and drawing thence the large earth-worms, which he swallowed at a gulp. Mr. Albert Vigne shot a fine female at Tygerhoek, on the River Zonder End, near Caledon, in February, 1866.

The Fourth Family, SCOLOPACIDÆ, or Snipes,

have the bill generally long, slender, curved, or straight throughout its length, with the sides compressed and grooved to the tip, which is obtuse ; the nostrils basal, longitudinal, closed by a membrane, and placed in the groove ; the wings long and pointed ; the tail usually short and even ; the tarsi more or less long and slender ; the toes generally long and slender, the lateral ones sometimes much united to the middle toe by a membrane ; the hind toe short, resting on the ground, or entirely wanting.

The Sub-Family, LIMOSINÆ, or Godwits,

have the bill generally long, slender, entirely curved, or straight from the base, with the sides more or less grooved to the tip, which usually overlaps that of the lower mandible, and is obtuse ; nostrils basal, placed longitudinally in the lateral groove ; wings long and pointed ; the tail usually short and even ; the tarsi lengthened and slender ; the toes long, the lateral ones united at the base of the middle toe ; the hind toe short, sometimes wanting.

Genus NUMENIUS, Latham.

Bill more or less long, slender, and curved from the base, with the sides compressed and grooved for nearly its whole length ; the tip of the upper mandible projecting over that of the lower, and rather obtuse ; the nostrils basal, lateral, and placed in a lateral groove, with the opening longitudinal and covered by a membrane ; wings long and pointed, with the first quill the longest ; tail short and even ; tarsi longer than the middle toe, slender, and covered in front with narrow transverse scales ; toes moderate, the lateral toes unequal, and united at their bases ; the hind toe long, slender, and partly resting on the ground ; the claws short and obtuse.

607. Numenius Arquatus. (Lath.) Pl. Enl.
818 ; *The Curlew*, Bewick's Brit. Birds, p. 68.

PLUMAGE of the head, neck, and fore part of the back, light reddish-grey, streaked longitudinally with blackish-brown ; hind part of the back white, with narrow longitudinal dark marks ; tail white, with twelve dark-brown bars ; the breast, sides, and abdomen, white, with lanceolate dark-brown spots and bars ; bill very long (six to seven inches) and curved downwards. Total length (of male), 25″ ; wing, 12″ 6‴ ; tail, 4″ 6‴. The female is somewhat smaller.

The Curlew is not uncommon on our sea-border throughout its whole extent. It maintains its characteristic shyness, and rarely falls to the gun of the sportsman. I never heard of its breeding in the colony, though it is found here throughout the year. I met with it up the whole East Coast, as far as the Line. Schlegel separates our South African species from the European bird on account of its size, and calls it *Numenius Major*.

608. Numenius Phœopus. (Linn.) Pl. Enl.
842 ; *The Whimbrel*, or *Little Curlew*, Bewick's Brit. Birds, p. 70.

UPPER part of the head, dark-brown, with a medial and two lateral whitish bands, streaked with dusky ; neck, all round, pale brownish-grey, streaked with brown ; feathers of the fore part of the back, scapulars, and wing-coverts, dusky, with marginal whitish spots ; hind part of back white ; tail and coverts, greyish-white, barred with dusky; chin, hind part of breast, and abdomen, white; fore part of breast streaked; sides and lower wing-coverts barred with dusky. Length, 17″, wing, 9″ 9‴; tail, 4″; bill, 3″ 2‴.

The Whimbrel is a rare bird in South Africa, only three specimens having reached my hands: one of these was procured in the neighbourhood of Walwich Bay by the Messrs. Chapman; the other two were shot near Cape Town. one by my son, the other by Mr. Butler, the Taxidermist of the Museum. From the statements of the two latter, it appears the birds were feeding in company with the common curlew, and may be others of the same species were with them, as they were killed by chance, and the difference not observed till submitted to me for inspection.

Genus LIMOSA, Brisson.

Bill long, rather slender, and more or less inclined upwards towards the tip, with the sides compressed and grooved on both mandibles to nearly their entire length; the nostrils lateral, basal, and placed in the groove, with the opening longitudinal, and closed with a membrane; wings long and pointed, with the first quill the longest; tail short and even; tarsi longer than the middle toe, rather slender, and covered in front with narrow transverse scales; toes long, the outer toe united to the middle toe by a membrane as far as the first joint; the inner toe slightly united; the hind toe long, slender, and partly resting upon the ground; the claws short and obtuse.

609. Limosa Cinerea. (Güld.) *Terekia Cinerea; Scolopax Cinerea*, Gmel.; *Sc. Terek*, Lath.; *Sc. Sumatrana*, Raffles; *Limosa Recurvirostra*, Pallas; "*Terek Sandpiper.*"

UPPER parts, brown-grey, shot with greenish, and with a slender black streak along the shaft of each feather; forehead, chin, throat, and chest, white, mottled finely with grey streaks and spots; under parts, and bar along the wing, pure white. Length, 9″ (of which the bill is 1″ 9″, and curved upward); wing, 4″ 9″; tail, 2″ 3‴.

Mr. Ayres records the shooting of a specimen at Natal "from a flock of four or five amongst the mangroves in the bay;" and adds: " they are scarce birds here." I have not met with it.

The Sub-Family, TOTANINÆ, or Longshanks,

have the bill long, slender, compressed, and grooved on the sides, with the culmen more or less straight, curved, and acute at the tip; the nostrils linear, and placed in the lateral

groove, which extends to or beyond the middle of the bill;
the wings long and pointed; the tail moderate and even, or
rounded; the tarsi more or less long, slender, and covered in
front with narrow transverse scales; the toes usually long
and slender; the lateral ones unequal, and generally united
at the base to the middle toe, especially the outer: the hind
toe long, slender, and hardly touching the ground.

Genus TOTANUS, Bechstein.

Bill more or less long and strong, with the culmen straight
or slightly curved, and the sides compressed to the tip, which
is slightly curved and acute; the gonys long, and slightly
curved upwards; the nostrils linear, placed in a membranous groove, which does not extend beyond half the
length of the bill; wings reaching beyond the end of
the tail, and pointed, with the first quill the longest;
tail moderate, and nearly even; tarsi as long as or longer
than, the middle toe, more or less slender, and covered in
front with numerous very narrow scales; toes long and
slender, the anterior toes united by a membrane, especially
the outer; the hind toe slender, elevated, and hardly touching the ground.

610. Totanus Stagnatilis, Bechst.; *Scolopax Totanus*, Linn., Pl. Enl., 876; *Tringa Guinetta*, Pall.

ABOVE, pale-cinereous; the margins of the feathers whitish;
eyebrows, cheeks, back, and under part, pure white; neck
and breast marked with brown; tail white, barred with
brown. Length, 8" 6'''; wing, 5" 1'''; tail, 2" 5'''.

Procured at Natal by Mr. Ayres; but I have not seen it from the
Cape Colony.

611. Totanus Calidris, Bechstein, Pl. Enl. 827, 845; *Scolopax Calidris*, Linn.; *Tringa Gambetta*, Gmel., and *Tringa Striata*, Lath.; *The Redshank*, Bewick's Brit. Birds, p. 96.

PLUMAGE in winter: upper parts of the head and hind-neck,
brownish-grey; fore part of the back, greyish-brown, streaked
with dusky; its hind part white; the tail and its coverts
barred with white and dusky-grey; the fore part and sides
of the neck pale-grey, streaked with dusky, as are the sides;
the rest of the lower parts white. In summer: the lower

parts are white, or less shaded with grey, and marked with decided dusky streaks, spots, and angular bars; feet and legs, orange-red in life. Length, 11" 3'''; wing, 6" 6'''; tail, 2" 10'''.

The "*Red-Shank*" and "*Green Sandpiper*" are found sparingly about the Knysna estuary, and the mouth of the Salt River, near Cape Town, in both of which places I have shot them. I also detected them among some birds shot by Messrs. Chapman on Lake N'Gami. At Zoetendals Vley they were abundant in November.

612. Totanus Ochropus. (Linn.) Temminck, Pl. Enl. 843; *Totanus Leucurus*, Gray; *The Green Sandpiper*, Bewick's Brit. Birds, p. 99.

TAIL, nearly even, white; the four middle feathers with three blackish-brown bars toward the end, the outermost feather plain; upper part of head and hind-neck, brownish-grey; back, scapulars, and inner secondaries, greenish-brown, with marginal whitish and dusky spots; neck greyish-white, with longitudinal dusky lines; the breast and abdomen white; lower wing-coverts, axillars, and some of the hypochondrial feathers, angularly barred with brown; bill dusky above, reddish beneath; feet greyish-blue, tinged with green. Length, 11" 6'''; wing, 5" 11'''; tail, 2" 6'''.

Several examples of this bird have been received from Mr. Arnot, procured near Colesberg; also common at Zoetendals Vley in November, 1865, and at the Knysna.

613. Totanus Glottis. (Linn.) Bechstein; *Limosa Grisea*, Briss.; *Scolopax Glottis*, Linn.; *The Green-Shank*, Bewick's Brit. Birds, p. 108.

ADULT, in winter, with the bill greenish-brown at the base, black towards the end; the feet greenish-grey; the head, hind part and sides of neck, greyish-white, streaked with brown; the fore part of the back, scapulars, and wing-coverts, greyish-brown, the feathers edged with whitish; hind part of the back, the fore part of the face, and all the lower parts, white, but with faint grey markings on the fore part of the sides, and on the lower wing-coverts; tail white, barred with greyish-brown. Adult, in summer, with the fore-neck and breast marked with oblong black spots; the fore part of the back and the scapulars, black, the feathers margined with whitish. Length, 14" 3'''; wing, 7" 6'''; tail, 3" 2'''.

Common with the preceding species, and found also on all vleys throughout the country.

*o

614. Totanus Glareola. (Linn.) Temm.; *Tringa Littoralis*, Linn. ; *Totanus Affinis*, Horsf. ; *The Wood Sandpiper*, Bewick's Brit. Birds, p. 101.

TAIL, doubly emarginate, white, all the feathers barred to the base with blackish-brown : the outermost feather, with the inner web, plain ; upper part of head greyish-brown ; back, scapulars, and inner secondaries, dark purplish-brown, with marginal whitish and dusky spots ; neck greyish-white, with longitudinal dusky lines ; the breast and abdomen white ; lower wing-coverts dusky, edged with white ; axillar feathers, and some of those of the sides, narrowly and irregularly barred with brown ; feet greenish-grey. Length, 9" ; wing, 5" 2''' ; tail, 2" 2'''.

Mr. Ayers found this species at Natal. It occurred to me about Zoetendals Vley, in November, 1865 ; but this is the only place where I have seen it.

615. Totanus Fuscus. (Briss.) Gould's Birds of Eur., Pl. 309 ; *Scolopax Fusca*, Linn. ; Bewick's Brit. Birds, p. 94.

ADULT, with basal half of bill red, the other half brownish-black ; the feet orange-red ; the secondary quills black, barred with white. In winter the upper part of the head, and hind neck, brownish-grey ; the fore part of the back, greyish-brown, streaked with dusky, its hind part white ; the tail and its coverts barred with white, and dusky ; the cheeks and fore-neck, greyish-white, faintly streaked with grey ; the rest of the lower parts white. In summer the head, neck, and lower parts greyish-black, the feathers slightly edged with whitish ; those of the fore part of the back and wing-coverts blackish-brown, marginally spotted with black and white ; the other parts as in winter. Length, 12" ; wing, 6" 9''' ; tail, 3'''.

A single specimen, procured by myself at the Knysna, is the only example that has occurred to me.

Genus TRINGOIDES, Pr. Bonap.

Bill as long as or rather longer than the head, with the culmen straight, and the sides compressed towards the tip, which is curved and acute ; the nostrils linear, and placed in a lateral membranous groove, which extends to near the tip on both sides ; wings reaching beyond the middle of the tail, and pointed, with the first quill the longest ; tail long, broad, aud much rounded; tarsi long, moderately strong, and covered in front with narrow transverse scales; toes long, with the outer one united at the base by a membrane to the middle toe, the inner one free, the hind one moderate, elevated; the claws short and acute.

616. Tringoides Hypoleuca. (Linn.) Pl. Enl.

850; *Tringa Hypoleucos*, Linn.; *Common Sandpiper*, Bewick's Brit. Birds, p. 103; *Tringa Leucoptera*, Pall.

BILL dusky; feet greenish-grey; upper parts glossy greenish-brown, transversely banded and undulated with dark-brown; lower parts white, excepting the fore part, and sides of the neck, which are greyish, with faint dusky lines. Young, with upper parts lighter, the feathers margined with dusky and reddish; iris hazel. Length, 8″ 6‴; wing, 4″ 9‴; tail, 2″ 5‴.

The *Sandpiper* is rather a rare bird in this colony, and but very few specimens have been seen. The first was forwarded to me in March, 1864, by Mr. Cowen, of the Paarl; another was shot about the same time on the Cape Flats; and my son procured a third on the rocks at Green-Point, near Cape Town. I saw a specimen or two at Zoetendals Vley in November, 1865.

The Sub-Family, RECURVIROSTRINÆ, or Avocets,

have the bill lengthened and slender, with the sides grooved to the middle, and compressed towards the tip, which is acute; the nostrils linear and membranous; the wings long and pointed; the tail rather short and rounded; the tarsi long, slender, and covered in front with reticulated scales; the tibia long, and denuded of feathers for some distance above the knee; the toes moderate, sometimes free, and sometimes entirely united by a web to the ends of the toes, the hind toe extremely short or wanting; the claws short.

Genus RECURVIROSTRA, Linnæus.

Bill very long and slender, with the culmen slightly depressed at the base, the sides grooved to the middle, and compressed to the tip, which is gradually pointed; the nostrils lateral and placed in the groove, with the opening linear and membranous; wings long and pointed, with the first quill the longest; tail short and rounded; tarsi much longer than the middle toe, rather compressed, and covered in front with reticulated scales; toes united together by an indented web; the outer toe rather longer than the inner; the hind toe extremely short; the claws short, compressed, and acute.

617. Recurvirostra Avocetta, Linn.; Pl. Enl.,
353 ; " *The Avoset*," Bewick's Brit. Birds, p. 58.

UPPER part of head and hind-neck, for half its length, black ; the rest white, excepting some of the scapulars, the smaller wing-coverts, and the primary quills and coverts, which are black ; legs and feet black, the latter partially webbed ; bill black, very long, slender, and turned considerably upwards. Length, 18" ; wing, 9" 3‴ ; tail, 3" 3‴ ; bill, 3" 3‴.

Occurs periodically in the colony in small flocks. It does not appear to be very shy, as many have been yearly procured on Zeekoe Vley, on the Simon's Town and Wynberg road; among them several specimens in very young plumage. I cannot, however, hear of their breeding here. Mr. Henry Jackson killed two or three specimens near Nel's Poort in February, 1866; at the same time, Mr. W. Keal procured some near Beaufort. What they could have been doing in a country so far from the sea and so badly watered, is a mystery. They only remained a few days in that locality, so were probably migrating.

Genus HIMANTOPUS, Brisson.

Bill much longer than the head, very slender and straight, with the sides grooved to the middle, and compressed towards the tip, which is acute ; the nostrils basal, and placed in the groove, with the opening long, linear, and closed by a membrane ; wings long and pointed, with the first quill the longest ; tail short, and nearly even ; tarsi very long, slender, and covered in front with reticulated scales ; toes moderate, and united at the base by a small membrane, especially the outer toe ; the hind toe wanting ; the claws small, compressed, and acute.

618. Himantopus Candidus, Bonn.; Pl. Enl.,
878 ; *Himantopus Melanopterus*, Temm. ; *Charadrius Himantopus*, Linn. ; *The Black-winged Stiltplover*, Bewick's Brit. Birds, Vol. 2, p. 6 ; *H. Albicollis*, Vieil.

GENERAL colour, pure white; wings black; back with a small brown patch near the shoulders; tail tinged with brown; in some phases of plumage the head and neck are also tinged with the same. This bird may at once be distinguished from all our waders by the disproportionate length and slenderness of its legs, which are of a bright-red colour. Length, 15"; wing, 8" 9‴; tail, 3" 6‴; legs, 10".

But one specimen has been obtained within the colony. It was killed on the Cape Flats by Mr. Dumbleton, to whom I am indebted for many rare birds, shot by his own hand. A pair were seen, but the other escaped. I have seen this species very abundantly in Ceylon, in the tanks scattered throughout the jungle. It is usually in small companies, wading up to the extent of its long legs, and even swimming across small holes or depressions which it may encounter while wading. It feeds on small crustaceans, worms, and shells, and is easy of approach, till it learns to fear the presence of man. On being flushed, the flock flies round and round, each bird uttering a loud 'cry of "wheet-wheet-wheet," in a different key, the whole forming a pleasing music, as agreeable to the ear of the naturalist as the cry of a pack of hounds is to that of a fox-hunter.

The Sub-Family, TRINGINÆ, or Sandpipers,

have the bill generally longer than or as long as the head, slender, and compressed on the sides, with the culmen near the tip slightly depressed and enlarged; the nostrils basal, and placed in a nasal groove, that extends for two-thirds of the bill; the wings long and pointed; the tail moderate and rounded; the tarsi usually long and slender; the toes more or less long, and united at the base.

Genus PHILOMACHUS, Mœhring.

Bill as long as the head, straight, rather slender, the sides compressed and grooved to near the tip, which is rather dilated; the nostrils basal, lateral, and placed in a basal groove; wings long and pointed, with the first and second quills equal and longest; tail rather short, and nearly even; tarsi long, slender, and covered in front with transverse scales; toes moderate, the lateral ones unequal, with the outer united at the base to the middle one as far as the first joint, and the base of the inner toe free; the hind toe elevated and short.

619. Philomachus Pugnax. (Linn.) Pl. Enl.

300, 305, 306; *Tringa Pugnax*, Linn.; *T. Variegata*, Brün.; *Machetes Pugnax*, Cuv.; The "*Ruff*," ♂, "*Reeve*," ♀, Bewick's Brit. Birds, p. 87.

MALE, in winter, with the bill brown; the feet greenish-yellow; the throat and abdomen white; the fore-neck and part of the breast, pale reddish-brown, spotted with dark-brown; the upper parts variegated with brownish-black and light-red. Female similar, but with the upper parts lighter, the lower more grey; the bill and feet dusky. Male, in summer, with numerous fleshy tubercles on the face, two occipital tufts, and a very large ruff of elongated feathers on the neck; the colour of the plumage varying in different individuals, and even in the same at different periods of the same season ; the bill and feet orange or yellow ; iris hazel. Length, 12″ ; wing, 7″ 6‴ ; tail, 3″.

The habits of this bird have been so well described by numerous authors, that they need not be alluded to here—the more so as it does not assume its nuptial dress and pugnaceous feelings in this country, only visiting us during the non-breeding period. It seems generally distributed. Mr. Ayres has found it in Natal ; and I have received it from Colesberg, the Knysna, Traka ; and have shot it myself on the Cape Flats.

Genus TRINGA, Linn.

Bill as long as or longer than the head, straight, slender, with the sides compressed at the base, and rather dilated and depressed at the tip ; the nostrils placed in a nasal groove, which extends to near the tip, basal, lateral, and longitudinal ; wings moderate and pointed, with the first quill the longest ; tail rather short, and nearly even ; tarsi strong, rather long, and covered in front with transverse scales ; toes moderate, slightly united at the base of the outer toe, and all margined on the sides by a membrane ; the hind toe very small and elevated.

620. Tringa Canutus, Linn. ; Pl. Enl. 365, 366 ; *Tringa Cinerea*, Lath. ; *The " Knot,"* Bewick's Brit. Birds, p. 84 ; *T. Glareola*, Pall. ; Gould's Birds of Eur., Pl. 324.

In winter, the plumage is ash-grey above, white beneath ; the neck and sides streaked with dusky. In summer, the back glossy black, variegated with yellowish-red ; the fore-neck and breast, yellowish-red. The young : grey above, each feather on the back with a double marginal band of black and white. Length, 10″ ; wing, 6″ 9‴ ; tail, 2″ 9‴.

South Africa, teste Hartlaub (Orn. W. Af., p. 237).—Sed non vidi.

621. Tringa Subarquata, Gmel., Pl. Enl. 851 ; *Scolopax Africana*, Gmel. ; *Scolopax Subarquata*, Gmel. ; *Sc. Caffra*, Forst. ; *Sc. Pygmœa*, Gmel. ; *Numenius Africanus*, Lath. ; *Numenius Pygmœus*, Lath. ; *Pigmy Curlew*, Bewick's Brit. Birds, p. 73.

Bill half as long again as the head, considerably decurved at the end, and black. Plumage in winter : brownish-grey above, each feather with a dusky streak ; the sides and fore part of the neck, with a small portion of the breast, greyish-white, streaked with brownish-grey ; the throat, breast, and abdomen, white ; the tail-coverts white ; the tail brownish-grey. In summer: the upper parts of the body black, mottled and streaked with light-red ; the sides of the head, the fore part and sides of the neck, and the breast, bright yellowish-red ; the neck faintly streaked with dusky ; tail-coverts white, spotted with brown and red ; the tail brownish-grey, glossed with green. Length, 9″ ; wing, 5″ 6‴ ; tail, 2″.

The *Pigmy Curlew* is abundant along the shores in the winter or non-breeding plumage, and I have shot some now and then, just beginning to assume that of the summer phase ; but not a single specimen has ever occurred to me, showing the full fine red tints of that state. It lives in great flocks on Robben Island, and at the mouth of the Salt River.

During my cruise in H. M. S. *Castor*, on the East Coast of Africa, we landed at an Arab settlement called Fazy, within a degree and a half of the Line. Our boat entered the mud-creek on which the little town is built, and seeing a vast flock of birds on the ooze, Lieut. Suttee and myself stole upon them behind an embankment, and fired our four barrels into the mass. The dead and wounded might have been counted by hundreds. The sailors collected a *bread-bag* full ! ! and they dropped by twos and threes from the flock, as it disappeared in the distance. The majority consisted of the species under consideration with *T. Minuta, Calidris Arenaria, Totanus Fuscus, T. Glottis, Charadius Marginatus*, &c.

622. **Tringa Minuta,** Leist.; Gould's Birds of Eur., Pl. 332 ; *Tringa Pusilla,* Mey. and Wolf ; *Pelidna Minuta,* Bonap. ; *Tringa Cinclus,* Pall. ; Bewick's Brit. Birds, p. 83.

PLUMAGE in winter: brownish-grey above, each feather with a dusky streak ; the sides of the neck paler ; its lower part tinged with brown ; the throat, fore-neck, breast, and abdomen, white ; the middle tail-coverts dusky ; the lateral white ; the tail-feathers ash-grey, edged with white ; the two middle greyish-brown. In summer: the upper parts yellowish-red, spotted with brownish-black ; the throat, breast, and other lower parts, white ; the lower anterior and lateral parts of the neck reddish-grey, streaked with brown ; a dusky loral band, and a whitish streak over the eye. Length, 6″ ; wing, 4″ ; tail, 1″.

Common in all marshes and on the seaboard. Their flight is very beautiful when in large flocks ; various evolutions performed simultaneously show alternately the dark upper plumage and the white under sides like flashes of light, particularly when seen against a dull, lowering sky.

Genus CALIDRIS, Illiger.

Bill as long as the head, straight, slender, the sides at the base compressed, and the tip slightly dilated and smooth; the nostrils basal, lateral, longitudinal, and placed in a nasal groove; wings moderate and pointed, with the first quill the longest; tail moderate and even, with the middle feathers rather longer than the lateral ones; tarsi longer than the middle toe, strong, and covered in front with transverse scales; toes rather short, slender, slightly united at their bases by a small membrane; the sides margined; the hind toe wanting; the claws short and curved.

623. Calidris Arenaria, Illig.; Gould's B. of
Eur., Pl. 335; *Tringa Arenaria*, Linn.; *Charadrius Rubidus*, Lath.; *The Sanderling*, Bewick's Brit. Birds, p. 4.

In winter: the upper parts pale-grey, with dusky lines; the lower white; wings black and grey, with a white band. In summer: the upper parts variegated with black and red; the lower red anteriorly; white behind. Young: with the upper parts black, and variegated with yellow and white; lower white; iris hazel. Length, 7" 9'''; wing, 5" 2'''; tail, 2" 2'''.

Found in large flocks everywhere along our seaboard, where sand patches afford it a congenial home and abundance of food. I have not procured it in the full nuptial dress, and fancy it must migrate to some colder climate to breed.

The Sub-Family, SCOLOPACINÆ, or Snipes,

have the bill long, straight, rather slender, compressed on the sides, and rather depressed, and turned down near the tip, which is obtuse, and bent over that of the lower mandible; the nostrils placed in a narrow, longitudinal groove, with the opening linear, and covered by a membrane; the wings moderate and pointed; the tail short and rounded; the tarsi more or less long, and covered in front with narrow transverse scales; the toes long, rather slender, with the hind toe short, elevated, and reaching to the ground.

Genus GALLINAGO, Leach.

Bill long, straight, grooved, and compressed on the sides, and the culmen rather depressed near the tip, which is obtuse, and curved over that of the lower mandible; the nostrils basal, placed in the groove, with the opening oblong and exposed; wings moderate and pointed, with the first and second 'quills equal and longest; tail short and rounded; tarsi moderate, shorter than the middle toe, strong, and covered in front with narrow transverse scales; the tibia bare for a short space above the knee; toes long, the inner toe shorter than the outer, and free at its base; the hind toe moderate and elevated, with the claw long and curved.

624. Gallinago Æquatorialis, Rüppell; *G. Nigripennis* et *Macrodactyla*, Bp.; *G. Scolopax Major* var., Gray; *G. Longirostris*, Licht. nec Cuv.

TAIL rounded, of sixteen feathers; bill twice the length of the head; two longitudinal black bands on the head, separated by a narrow medial, reddish-white band, and on each side a yellowish-white band; the upper parts variegated with black and light-red, with four longitudinal yellowish-white bands; wing-coverts tipped with white; three lateral tail-feathers on each side, white; sides transversely barred with dusky; axillar feathers white, barred with greyish-black. Length, 12″; wing, 5″ 6‴; tail, 2″ 6‴.

This fine Snipe may be easily distinguished from the "common snipe," which is said by some sportsmen* to be likewise an inhabitant of this colony, by the greater number of feathers in the tail, which in the latter species only amount to fourteen; it is also more brilliant in colouring, and exceeds it in size. One specimen forwarded to the Museum by the Hon'ble Mr. Southey weighed 9 oz. (avoir.)

It is distributed throughout the colony, migrating from place to place, according as the waters dry up. It prefers muddy swamps to clear streams, crouching amid the rank herbage. Its flight, compared to that of the common snipe, is slow and heavy, but is sufficiently rapid to puzzle the Boer with his long flint gun, who never dreams of firing a charge of shot at so insignificant a bird, which he is almost sure to miss! Both this and the Painted Snipe breed in the marsh below the Observatory near Cape Town. My son has captured the young birds just excluded from the egg in November and December.

Genus RHYNCHÆA, Cuvier.

Bill moderate, rather curved, compressed and grooved on the sides to the tip, which is much curved, and slightly hooked over that of the lower mandible; the nostrils basal, placed in the lateral groove, with the opening linear and exposed; wings moderate, with the first three quills equal and longest; the tertials as long as the quills; tail very short; tarsi as long as the middle toe, and covered in front

---

* Shaw, Vol. XII., p. 57, also says that it is found at the Cape of Good Hope; but as he does not give that as the habitat for the very plentiful *G. Æquatorialis*, I presume he has mistaken the one for the other.

*P

with narrow scales; the tibia bare of plumes for a short space above the knee; toes long and slender, with the inner toe shorter than the outer; the hind toe long, slender, and elevated.

625. Rhynchæa Capensis. (Linn.) Pl. Enl. 270;
Scolopax Capensis, Linn.; *Sc. Bengalensis,* Gmel.; *Rh. Variegata,* Vieil.; *The Painted Snipe,* Latham's Gen. Syn., Pl. 81.

MALE : head brown, slightly variegated with white; a yellow stripe passes from the base of the bill over the centre of the head, to the back of the neck; a similar stripe rises imme- diately in front of each eye, passing to the back of the head; two similar stripes, bordered by black, extends from the shoulders, down the back, to the root of the tail; neck greyish-brown, barred finely on the back portion with dark- brown, on the anterior portion more coarsely with the same colour; it is likewise here bisected by a white collar; under parts all white; the back beautifully marked with transverse black, white, and grey bars; the wings with numerous yellow semiocelli, edged with black. When extended, they expose numerous yellow or white ocelli, also edged with black, and also numerous black wavey transverse bars, chiefly on a slate-coloured ground; the tail is similarly marked. The female, reversing the ordinary rules, is more brilliantly marked than the male. The eye-stripe is pure-white; the chin dirty-white; neck and breast deep dull-rufous, succeeded by a broad black collar, and the pure white of the under parts broken only by a patch of black on each side. The plumage of the back is much darker, and the wings, when closed, instead of showing the black and yellow markings of the male, are a shining-green, transversely and minutely barred with black, all lustrously shot with metallic-green; legs green; bill red at the tip; iris hazel. Length, 10″; wing, 5″ 2‴; tail, 2″.

The " *Painted Snipe* " frequents the same localities as the preceding species, and is very common at certain seasons, breeding with us in the marshy ground near Cape Town.

The Fifth Family, PALEMEDEIDÆ, or Screamers,

have the bill generally long and slender, with the culmen rather depressed, straight at the base; the apical part vaulted, and the tip overhanging that of the lower mandible; the gonys short, and sometimes angulated; the nostrils lateral, generally placed near the middle of the bill, and longitudinal; the wings long, and generally armed at the shoulder with an acute spine, or blunt tubercle; the tail generally short, and slightly rounded; the tarsi long and slender; the toes very long, slender, and furnished with long and straight, or short, and slightly curved claws ; the hind toe long, furnished with a more or less long and acute claw.

The Sub-Family, PARRINÆ, or Jacanas,

have the bill lengthened, slender, with the culmen straight at the base, and vaulted at the tip, which is entire ; the nasal groove long and narrow, and the opening of the nostrils, placed in the middle of the bill, small and oval; the wings more or less long and pointed ; the tail generally short; the tarsi long and slender; the toes very long, slender, and armed with long slender claws, especially the hind toe.

· Genus PARRA, Linnæus.

Bill as long as the head, straight, slender, with the culmen straight from the base, and slightly vaulted and curved to the tip, which is entire ; the sides compressed ; the lateral margins straight, and the gonys short, ascending, and angulated ; the nostrils small, oval, medial, and placed in a lengthened groove ; wings long, with the first quill rather shorter than the second, the third the longest ; tail very short, and partly concealed by the coverts ; tarsi as long as the middle toe, without the claw, rather slender, and covered with transverse scales ; toes very long and slender, the outer rather longer than the inner toe ; the hind toe long; the claws long, especially that of the hind toe, which is extremely acute ; the base of bill, and more or less of the head, denuded, and furnished with caruncles.

626. Parra Africana, Gmel., Lath., Gen. Syn., V.
Pl. 17 ; Cuv., Vol. 3, p. 397 ; *Metopidius Africanus,* Wagler.

GENERAL colour, rich cinnamon-brown ; back of head and neck black, tinged with green ; chin and front of neck white, darkening into shining-yellow on the chest, which is divided from the cinnamon of the body by a narrow dark line ; front of head bare, which together with the basal half of the bill is blue in life ; the tip is horn-coloured ; legs dark-green ; claws horn-coloured. Length, 10″ ; wings, 5½″ ; tail, 1¾″.

This Jacana is decidedly rare in the colony, though I have received specimens from several different and widespread localities. It appears in Mr. Chapman's collections, and is cited as from Natal by Mr. Gurney.

627. Parra Capensis, Smith, Zool. S. Af., Pl. 32.

FOREHEAD, yellow; eyebrow, white; crown of the head, a narrow bar on each side of the base of the neck, sides of the body, and tail, deep reddish-orange; back of the neck black; sides of the neck and breast, pale greenish-yellow; chin, throat, breast, belly, and under tail-coverts, white; wings and back, grey-brown; bill yellowish-brown; legs and feet greenish-yellow. Length, 7″ 3‴; wing, 3″ 9‴; tail, 1″ 10‴.

Dr. Smith, loc. cit., mentions a single specimen killed near Algoa Bay. It has also been procured in Natal. No specimen has, however, come under my notice.

The Sixth Family, RALLIDÆ, or Rails,

have the bill more or less long, with the culmen sometimes advancing on the forehead, but always curved at the apex, and the sides compressed; the gonys usually. ascending; the nostrils lateral, and placed in a membranous groove; the wings moderate, and generally rounded; the tail more or less long and rounded; the tarsi long and rather slender; the toes mostly lengthened and slender; the claws short, compressed, and acute.

The Sub-Family, RALLINÆ, or Rails,

have the bill more or less slender and long, with the culmen keeled and curved at the apex, and the sides compressed; the nostrils lateral, with the opening placed in a membranous groove; the wings moderate, with the first quills usually graduated; the tail mostly short, and rounded; the tarsi long and slender; the toes more or less long and slender, and free at their base.

Genus RALLUS, Linnæus.

Bill longer than the head, slender and straight, with the culmen slightly curved from the front of the nostrils, and the sides compressed to the tip, which is obtuse, and slightly emarginated; the gonys long and slightly curved upwards; the nostrils placed in a membranous groove, which extends for two-thirds the length of the bill, with the opening exposed and linear; wings short, with the second and third. quills equal and longest; tail short and rounded; tarsi moderate, shorter than the middle toe, and covered with transverse scales; toes long and rather slender, the inner toe shorter than the outer, both free at their bases; the hind toe short and slender; the claws short, compressed, and very acute.

628. Rallus Superciliosus, Swains., An. in Menag., p. 335.

PLUMAGE above, brown; beneath, cinereous; front with a lateral white stripe before the eye; chin white; belly tinged with ferruginous; flank banded black and white; lower part of back and tail black; feet and bill red. Length, 12″.

Inhabits South Africa, Swain. loc. cit.—Non vidi.

629. Rallus Cœrulescens, Gmel.; Cuv., Vol. 3, p. 402; Shaw, Vol. 12, p. 211; R. Caffer, Forst. Descr. Anim., p. 50.

UPPER parts of head, neck, and body, reddish-brown; the chin, fore part of neck and breast, pale slate-colour; the flanks transversely striped with black and white. Length, 8″; wing, 4½″; tail, 1″.

This Rail has reached me from several different places in the colony; but seems to be nowhere plentiful. Mr. Cairncross has sent it from Swellendam; Mr. Jackson from Nel's Poort; Mr. Arnot from Colesberg.

630. Rallus Madagascariensis, Smith, S. Af. Quar. Jour. 1, p. 80.

HEAD and chin grey; neck, breast, and belly, rusty grey-brown; back rusty-olive, blotched with black; vent barred transversely with reddish white; under tail-coverts white; point of bill and upper surface of upper mandible, blackish-brown, the other parts of the bill yellowish. Length, 8½″

Inhabits South Africa, teste Dr. A. Smith.—Sed non vidi.

631. Rallus Aquaticus, Linn.; Pl. Enl. 749; Rallus Sericeus, Leach.; Scolopax Obscurus, Gmel.; Gould's B. of Eur., Pl. 339; Bewick's Brit. Birds, p. 137.

THROAT whitish; sides of head, neck, breast, and belly, ashy lead-colour; the feathers on the upper parts of the body reddish-brown, with a deep black mark in the centre of each; flanks black, transversely rayed with white bars; under tail-coverts white; bill red, shaded with brown at the tip; irides orange. Length about 12″.

Mr. Gurney (Ibis Vol. 1859) quotes this species from Natal. I have not met with it.

Genus ORTYGOMETRA, Linnæus.

Bill shorter than the head, and more or less strong, with the culmen keeled, slightly curved, and the sides compressed to the tip, which is slightly emarginated ; the gonys short and ascending ; the nostrils lateral, and placed in a membranous groove, with the opening exposed, linear, and near the middle ; wings moderate, with the second and third quills equal and longest; tail short and graduated ; tarsi rather robust ; toes more or less long and slender, with the inner toe rather shorter than the outer ; the hind toe very slender, and rather short ; the claws moderate, compressed, and acute.

632. Ortygometra Crex. (Gmel.) Pl. Enl. 750 ;
Crex Pratensis, Bechst. ; *Rallus Crex,* Linn. ; *Gallinula Crex,* Lath. ; *The Landrail,* or *Corn-Crake,* Bewick's Brit. Birds, p. 141.

ALL the feathers on the upper parts of the plumage, dark-brown, edged with pale-ferruginous ; wing-coverts and quills, deep-chestnut ; fore parts of neck and breast, pale-cinereous ; belly yellowish-white ; sides, thighs, and vent, rufous, marked with white bars.   Length, $9\frac{3}{4}''$; wing, $5\frac{1}{2}''$; tail, $1''$ $5'''$.

The "*Landrail*" or "*Corn-Crake*" seems to be not uncommon in Natal, where, according to Mr. Ayres (Ibis 1863, p. 331), they are plentiful inland during the summer months.   Only one single specimen, however, has occurred in the Cape Colony, and this was killed on the Cape Flats, near Wynberg, by Mr. H. Dumbleton in 1864.

633. Ortygometra Minuta. (Pallas.) *Rallus Pusillus,* Gmel. ; *Rallus Parvus,* Scop. ; *Gallinula Minuta,* Mont. ; Gould's Birds of Eur., Pl. 345 ; Bewick's Brit. Birds, p. 145.

UPPER parts brown, variegated on the back with black and white ; chin whitish ; sides of the head, neck, and chest, light slate-colour ; flanks, thighs, and belly, brown-black, crossed with white bars; irides red : according to Mr. Atmore, dark-brown.   Length, $7''$; wing, $3\frac{1}{2}''$; tail, $1''$.

Common in a few favoured localities.   At Swellendam I shot several in one small vley near the Buffeljagts River.   They are very difficult to flush, and fly heavily when on the wing.   A pointer will stand to them as to snipe, and the bird often rises from under his nose.   They feed on minute crustaceans, a little *Succinea* (*S. Delalandi*), which abounds in its favourite haunts, water-weeds, and worms.

Genus CORETHRURA, Reichenbach.

Bill shorter than the head, rather slender, and straight, with the culmen rather elevated at the base, and curved before the nasal groove, and the sides much compressed to the tip, which is slightly emarginated ; the gonys short and ascending ; the nostrils lateral, and placed in a membranous groove, with the opening linear and in the middle ; wings moderate, with the first quill much shorter than the second, which is also much shorter than the third and fourth, which are equal and longest ; tail short, rather graduated ; tarsi as long as or shorter than the middle toe, rather slender, and the front covered with transverse scales ; toes lengthened and slender, with the inner shorter than the outer; the hind toe long, and very slender ; the claws moderate, weak, and acute.

634. Corethrura Pulchra. (Gray.) Griff. An.

Kingd. III., p. 140 ; *Crex Pulchra*, J. E. Gray ; *Gallinula Pulchra*, Swain., W. Af. 11, p. 243 ; *Ortygometra Pulchra*, G. R. Gray ; *Gallinula Elegans*, Smith, Zool. S. Af., Pl. 22 ; *Rallus Cinnamoneus*, Less., Rev. Zool. 1840, p. 99 ; *Rufous-headed Rail*, Lath., Genl. Hist., Vol. IX., p. 379.

HEAD, neck, breast, and tail, bright-rufous ; the rest of the plumage above and beneath, black, spotted with white. Length, 6″ ; wing, 3″ 3‴ ; tail, 1″ 7‴.

A single specimen of this little species was sent to me by the late Mr. G. Rex from the Knysna. It must be very rare, as it has not since occurred to me in twelve years' observation.

635. Corethrura Dimidiata. (Temm.) *Gallinula Dimidiata*, Smith, Zool. S. Af., Pl. 20.

MALE : head, neck, breast, and shoulders, chestnut; body black, streaked with white ; plumage very lax. Length, 6″ 6‴; wing, 3″; tail, 2″.

The Female is black-brown, minutely spotted with light-buff spots ; chin white ; centre of breast, dirty-buff.

This elegant little Gallinule is not unfrequent among reeds, and appears widely distributed. It is difficult to flush, preferring to skulk among the dense rushes and tangled herbage ; if, however, it does take wing, it quickly relinquishes its weak, butterfly-looking flight, and drops into the grass again, from which it will suffer itself to be captured by the hand rather than rise. Mr. Atmore sent me an egg taken from the abdomen of a female shot by him, which is pure white : axis, 13″; diam., 10″.

636. Corethrura Ruficollis. (Gray.) Zool. Misc.,

p. 13 ; *Gallinula Jardinii*, Smith, Zool. S. Af., Pl. 21 ; *Alecthelia Lineata*, Swain., ♀, An. in Menag., p. 338.

♂.—Head, posterior portion of neck and tail, bright-rufous ; chin white ; the rest of the body dark grey-brown, variegated with numerous white stripes; plumage very lax. Length, 5" 6"'; wing, 2" 9"'; tail, 1" 8"'.

♀.—Above, blackish, varied with numerous undulated, transverse lines of light fulvous-brown, which assume the appearance of stripes on the head and neck; beneath whitish, with black stripes on the neck and breast, and bands on the flanks and body; tail and tail-coverts black; banded with ferruginous lines.

Of this rare little Rail I have only seen one single specimen, procured at the Knysna by Mr. G. Rex. The female has been described as a distinct species by Mr. Swainson, loc. cit. His type specimen is in the Cambridge Museum.

637. Corethrura Capensis. (Linn.) *Rallus Capensis*, Linn.; Brown's Ill., Pl. 38 ?; Cuv., Vol. 3, p. 405; Shaw, Vol. 12, p. 216.

HEAD, neck, back, and lower parts of the neck, ferruginous; the lower part of the breast, the belly, thighs, vent, quills, and tail, undulated with black and white; the two middle tail-feathers ferruginous; bill black; legs blood-red. Size of the common Crake.

" Inhabits the Cape of Good Hope."—Shaw, loc. cit.—Non vidi.

The Sub-Family, GALLINULINÆ, or Gallinules,

have the bill short, with the culmen more or less elevated and advancing on the forehead, and the sides compressed; the wings short and rounded; the tail short and rounded; the tarsi long and slender; the toes very long, slender, more or less margined on their sides, and the hind toe long; the claws generally long and slightly curved.

Genus PORPHYRIO, Brisson.

Bill short, very much elevated at the base, which is flat, and broadly dilated on the forehead; the culmen much arched to the tip, and the sides much compressed; the nostrils placed in a small basal groove, and rounded; wings moderate, with the second, third, and fourth quills nearly equal and longest; tail short and rounded; tarsi long, shorter than the middle toe, and scutellated with broad transverse scales; toes very long, slender, and free at their base, with the lateral ones unequal, the outer longest; the claws long, slender, and somewhat curved.

638. Porphyrio Pulverulentus, Temm., Pl. Col. 405 ; Cuv., Vol. 3, p. 414.

BLUISH-GREY ; back, rump, and secondaries, olive-brown ; vent white ; frontal plate produced behind the eyes ; bill and feet red ; wings ashy-blue."

South Africa.—Griffiths, Cuvier, loc. cit. I have not seen it from South Africa, but it is a well-known Indian species ; and I fancy Cuvier has mistaken the next species for it.

639. Porphyrio Erythropus, Shaw, Vol. 12, p. 255 ; *Fulica Porphyrio*, Linn ; *Gallinula Porphyrio*, Lath. ; *P. Madagascariensis*, Lath. ; *P. Smaragnotus*, Temm. ; *G. Poliocephala*, Lath.

HEAD, hind part of neck, and wing-feathers, glossy-violet ; back and rump, dull glossy-green ; cheeks, throat, fore part of neck, and under part of body, violet-blue ; tail dull-green ; vent-feathers pure white ; legs, bill, and frontal shield, bright-crimson. Length, 17″ ; wing, 9″ ; tail, 3″.

The "Blue Gallinule" is generally distributed throughout the colony, frequenting vleys and large ponds. It breeds in such places among reeds, forming a large nest of sedge, and depositing from six to ten eggs, of a ruddy-brown, spotted with dark purple-brown : axis, 2″ 2‴ ; diam., 1″ 6‴.

Genus GALLINULA, Brisson.

Bill moderate, with the base of the culmen more or less extending on the forehead, and suddenly curved at the tip ; the nostrils placed in a large nasal groove, with the opening near the middle of the bill ; wings moderate, with the second, third, and fourth quills nearly equal and longest ; tail short and rounded ; tarsi strong, shorter than the middle toe, and broadly scutellated in front ; toes very long, divided at the base, margined throughout their length by a membrane, and the outer longer than the inner.

640. Gallinula Chloropus. (Linn.) Lath., Pl. Enl. 877 ; *Fulica Fusca*, Linn. ; *Moorhen* or *Waterhen*, Bewick's Brit. Birds, p. 148 ; Gould's Birds of Eur., Pl. 342.

UPPER parts, dull greenish-brown ; under parts, dark slate-colour ; head and neck darkest ; vent white ; as are also some stray feathers on the flanks and in the centre of the belly. In some specimens the slate-coloured feathers of the

*Q

under parts are edged with dirty-white ; the throat is also
spotted with it ; tip of bill rich-yellow ; base and frontal
shield, bright-red ; legs green, with a red and yellow garter.
Length, 11″ ; wing, 6″ 9‴ ; tail, 3‴.

The " *Waterhen* " or " *Moorhen* " is not uncommon in the colony,
frequenting the same places as the Crested Coot. Most of the specimens received by me came from Mr. Atmore, who shot them about
Swellendam. It is common at Zoetendals Vley, where I met with it
in November.

641. Gallinula Angulata, Sund., Œfv. K. V. A.
Handl., 1850 ; *G. Pumila*, Sclater, Ibis 1859, p.
249, Pl. VII.

UPPER parts, dark olive-green ; wings cinereous, the outer
edge white, the inner rufous ; chin and centre of belly, light-
cinereous, almost white ; the flanks and chest darker ; outermost tail-coverts of the under side, white ; the inner black ;
a few white feathers are scattered along the flanks ; frontal
shield and top of the bill near the tip, bright-crimson ; the
rest bright greenish-yellow ; legs and feet the same. Length,
8″ ; wing, 5″ ; tail, 2½″.

A single specimen of this elegant little Waterhen, in superb
plumage, was sent to me by Mr. W. Cairncross, of Swellendam, who
procured it in that neighbourhood.

642. Gallinula Niger ; *Gallinula Flavirostris*,
Swain., Orn. W. Af., Vol. 2, p. 244 ; *Crex. Nigra*,
Licht. ; Cuv., Vol. 3, p. 409 ; *Rallus Niger*, Linn.,
Shaw, Vol. 12, p. 210.

GENERAL colour, deep-black, shining-brown if held in certain
lights ; bill bright-yellow ; eyelids and feet red, changing
in death, the first to a dirty-green, the latter to a light-brown ;
irides dark-crimson. Length, 7¼″ ; wing, 4¼″ ; tail, 1¾″.

Not uncommon in vleys and among the sedges along river banks.
The Museum taxidermist, Mr. Butler, shot a fine specimen, from
which the above description was taken, close to Cape Town, in the
Black River.

Genus FULICA, Linnæus.

Bill shorter than the head, strong, straight, more elevated
than broad, with the culmen straight, advancing on the
forehead, forming a broad shield, and curved near the tip ;
the nostrils placed in a nasal groove, linear, and near
the middle of the bill ; wings short, with the second and
third quills the longest ; tail very short and rounded ; tarsi

moderate, shorter than the middle toe, and covered with transverse scales; toes long and united at the base, lobated on the sides, especially on the inner; the inner toe with two, the middle with three, and the outer with four decided rounded membranes; the hind toe long, and lobated in its entire length.

643. Fulica Cristata, Gmel., Pl. Enl., 797 ; Vieil. Gal. des Ois., t. 269 ; *The Crested Coot.*

THE entire plumage (with the exception of the head and neck, which are quite black) is a dull, dark slate-colour; bill livid; frontal shield pure-white, and surmounted by two singular knobs of a deep reddish-chestnut colour, which shrivel up and become quite brown in death; irides red; legs green, with a pink garter just under the feathers of the thigh. Length, 16"; wing, $9\frac{1}{2}$"; tail, $2\frac{1}{2}$".

The Crested Coot is abundant on all ponds and sheets of water. It also frequents deep holes and still reaches in rivers, concealing itself amid the herbage during the day, and feeding morning and evening about the banks. It wanders at these times several hundred yards from its lair, and on being alarmed scuttles away on its feet, aided by its wings, until, a sufficient impetus being attained, it rises in the air, and will not unfrequently fly a very long distance. It swims well and boldly, and I have seen it alight on the waters of Table Bay and breast a considerable sea. It constructs a nest of sedge usually floating among the rushes, and lays seven eggs, of a cream-coloured ground, covered with dark-brown spots: axis, 2" 1'''; diam., 1" 6'''.

The young when first hatched are little black balls of soft down, but they swim and dive almost from the moment of leaving the egg.

Order VIII. ANSERES. Linnæus.

THIS division embraces many species of Birds, which are peculiar for having their tarsi usually very short, compressed, and placed more posteriorly than is the case with those that compose the other orders; the toes are connected together by a membrane, which sometimes extends to the end of each toe; the hind toe is sometimes free, though more or less webbed, and thus their feet are well adapted to assist them in their progress on or in the water.

The First Family, ANATIDÆ, or Ducks,

have the bill generally depressed, broad, and always laminated on the sides; the lamination being more prominent in some species than in others.

The First Sub-Family, PHŒNICOPTERINÆ, or Flamingoes,

have the bill large, compressed, suddenly bent downwards in the middle, and the lateral margins laminated; the tarsi very long, slender, with the tibia also lengthened and naked; the toes short; the anterior ones united by a membranous web.

Genus PHŒNICOPTERUS, Linn.

Bill larger than the head, higher than broad at the base, with the culmen flattened, and suddenly bent downwards in the middle; the sides narrowing, and rather obtuse at the tip; the lower mandible narrow at the base, widening in the middle, and then narrowing to the tip; the lateral margins curved, and finely laminated; the nostrils placed in a groove, linear, and covered by a membrane; wings moderate, with the first and second quills nearly equal and longest; tail short; tarsi very long, slender, slightly compressed, and

covered in front with transverse scales ; the tibia lengthened
and exposed ; toes short, the anterior ones united together by
a membrane ; the hind toe free, very short, and almost
touching the ground ; the claws short.

644. Phœnicopterus Erythræus, Verr.

GENERAL colour, pure-white ; the wings black and crimson ;
base of the bill crimson ; the tip black ; legs light-crimson.
Length, 4' 6" ; wing, 17½" ; tail, 8" ; leg, 2' 2".

Our large Flamingo has a wide range throughout South Africa, and
has been killed, both in mature and young plumage, at the mouth of
Salt River, close to Cape Town. I am informed that it is very
abundant at Verloren Vley at certain seasons ; and many persons have
assured me that it breeds there.

Mr. Chapman also informs me that both the Flamingos found in
South Africa breed on Lake N'Gami, forming a large elevated nest of
rushes, amid the reeds that surround the lake. I saw it in great numbers at the mouth of the river flowing out of Zoetendals Vley in
November. An egg, said to belong to this species, was presented to
the Museum by Miss Boonzaier, of Hoedtje's Bay. It is pure white,
finely granulated ; shell extremely thin : axis, 3" 9" ; diam., 2" 7"'.

645. Phœnicopterus Minor, Geoffr., Pl. Enl.

419 ; *Ph. Parvus*, Vieil., Ann. d'Ornith. Gal. Des.
Ois., t. 273.

GENERAL colour, pale-pink ; wings black ; outside edges deep-
pink, or crimson ; bill purple-pink. Length, 3' 3" ; wing,
14" ; tail 6" ; leg, 16".

Several specimens of this Flamingo have reached my hands from
widely different localities ; and Mr. Piers informs me that they approach
the end of the peninsula as far as Vogel Vley, near Tulbagh.

The Sub-Family, PLECTROPTERINÆ, or Spur-winged Geese,

have the bill long, generally of equal width throughout, and
armed at the tip with a strong broad nail ; the apical part of
the tibia and knee naked ; the tarsi lengthened, compressed, and covered with small subquadrate scales ; the toes
moderate, and the anterior ones more or less united by a
membrane ; the hind toe long, simple, and generally elevated.

Genus PLECTROPTERUS, Leach.

Bill lengthened, the base as broad as high, and the culmen
gradually sloping to the tip, which is armed with a large

broad nail ; the nostrils oval, and placed near the middle and culmen ; wings lengthened, with the second, third, and fourth quills equal and longest, and the bend of the wing armed with a strong spur ; tail short and rounded ; tarsi long, a trifle shorter than the middle toe, strong, and covered with small quadrate scales, those of the front the largest ; toes long, the anterior ones united by an indented web ; the hind toe elevated and simple ; the base of the culmen furnished with a naked protubrance ; the cheeks and a portion of the neck of the adult denuded of feathers.

646. Plectropterus Gambensis. (Linnæus.)

Steph., Lath., Syn. VI., Pl. 102 ; *Anas Gambensis*, Linn. ; *Anser Spinosa*, Bonn. ; *Anser Gambensis*, Bonn. ; *Cygnus Gambensis*, Rüpp.

UPPER parts, black, shot with bronze and green ; wings mottled with white ; lower parts white, with patches of black behind the thighs ; the front of the head is bare (more so in the ♂ than in the ♀), granulated, and reddish ; this is succeeded by a white patch, which does not extend to the top of the head, but occupies chin and part of the front of the throat ; wings armed with a strong powerful spur. ♂ : Length, 3' 6''; wing, 21''; tail, 7''. ♀ : Length, 3''; wing, 19''; tail, 6''.

This is the largest of our Geese, and is easily distinguished by the strong spur on the edge of its wing. Hartlaub (Orn. W. Af., p. 246,) includes it among the birds of South Africa. My descriptions are taken from a fine pair, male and female, brought by Mr. Chapman from the Zambezi. I have also received it from Mr. Arnot, who procured it in Mahura's country.

Genus SARKIDIORNIS, Eyton.

Bill moderate, of equal length throughout, more elevated at the base than broad, and the tip armed with a broad strong nail ; the nostrils large, oval, placed near the middle and culmen ; wings lengthened, with the first and second quills nearly equal and longest, each shoulder armed with a blunt tubercle ; tail moderate and rounded ; tarsi moderate, as long as the middle toe without the claw, and covered with subquadrate scales ; toes long ; the anterior ones united by a full-web ; the hind toe moderate, much elevated and simple. The male furnished with a large rounded compressed caruncle on the top of the culmen.

647. Sarkidiornis Africana, Eyton, Monogr. Anat., p. 103.

UPPER parts, black-brown, shining-coppery on the back, and brilliant-green on the wing; rump brownish-grey; under parts whitish; flanks grey; head and neck white, covered with black spots, which form a broadish line from the top of the head, down the back of the neck. The male, in the breeding season, has a large elevated flattened black caruncle extending down the centre of the upper mandible. Length, 24"; wing, 15"; tail, 7".

This Goose may be at once recognised in the breeding season by the curious flattened knob on the bill. I have received it from Mr. Arnot, and Mr. Chapman, who tells me it is common on the rivers to the North.

Genus CHENALOPEX, Stephens.

Bill rather short, strong, with the base much elevated, and the culmen gradually sloping to the tip, which is armed with a broad strong nail; the nostrils large, rather rounded, and placed near the middle and culmen; wings long, with the second and third quills nearly equal and longest; the bend of the wings armed with a blunt spur; tail moderate, and slightly rounded; tarsi longer than the middle toe, and covered with sub-quadrate scales; toe moderate, the anterior ones united by a full web; the hind toe long, elevated, and slightly lobed.

648. Chenalopex Ægyptiacus. (Linn.) Pl.

Enl., 379; Shaw, Vol. 12; *Anas Montana*, Lath.; *Anser Varius*, Schn.; "*Berg-Gans*," or *Mountain-Goose* of Colonists.

UPPER parts, ruddy-grey; between the shoulders, finely variegated with black, deepening into pure black on the rump, and brick-red on the wings; on the closed wing, a broad white patch commences at the shoulder, and extends half-way, when it is met by a brilliant green patch; near this junction the white is interrupted by a narrow black line, which extends across the white; under parts ruddy-grey, finely mottled; the centre of the belly least so; in the middle of the posterior part of the breast is a patch of deep rufous; top of the head, chin, and cheeks, dirty-white; base of the bill, a patch round the eye, back of the neck, and ring round the lower part, ruddy. Length, 28"; wing, 16"; tail, 5".

This is the commonest of our Geese, and is found in considerable abundance throughout the colony. It congregates, at particular seasons of the year, in vast numbers on certain well-known vleys, such as Vogel Vley (Bird Lake), Verloren Vley (Lonely Lake), Bot River Mouth, &c., &c. In these places it sheds the wing-feathers and becomes unable to fly. It is then much sought after, being very fat; but it dives with such rapidity that even in this condition it is very difficult to kill. It breeds in the rushes in these vleys; but in other places, such as the Keurboom River, in the Knysna division, it breeds in holes of rocks or hollow trees near the water, and is said to bring down its young on its back. The goslings dive as actively as their parents, which show much solicitude about them, keeping with them until they are fully grown; in fact, the little flocks in which they are generally found consist usually of an old pair and their brood, from ten to fifteen in number.

They are easily tamed, and a great ornament to a poultry-yard; but sad plagues to the rest of the inhabitants, over which they domineer. They have a loud, harsh cry, which, however, sounds well amid the wild rocky scenes where they love to dwell. They are strong on the wing, and require a hard-hitting gun and a charge of buck-shot, well aimed, to bring them down. Their eggs are white, and small for the size of the bird, being: axis, 2″ 2‴; diam., 1″ 7‴.

The Sub-Family, ANSERINÆ, or Geese,

have the bill as long as or shorter than the head, the culmen much elevated at the base, and gradually sloping to the tip, which is armed with a large broad nail; the sides compressed, and the marginal laminæ more or less apparent; the wings lengthened; the tarsi longer than the middle toe, and the knee naked; the toes short, and strongly webbed, with the hind toe short, elevated, and scarcely lobed.

Genus NETTAPUS, Brandt.

Bill small, more elevated at the base than broad, the culmen gradually sloping to the tip, which is armed with a large nail; the lamellæ not exposed, but short and widely set; the nostrils somewhat basal, placed near the culmen, with the opening rounded; wings moderate, pointed, and the first two quills the longest; tail short and rounded; tarsi shorter than the middle toe; toes long, and united by a full web, the lateral one unequal, and the hind toe very short, and more or less lobed; the claws short and curved.

649. Nettapus Madagascariensis. (Gmel.)

Anas Madagascariensis, Gmel.; *A. Aurita*, Bodd., Pl. Enl. 770.

UPPER parts, shining black-green, with a longitudinal white stripe along the wing; under parts white; chest, flanks, and sides rufous, the first transversely marked with narrow black bars; front of head, cheeks, and throat, white; top of head shining black-green; on each side of the neck a large dead-green patch, surrounded by the black-green of the head. Length, 14″; wing, 6″; tail, 2″ ‘ ‴.

This lovely little Goose may at once be known by its brilliant colouring, particularly the green patch on each side of the neck. It is abundant at Natal, where a friend informs me it swims in such large flocks, that he has killed from ten to fourteen in a single discharge of a common fowling-piece. Mr. Chapman found it equally abundant in his travels to the northward; but as yet I have not heard of its being procured in the colony.

The Sub-Family, ANATINÆ, or River-Ducks,

have the bill lengthened, more or less broad, depressed towards the tip, which is furnished with a hard nail, and the inner portion of the lateral margins more or less lamellated; the tarsi compressed, and generally the length of the inner toe; the hind toe lengthened, and slightly bordered with a membranous lobe from the base to the tip.

Genus DENDROCYGNA, Swainson.

Bill long, higher at the base than broad, with the culmen sloping to the tip, which is armed with a strong broad nail, and the lateral margins straight; the lamellæ of the upper mandible advancing below the lateral margins, slender, and set widely apart; the nostrils large, oval, and placed near the base and culmen; wings short and rounded, with the second, third, and fourth quills the longest; the first quill with a deep notch in the middle, and the secondaries nearly as long as the quills; tail moderate, and rounded at its end; tarsi slightly shorter than the middle toe, robust; toes long, the lateral ones united to the middle one by an indented membrane; the hind toe very long, elevated, and simple.

650. Dendrocygna Viduata. (Linn.) Eyton.,
Pl. Enl. 808; *D. Personata,* Herz. V., Würtemb.; Penn. Gen. Birds, Pl. 13.

ABOVE, brown; the feathers broadly margined with dirty-fulvous; wings and rump blackish-brown; in the centre of the back a rufous patch; centre of under parts immaculate black-brown; flanks and sides transversely barred with fulvous; fore part of the head, chin, and spot on the middle of the throat, white, more or less tinged with fulvous; back of head and neck black; front of neck and chest, deep-ruddy. Length, 17″; wing, 8½″; tail, 3″.

This curiously marked Tree-Duck (which is easily distinguished from all our other species by its half-white face) is occasionally met with in

*R

Natal, to which place it doubtless wanders from the Zambezi, where it is common. I found it also abundant in Madagascar, in the swamps and rice-fields, and very easy of approach. The habit of perching on trees resembles that of the allied *D. Arcuata* of India.

Genus CASARKA, Pr. Bonap.

Bill as long as the head, nearly straight, the width equalling the height at the base, the anterior half depressed, and scarcely curved upwards at the tip, which is armed with a strong broad nail; the basal part of the lateral margin straight, and the apical part slightly curved upwards; the lamellæ of the upper mandible prominent below, the lateral margins slender, and set rather widely apart; the nostrils suboval near the base and culmen; wings moderate, with the second quill the longest; tail short and rounded; tarsi robust, shorter than the middle toe; toes long, and united by a full web; the hind toe long, elevated, and lobed.

651. Casarca Rutila, Pall., Hist. D'Egypt. Ois., t. 10, f. 1; *Anas Casarca*, Linn.; *Tadorna Rutila*, Temm., Man. d'Orn.; *Anas Cana*, Gmel.; Latham's Ind. Orn.; *A. Rubra*, Gmel.; *Berg-Eendt* of Colonists.

GENERAL colour, rufous, inclining to yellow (fulvous) on the chest and vent, and mottled with minute wavy black lines on the back; rump and lower part of back, very dark-grey; tail black, shining-green; shoulder pure-white; green patch on the wing very large; head and neck ash-coloured, with a dark-rufous collar. Length, 2′ 3″; wing, 15″; tail, 6½″.

This Goose is readily distinguished by its rufous colour and grey head. It is rare in the colony, appearing at odd times in different places. I am told it breeds among rocks, and, like the previous species, conveys its young down to the water on its back.

Genus MARECA, Stephens.

Bill shorter than the head, higher than broad at the base, of equal breadth, but depressed, and much rounded at the tip, which is armed with a strong broad nail; the lamellæ of the upper mandible prominent, especially in the middle of the lateral margins, slender, and rather widely set; the nostrils situated near the base and culmen, lateral, and oval; wings long and pointed, with the first and second quills the longest; tail moderate, wedge-shaped; tarsi shorter than the middle toe; toe united by a full web; the hind toe moderate and lobed.

652. Mareca Capensis. (Gmel.) L.; *Cape Wigeon*,
Lath. Gen. Syn., 6, 519.

HEAD, ash-grey, profusely streaked with blackish dots; lower part of neck and breast, ash-grey, profusely variegated with reddish-brown broken bars, giving the plumage a scaled appearance; feathers of the back, dark reddish-brown, each feather edged with lighter; speculum of the wing, bright-green, edged with white and black; shoulders dark-ash; legs reddish; webs dusky; claws black; bill red, the base black. Length, 19''; wing, 8½''; tail, 2¾''. Female smaller than the male.

Usually confounded here with the *Smee Eendtje*, in company with which I have shot it at Beaufort, the Knysna, on the Cape Flats, and at Vogel Vley. At this latter place I killed one near a nest which contained a single egg; it was probably one of this species, and is of a dirty greenish-white throughout: axis, 1'' 10'''; diam., 1'' 6'''.

Genus ANAS, Linnæus.

Bill longer than the head, higher than broad at the base, nearly of equal breadth throughout; the culmen nearly straight, and depressed to the tip, which is armed with a strong, broad nail; the lamellæ of the upper mandible hardly visible beyond the lateral margin, strong and widely set, especially near the middle; the nostrils placed near the base of the culmen, lateral, and oval; wings moderate and pointed, with the tertials lengthened and acute, and the first quill the longest; tail short and wedge-shaped; tarsi shorter than the middle toe, and compressed; toes united by a full web, and the hind toe small, and somewhat lobed.

653. Anas Erythrorhyncha, Gmel. Sys. Nat.;
Tadorna Erythrorhyncha, Shaw, Vol. 12, p. 75; *Pœcilonitta Erythrorhyncha*, Smith, Ill. S. Af. Zool., Pl. 104; *Smee Eendtje* of Colonists.

UPPER parts, brown, tinged faintly with green, each feather margined with pale-pinkish; below brown, each feather so broadly margined with white, as to cause that colour to predominate; head and neck dark-brown, the latter minutely mottled with dirty-white; chin and lower parts of the cheeks below the eye, white; on the wings a broad bar of pink, crossed at the upper side by a narrow green line; centre of the bill brown, the rest pink; iris hazel. Length, 18½''; wing, 8¼''; tail, 3''.

The "Red-billed Teal" is common and very generally distributed. It is usually found in little flocks of six or eight individuals, frequenting tanks and rivers. It breeds among rushes, forming a large flat nest of sedge, and lays from six to ten eggs, similar in colour and size to that of the preceding.

654. Anas Sparsa, Smith, Zool. S. Africa, Pl. 97; Eyton's Anat., p. 142; *Black-Duck* of Sportsmen.

THROUGHOUT of a very dark-brown, finely mottled on the head and neck with dirty-white, and marked on the wings and tail with large pure white spots; vent edged with white; across the wing is a bright-green bar, edged with black and white; the black nearest the green. Length, 22"; wing, 10"; tail, 5".

The Black-Duck is solitary and shy in its habits, frequenting rivers, under the banks of which it conceals itself during the day. It is scarce, but generally distributed. I have received specimens from Colesberg and Traka, and heard of it at Middelburg, where my late friend, Mr. Jackson, shot several specimens. It may be distinguished from all our other ducks by its dark plumage and abrupt white spots. Mr. Atmore writes : " The common duck of the Oliphants River— very shy—delicious eating."

655. Anas Flavirostris, Smith Zool. S. Af., Pl. 96; Eyton's Anat., p. 141; *A. Xanthorhyncha*, Forst. Desc. Ani., p. 45; *Geelbec* of Colonists and Sportsmen.

THROUGHOUT a light-brown, each feather broadly edged with white, giving the whole a scaled appearance ; head and neck minutely mottled with white; a broad green band, narrowly edged with black, and again by white, extends partially across the wing ; bill bright-yellow, with the tip and centre of upper mandible black. Length, 22"; wing, 9"; tail, 4".

This is our commonest Duck, being abundant all over the colony and South Africa generally. It is migratory, as indeed are all our ducks, coming and going with the waters. It is usually seen in pairs, male and female, and exhibits all the wariness and caution of its race. They breed in considerable numbers at Vogel Vley, among the rushes and rocks scattered over that lake. I am told that at one season of the year the farmers in that neighbourhood assemble for a grand hunt after these birds and the *A. Erythrorhyncha* The method of hunting is as follows : The shooters are posted in different parts of this long sheet of water, hidden among the rushes and the bush-covered rocks which jut out here and there in the shallows. Men are then sent about with the long wagon whips, and with these they beat the rushes and keep up a continual cracking (and the crack of a Cape whip is nearly equal to that of a gun); the wretched birds fly backwards and forwards (having no other water within many miles), and as they pass the ambuscades, are shot down ; when the day's butchery is over, the dead and wounded are sought for, and usually fill many sacks.

They construct their nests in the dry veldt at a distance from the water, generally in a dense bush; the female sits so close that unless hunted for she will scarcely rise. Her eggs, usually six in number, are cream-coloured : axis, 2″ 3‴; diam., 1″ 9‴.

656. Anas Dominicana, Gmel.; Shaw, Vol. 12, p. 109.

SIZE of wild-duck; bill black; chin and throat white; from the bill through the eye a streak of black, ending in an angle behind; hind-head, neck, and breast, black; back and lesser wing-coverts, deep cinereous-grey, crossed with two bands of very pale-grey; bill and vent pale-grey; legs black.

"Inhabits South Africa."—Shaw, loc. cit.; but is in reality a South American species.

Genus QUERQUEDULA, Stephens.

Bill as long as the head, straight, the height equalling the breadth at the base, of equal width throughout, depressed towards the tip, which is armed with a small narrow hooked nail; the lamellæ of the upper mandible hardly visible, broad, and rather widely set; the nostrils situated near the base and culmen, lateral, and oval; wings moderate and pointed, with the second quill the longest, and the secondaries lengthened and pointed; tail moderate, and wedge-shaped; tarsi rather shorter than the middle toe; toes united by a full web, and the hind too short, and slightly lobed.

657. Querquedula Hottentotta, Smith, Zool.

S. Af., Pl. 105; Eyton's Anat., p. 129; *Hottentot Teal.*

MALE : umber-brown, edged with lighter; crown, occiput, and quills, dark-brown; chin, throat, cheeks, rump, and under tail-coverts, light-brown; the last narrowly edged with darker; breast, lower part of neck, and under surface, darker than on the rump; abdomen barred with black; speculum and secondaries, bright brassy-green; the former edged behind with black, then with white; coverts brown, slightly glossed with brassy; tail dark-brown. Female like the male, but markings less distinct; under plumage lighter; legs in both sexes brown; bill lead-coloured; nail horn-brown. Length, 13″ 9‴; wing, 5″ 8‴; tail, 2″ 10‴.

Dr. Andrew Smith, whose description I quote, states "that the only specimens of this bird seen by him, two in number, were procured on a large rley about one hundred miles north-west of Cape Town. This was probably Verloren Vley. I have not seen it from the colony; but a mutilated specimen appears in Mr. Chapman's collection; and Mr. Andersson procured it in Damaraland."

Genus SPATULA, Boie.

Bill longer than the head, narrowed at the base; the culmen straight, depressed, and the side much dilated for nearly half its length from the tip, which is furnished with a small hooked nail; the lamellæ of the upper mandible very slender and long, especially near the middle, those of the lower mandible concealed by the lateral margins of the upper, but also very slender and lengthened; the nostrils placed near the base, small and oval; wings lengthened and pointed, with the first two quills of nearly equal length; tail moderate, and somewhat pointed; tarsi shorter than the middle toe; toes united by a full web; and the hind toe short, and slightly lobed.

658. Rhynchapsis Capensis, Smith., Zool. S.
Af., Pl. 98 ; Eyton's Anat., p. 135 ; *The Cape Shoveler.*

GENERAL colour above and below, dark-brown, each feather being bordered with dirty-white, thus imparting to the bird a scaled appearance; shoulder blue, succeeded by a bar of white, and then by another of shining-green, on the inner side of which there is a patch of blue; head grey, finely mottled with dark-brown speckles. The female is much less brilliantly coloured. Length, 21″; wing, 9½″; tail, 3¾″.

The "Cape Shoveler" is rarely seen in the neighbourhood of Cape Town; but when the *E. Maccoa* and *N. Brunnea,* hereafter mentioned, appeared there in such numbers, the Shoveler accompanied them. Its usual habitat is Verloren Vley. Hartlaub states (Orn. W. Af., p. 248), that the European Shoveler has been found in South Africa. Surely this must be the species alluded to. My son captured a wounded bird near Salt River in November, 1865.

The Sub-Family, FULIGULINÆ, or Sea-Ducks,

have the bill of various lengths, elevated at the base, and more or less broad and depressed towards the tip, which is armed with a broad strong nail; wings moderate and pointed; tail generally short, more or less wedge shaped; the tarsi much shorter than the middle toe, and compressed; toes long, and united by a full web; the outer as long as the middle toe, the hind one short, and deeply margined with a broad membranous web.

Genus NYROCA, Fleming.

Bill as long as the head, higher at the base than broad; the culmen gradually sloping towards the tip, which is depressed, slightly dilated, and armed with a strong nail; the lamellæ of the upper mandible not prominent, and the nostrils oval, and placed near the base; wings lengthened and pointed, with the two first quills the longest; tail short and rounded; tarsi half the length of the middle toe, and compressed; toes lengthened, and united by a full web.

659. Nyroca Brunnea, Eyton's Anat., Pl. p. 161.

♂.—General colour above, deep brown, minutely variegated with grey; below, deep brown, tinged with rufous, more especially on the flanks and shoulders; lower part of the neck and breast approaching to black, tinged with faintish purple; checks and sides of upper part of neck, rich dark-chestnut; small spot on the chin and bar on the wing, white.

♀.—Throughout a lighter-brown, approaching to white on the under parts, and all tinged and blotched with dirty-rufous; chin, anterior portion of throat, base of the bill, and stripe through the eye, white; irides, bright reddish-crimson. Length, 19″; wing, 9½″; tail, 3″.

A periodical visitant to this end of the colony. I saw a pair in Zoetendals Vley in November, 1865, apparently breeding in the vast bed of rushes at the south end of the vley.

The Sub-Family, ERISMATURINÆ, or Spiny-tailed Ducks,

have the bill elevated at the base, and the anterior half much depressed to the tip, which is furnished with a nail; the wings short and concave, with the ends of the quills incurved; the tail lengthened, and composed of narrow, rigid feathers, which are but slightly protected with coverts both above and below; the tarsi shorter than the middle toe, and compressed; the toes lengthened, the anterior ones united by a full web, and the hind toe long, and furnished with a broad web.

Genus THALASSORNIS, Eyton.

Bill nearly the length of the head, more elevated at the base than broad; the culmen sloping to near the tip, and then depressed, and armed with a strong, broad, hooked nail; the width of the upper mandible nearly equal throughout, and the sides somewhat compressed; the nostrils small, oval, and placed in the middle of the bill; wings short, with the second and third quills longest; tail rounded, and composed of slightly rigid feathers; tarsi much shorter than the middle toe; toes lengthened, the outer nearly as long as the middle toe, and all the anterior ones united by a full web; the hind toe moderate, and strongly lobed.

660. Thalassornis Leuconota. (A. Smith.)

Eyton's Monograph Anat., I., 168; *Clangula Leuconota*, Smith's Zool. S. Af., p. 107.

BACK white; all the rest of the upper parts variegated with fulvous, black, and rufous; under parts fulvous, transversely striped with dark-brown, least so in the centre of the breast and belly; head and back of neck fulvous, profusely mottled with black round spots; front and sides of lower part of neck bright-fulvous; chin and spot on each side at the base of the bill, white. Length, 18″; wing, 8″; tail, 2½″.

This singular Duck makes periodical visits to the vleys in the neighbourhood of Cape Town. It is found in abundance to the Northward, and was included in Mr. Chapman's collection. Dr. Smith says it inhabits Verloren Vley, that great haunt of all waterfowl, and other lakes on the Western Coast of South Africa, and that it prefers diving to seeking safety by flight. I fancied I made out a pair of these birds at Zoetendals Vley. I examined them for some time with my binoculars, but they would not allow of a near approach; still the yellow patch on the throat was very visible.

Genus ERISMATURA, Pr. Bonap.

Bill nearly as long as the head, higher at the base than broad, the culmen suddenly curved to the front of the nostrils, and then depressed, straight, and the sides somewhat dilated near the tip, which is armed with a narrow nail, enlarged and hooked beneath; the nostrils oval and placed nearly in the middle of the bill; wings short and concave, with the first two quills the longest; tail long, wedge-shaped, and composed of narrow stiff feathers; tarsi half the length of the middle toe, and compressed; toes lengthened, the middle and outer ones of equal length, and the three anterior ones united by a full web; the hind toe long, elevated, and margined by a lobed membrane; the claws short, curved, and acute.

661. Erismatura Maccoa. (A. Smith.) Eyton's

Monogr. Anat., p. 169; *Oxyura Maccoa*, Smith's Zool., p. 108.

UPPER parts, chestnut-brown; under parts and wings, brown; head and upper parts of neck, black; bill blue; tail-feathers very narrow and rigid. Length, 17″; wing, 6″ 6‴; tail, 3⅓″.

The young male is beautifully mottled, and the female is generally of umber-brown, with the chin and sides of the head clear white.

The singular, rigid tail-feathers of this Duck will serve to distinguish it from any of the other species inhabiting South Africa. It is found at Verloren Vley ; and Mr. Dumbleton informs me he shot a specimen at Victoria. Dr. Smith states that it dives with great facility.

In the winter of 1858, great flights of these birds, and of *Fuligula Brunnea*, appeared in the neighbourhood of Cape Town, and were shot in considerable numbers on the Cape Flats, and on all the vleys of the neighbourhood. Since that date only a few stray specimens have made their appearance at odd times ; and it may be reckoned as one of our scarcest ducks.

The Fourth Family, PROCELLARIDÆ, or Petrels,

have the bill more or less lengthened, straight, more or less compressed and grooved, as if composed of several pieces, with the tip strong, arched, suddenly hooked and acute, and the nostrils tubular and exposed.

The Sub-Family, PROCELLARINÆ, or Petrels Proper,

have the nostrils placed in the basal portion of the culmen, tubular, and generally opened in front.

Genus PUFFINUS, Brisson. ·

Bill as long as or shorter than the head, slender, and much compressed, and grooved obliquely on the sides ; the tip lengthened, arched, suddenly hooked, and acute ; the lower mandible somewhat shorter than the upper, with the apical margin and gonys equally curved with the upper, the latter angulated beneath, and the sides longitudinally grooved ; the nostrils basal, elevated above the culmen, opening obliquely in two tubes placed side by side ; wings long, slender, somewhat acute, with the first quills the longest ; tail moderate and rounded, composed of twelve feathers ; legs moderate, with the apical part of the tibia naked ; tarsi compressed, and equal in length to the middle toe ; all the toes long, the outer equal with the middle one, the inner shortest, and the lateral toes margined exteriorly by a narrow membrane.

*s

662. Puffinus Cinereus, Gmel.; *Procellaria Puffinus et Cinerea*, Linn., Zool. S. Af., Pl. 56.

ABOVE fuscus, tinged with yellowish-brown; below cinereous-brown; wing and tail-feathers, and lower tail-coverts, brown; bill livid-brown; at the tip, yellowish-brown; front of the tarsi yellow, brown behind, and external!y reddish-brown; eyes brown. Length, 17″; wing, 11″ 9‴; tail, 3″ 10‴.

The above description by Dr. Smith is taken from a young specimen. He states they are common in the Cape seas; but I never met with them.

Genus THALASSIDROMA, Vigors.

Bill shorter than the head, slender, weak, the sides much compressed, and slightly grooved, with the tip suddenly hooked and acute; the lower mandible shorter than the upper, the tip arched, with the gonys hardly angular beneath; the nostrils elevated above the culmen at its base, tubular, with a single aperture in front; wings long and pointed, with the first quill shorter than the third, and the second the longest; tail emarginated, or more or less forked; legs long, slender, with the naked space of the tibia extensive; tarsi longer than the middle toe; all the toes rather short, the outer one nearly equal with the middle, and the inner one the shortest; the hind toe only in the form of a triangular claw.

663. Thalassidroma Melanogaster, Gould.; *Th. Grallaria*, Licht., Cat. Dup. Berl. Mus., No. 764; *Pro. Oceanica*, Pr. Bonap. nec Temm., Pl. Enl. 993.

GENERAL colour throughout, brownish-black; belly white, divided down the centre by a black stripe; rump and sides behind the thighs, pure-white; legs, feet, and webs between the toe, black. Length, 8½″; wing, 6″ 5‴; tail, 3″.

Not uncommon off Cape L'Agullas and on the South-Eastern Coast. While cruising with Admiral Trotter in the old *Castor* frigate, the boats being always at my disposal, I obtained many specimens of this and other species of birds inhabiting the Cape seas. Indeed, most of my knowledge of our oceanic birds is due to the opportunities so kindly afforded me by my kind and noble old friend, now, alas! no more.

664. Thalassidroma Leucogaster, Gould; Birds of Australia.

RESEMBLES preceding species in size and colour, with the exception of being rather lighter, and the white of the belly *not* divided by a black stripe; the toes also are shorter.

Obtained by His Excellency Sir George Grey in the Cape seas. The only specimen that has fallen under my notice was kindly

presented to the South African Museum by Lieutenant-Commander
L. A. Beardslee, of U. S. Navy, and captured on board his ship,
the U. S. Steamer *Aroostook*, about 300 miles to the westward of the
Cape, May, 1867.

665. Thalassidroma Wilsoni, Pr. Bonap.

GENERAL colour throughout, brownish-black; rump and
thighs white; legs and feet black; webs between the toes
yellow, edged with black; irides black. Length, 7"; wing,
6"; tail, 2½".

The accompanying description is from a specimen killed in Table
Bay on the 29th April, 1865, by my son, who informs me that he saw
several more of the same species. I procured several specimens off
L'Agulhas Bank in 1856, since which time, until my son found it in
Table Bay, I had not seen them on the coast.

In habits they resemble the preceding species, with which they
freely consort, flitting over the waves and picking up odd bits flung
over the ships' sides. When taken in the hand they disgorge large
quantities of an oily matter, which quickly congeals, and assumes the
appearance of dirty lard.

666. Thalassidroma Oceanica, Temm.

GENERAL colour of plumage, soot-colour; quill-feathers approaching to black; edges of the secondaries whitish; rump
and vent white; four outer tail-feathers white on the inner
webs, at the base; bill and legs black; eye dark-brown.
Length, 6"; wing, 5"; tail, 1" 9"'.

A few of these Petrels appeared in Table Bay in company with
T. Wilsoni in the month of May, 1865. Specimens of each were procured by myself and my son. I had never previously seen either of
them so near inshore.

Genus PROCELLARIA, Linn.

Bill as long as or shorter than the head, more or less broad
at the base (varying in the sexes); the sides more or less
compressed towards the tip, which is more or less compressed,
much elevated and arched, lengthened, and acute; the lower
mandible shorter than the upper, with the tip and gonys
arched and acute; the nostrils tubular, horny, varying in
length from one-fourth to nearly two-thirds of the length of
the bill, with the aperture single, frontal, and crescent-
shaped; wings long, pointed, with the first quill the longest;
tail moderate, rounded, or wedge-shaped; legs with the
apical part of the thigh hardly naked; tarsi shorter than the
middle toe, laterally compressed, and covered with small
scales; toes long, with the outer as long as the middle one,
the inner shortest, and all united by a full web; the lateral
toes margined exteriorly, the hind toe in the shape of a large
subtriangular claw.

667. Procellaria Gigantea, Gmel.; Lath. Syn., t. 100.

GENERAL colour throughout, deep brown-black; bill livid. Length, 2′ 10″; wing, 21″; tail, 11″.

The Giant Petrel is common in Table Bay, and all along our coasts, throughout the year. Though very plentiful beyond the shipping in the bay, it never approaches the shore or wharves—certainly not from the fear of man, because when a boat cruises about the bay, it is sure to pay it a visit and examine its inmates closely. A white variety is common up the West Coast towards Walwich Bay. It is called "the Nelly" by sailors. Some of them have here and there a black feather. Eggs of this species measure: axis, 4″ 3‴; diam., 2″ 8‴; pure white, rough to the touch, and rather pointed at each end. They retain the strong smell of the bird for a long period.

668. Procellaria Æquinoctialis, Linn.; Ed.
Birds, Pl. 89; Smith, Zool. S. Af.; *Procellaria Fuliginosa*, Sol. M. S. Less., Vol. 2, p. 348; *Black White-throated Petrel*; *Cape Hen* of Sailors.

GENERAL colour throughout, deep brown-black, with chin, and more or less of the top of the throat, pure-white. Length, 21″; wing, 15½″; tail, 6½″.

The "Cape Hen" is a constant resident in Table Bay, though the majority leave us at one season of the year to breed. It cruises among the shipping and wharves, looking out for stray scraps washing about, on which it feeds. It extends far out to sea, and may be met with for some distance up both East and West Coasts.

The variety or species, *P. Larvata*, Lesson, *P. Conspicularis*, Gould, having a white band over the top of the head and round the check, never appears in South African waters, not coming so far north. I believe it to be quite distinct from *P. Æquinoctialis*. The eggs of the latter are pure white, rather rounded at each end: axis, 3″ 2‴; diam., 2″ 2.‴

669. Procellaria Macroptera, Smith, Zool. S. Af., Pl. 52.

THE whole plumage intermediate between brownish-red and liver-brown; several of the scapulars and interscapulars narrowly tipped with yellowish-brown; the feathers of the head, neck, and body, silvery-white towards their base, with a satin lustre; bill rich-black; tarsi, toes, and interdigital membrane, liver-brown, the two first tinted lake-red. Length, 17″; wing, 13″ 9‴; tail, 6″.

Dr. Smith states simply that "this is a rare bird in the Cape seas." I obtained a single specimen of it off the Cape in 1856, while cruising in H. M. Frigate *Castor*, Commodore Trotter.

670. Procellaria Glacialoides, Smith, Zool. S.
Af., Pl. 57 ; Forst. Icon. Ined., t. 91 ; *Pro. Tenuirostris*, Audub.

ABOVE, cinereous ; below white ; head and back of neck white, tinged with cinereous-yellow ; the coverts of the primaries, and the primaries and secondaries, outward, ruddy-brown, with the two last white ; tail pale-cinereous ; the sides of the body tinged with bluish-grey ; bill above, purplish-blue ; below, livid flesh-colour ; top of mandible, livid-black ; feet livid-grey. Length, 18″ 9‴ ; wing, 12″ 4‴ ; tail, 5″ 3‴.

South African Coast.—Dr. A. Smith (loc. cit.) ; non vidi.

671. Procellaria Capensis, Linn. ; Pl. Enl., 964 ;
Cape Pigeon ; Daption Capensis.

UPPER parts, mottled black and white ; under parts, pure white ; head black ; chin speckled. Length, 15″; wing, 10½″; tail, 4″ 9‴.

This bird, though common enough along the coast, rarely enters our harbours, preferring the open sea for its hunting-field. At one season of the year, about November and December, they disappear, and the voyager finds the sea duller and tamer than ever. I presume they go off to breed ; but where they select their nurseries I know not.

On a recent voyage to England, I left the Cape at the end of March : not a single Daption was visible throughout our trip ; nor indeed, after we left the land, until off Ushant, did we see more than half-a-dozen birds in all (1866). On my return in November, we fell in with one *three degrees north* of the Line ! ! I have never before heard of this species extending beyond the Equator.

672. Procellaria Turtur, Forst., A. Smith's Ill.
S. Af. Zool., Pl. 54.

ABOVE, blueish ash-coloured ; beneath white ; space before the eye, eyebrow, and lower eyelids, white; lesser wing-coverts, tips of the scapulars, webs of the primaries, the outer vanes of the four external quill-feathers, and the tip of the tail, dull reddish-brown ; base of tail cinereous ; bill greenish black ; upper mandible at the point, yellow-brown ; lower mandible livid ; feet reddish-brown ; eyes black-brown. Length, 10″ 9‴; wing, 7″ 6‴; tail, 4″.

Resembles in its habits *P. Fosteri*, and inhabits the Cape seas.

673. Procellaria Cœrulea, Gmel. ; *Procellaria Forsteri*, Smith, Zool. S. Af., Pl. 53.

UPPER parts, darkish pearl-grey ; under parts white ; scapulars grey, darkened by a distinct tint of broccoli-brown ; several of them *finely tipped with white* ; tail pearl-grey, the *tips of all the feathers, as well as their inner edges, white;* lateral feather on each side, dull white, with a freckling of grey towards its shaft; forehead, space in front of the eyes, eyebrows, and sides of head, white. Length, 10″ 5‴; wing, 8½″; tail, 3″ 10‴.

The white edges of the scapulars, and the tips and two outer tail-feathers distinguish this species at once from all the Blue Petrels inhabiting our seas. It is not uncommon along the coast, and is occasionally cast ashore after a gale of wind.

Genus PRION, Lacépède.

Bill the length of the head, very broad at the base, depressed above; culmen nearly straight, laterally swollen, but gradually compressed towards the tip, which is arched, elevated, compressed, and acute; the lateral margins dilated near the base, with a series of very fine laminæ running along the whole length internally, rather above the margin; the lower mandible broad at the base, gradually compressed towards the tip, which is much compressed, with the margin and gonys arched; the nostrils basal, tubular, elevated above the culmen, short, opening with two appertures in front; wings moderate, pointed, with the first quill nearly equalling the second, which is longest; tail moderate, broad, and rounded at the end; tarsi shorter than the middle toe, laterally compressed, and covered with small scales; the outer toe nearly as long as the middle toe, and the hind toe nearly in the form of a broad, short-pointed claw.

674. Prion Banksii. (Smith.) *Pachyptila Banksii,* Smith, Zool. S. Af., Pl. 55.

GENERAL colour above, pearl-grey; back of neck and inter-scapulars, tinged with brown; lesser wing-coverts, and the ends of the scapulars, ruddy-brown; tail grey, tinged with yellow-grey; the ends of the feathers dirty-brown; through the eye a stripe of purple, tinged with ruddy-brown; sides of the chest shining-grey; below white; eye brown; bill livid-brown; towards the tip of the upper mandible, clouded greenish-yellow; feet ruddy-brown. Length, 10″; wing, 7″ 9‴; tail, 3″ 9‴.

This bird occasionally frequents the bays of our coasts, and is constantly to be seen in the Cape seas. Mr. Ayres states that it has been thrown ashore at Natal, dead, in vast numbers. The same thing has occurred here several times. Many were thus cast away at Green-Point on the 21st of April, 1866; and a large flock perished in a vineyard at Constantia, belonging to Mr. Cloete.

The Sub-Family, DIOMEDEINÆ, or Albatrosses,

have the nostrils short, tubular, widest anteriorly, and placed near the base of the lateral groove.

Genus DIOMEDEA, Linn.

Bill longer than the head, very robust, straight; the sides compressed and longitudinally grooved, with the tip greatly curved and acute; the lateral margins dilated and curved; the culmen broad, convex, and rounded; the lower mandible' weak, compressed, with the tip truncated; the nostrils placed near the base in the lateral groove, covered by a tube which is short, widening and spreading anteriorly from the side of the bill, with the aperture somewhat rounded and open in front; wings very long, very narrow, with the second quill the longest; tail short and rounded; legs short, strong, with the tarsi one-fourth shorter than the middle toe, and the inner toe the shortest; the two lateral toes margined exteriorly by a narrow membrane; the web between the toes full and entire; the hind toe and claw entirely wanting; the claws short and obtuse.

675. Diomedea Exulans, Linn.; Pl. Enl. 237;

Less. Manuel D'Orn., Vol. 2, p. 351; *The Great Albatross*, or *Cape Sheep* of Sailors.

ADULT: general colour pure white, mottled minutely on the back with wavy black lines; wings black and white, mottled; tail spotted with black. Young: dull-brown, with white face and neck. Length, 3' 9"; wing, 2'; tail, 12".

The Albatross is a well-known inhabitant of our seas, and though it seldom ventures into the bays, it is sometimes caught by the fishermen between Robben Island and the mainland, together with the next species. It retires from our shores about November, and breeds on the Island of Tristan D'Achuna. An egg brought thence by Capt. Nolloth, of H. M. St. *Frolic*, and now in the South African Museum, is of a dull-white, minutely spotted with brown at the obtuse end: axis, 5"; diam., 3" 3‴.

I was on that Island at Christmas, twenty-one years ago; but I still retain a vivid recollection of the marvellous number of albatrosses of all kinds which we encountered on our arriving in that neighbourhood, and of the thousands that we found sailing about the singular peak, looking at that great attitude (9800 feet) like mere specks in the sky.

Eggs have been brought to me by Capt. Armson from the Crozette Islands. This gentleman informs me that they make no nest, but simply lay in a depression in the soil caused by their frequenting the same spot for many successive years. He tells me the young birds remain for several months in the nest, not eating anything, that he could ascertain, the parent birds having long left the islands. However, this statement, though made in full sincerity, may not represent the facts of the case. We know that a few hundred miles of flight would be nothing to a bird of such vast powers as the albatross. Nocturnal visits may therefore be made to the young, and entirely escape detection. This appears to me a likely solution of the difficulty.

Since the above was written, I have spoken to many of the sealers frequenting these islands. They declare the albatrosses *visit their young daily*, and turn them off the nests when the next breeding season comes round.

676. Diomedea Melanophrys, Temm., Pl. Col.
456 ; *The Small Albatross, or "Mollymaw."*

BACK, wings, and tail, brown-black, the latter with a grey tinge; head, neck, rump, and under parts, pure white; a blackish line passes over the eye. Length, 36"; wing, 21"; tail, 11".

Very abundant in all our bays, and on the coast generally. Hundreds are caught by the fishermen while out pursuing their usual trade, and are sold for three-pence each in the fishmarket. They are much eaten by the lower orders, and I am told are very palatable. They retire with the preceding to breed, and their eggs resemble theirs in every particular, save size, being: axis, 3" 9'''; diam., 2" 6'''.

This and the next species alight readily on the water to pick up garbage thrown from ships, and are easily caught with a hook and line in calm weather.

677. Diomedea Chlororhyncha. (Gmel.) Lath.
Syn., t. 94 ; Pl. Col., 468 ; *D. Profuga*, Banks; *D. Chrysostoma*, Forst. Icon. Ined., t. 100, 101.

WINGS, back, and tail, sooty black-brown; the shafts of the latter white; the rump and rest of the plumage, pure white, with the faintest tinge of black for an eyebrow; bill black, with broad yellow stripe commencing about 4''' of an inch from the feathers, and extending along the ridge of the upper mandible to the tip, which is reddish ; lower mandible black, with the exception of the extreme tip, which is livid. Length, 31"; wing, 19"; tail, 11".

Not uncommon off the South Coast, but does not approach so near to land as the preceding. All albatrosses feed much on squids. From the stomach of one I took a handful of their hard, horny, parrot-billed-shaped jaws.

678. Diomedea Culminata, Gould ; Proc. Z.
S., 1843, 107.

UPPER part of back, pale sooty-black; head and neck lighter; wing much darker, and tinged with brown; eyebrow, above dark, like the back, below white ; under parts, from the base of the neck, pure white. Length, 30''; wing, 19''; tail, 11''.

I obtained a single specimen of this Albatross off the South Coast, while cruising in H.M.S. *Castor*, and for some time confounded it with the preceding, with which it was consorting, and agrees perfectly in habits. I have never before heard of this species so far north. On the coasts of Australia it is a common species.

679. Diomedea Fuliginosa, Gmel., Pl. Col., 469 ; *D. Antarctica*, Banks, Icon. Ined., t. 26 ; *D. Palpebrata*, Forst., Icon. Ined., t. 102 ; *D. Fusca*, Audub. B. of Am., Pl. 407 ; *Blue-Bird* of Sailors.

THROUGHOUT of a dull sooty-black, darkest on the wings ; eye surrounded by a white eyelid ; bill jet-black, with a narrow, yellowish stripe along the lower mandible ; legs flesh-coloured ; irides brown ; tail wedge-shaped. Length, 33''; wing, 20''; tail, 13''.

The Black Albatross is common along our Southern Coast, at a little distance from land. It may at once be distinguished from *Procellaria Gigantea* by its wedge-shaped tail, and when close, by its white eyelids. It is fond of hovering over a ship's deck, and I have obtained specimens by watching my chance and sending a small rifle-bullet through them, thus killing them instantly ; if struck with shot, their feathers are so dense that they seldom receive a death-wound, but drift to seaward and perish miserably. I have never observed any of the other albatrosses hover over the deck in the manner that this bird constantly does : sometimes I have seen it almost touch a man on the royal-yard or on the lift of the spanker. It seems actuated by curiosity, turning its head from side to side, and scanning everything with its brilliant dark eye. Eggs of this species resemble those of *D. Exulans*, but are smaller : axis, 4'' 2'''; diam., 2'' 6''',

The Fifth Family, LARIDÆ, or Gulls,

have the bill of various forms, more or less straight and compressed on the sides; the nostrils lateral, generally longitudinal and submedial ; the wings lengthened and pointed ; the tail more or less long, and of various forms; the tarsi generally moderate, strong, and covered in front with transverse scales; the toes moderate, with the anterior ones united by a full web; the hind toe usually short and elevated.

The Sub-Family, LARINÆ, or Gulls,

have the bill more or less lengthened, straight, compressed on the sides, with the culmen straight at the base and curved to the tip, which is acute ; the nostrils lateral, submedial, and oblong ; the wings lengthened and pointed ; the tail

* T

moderate, and usually even ; tarsi moderate and strong; the toes moderate ; the fore toes united by a web, and the hind toe generally short and elevated.

Genus STERCORARIUS, Brisson.

Bill moderate, straight, and strong, with the culmen straight, rounded, and covered with a membranous or bony cere ; the apex curved, vaulted, and strong ; the gonys much angulated and ascending ; the nostrils placed in the fore part of the cere, narrow, and enlarging anteriorly ; wings length-ened and pointed, with the first quill the longest ; tail moderate, rounded, with the two centre feathers sometimes elongated ; tarsi longer than the middle toe, strong, and covered in front with strong scales ; toes moderate and strong, the anterior ones united by a full web ; the hind toe very small, and hardly elevated.

680. Stercorarius Catarractes. (Linnæus.)

Gould's B. of Eur., Pl. 439 ; *Larus Catarractes,* Linn. ; *Lestris Catarractes ; Common Skua,* Bewick's Brit. Birds, p. 216.

GENERAL colour of the upper parts, dark greyish-brown, streaked with brownish-yellow ; primary quills, brownish-black, with their shafts and basal parts white ; there being a conspicuous patch of that colour on the wing ; tail blackish-brown, white at the base, but that colour not apparent there. Length, 24″ ; wing, 16½″; tail, 7¼″. (McGillivray.)

Several specimens of this fine Skua Gull appeared in Table Bay on the 29th of April, 1865, chasing the Solan Geese *(Sula Melanura),* which flocked in to prey on the vast shoals of fish in the bay. My son fired at one a week previous to this. They were plainly recognis-able at a great distance, owing to the white spot on the wing, and their size ; and while the smaller *S. Parasiticus* attacked the little gulls and terns, they confined their attentions to the solan geese, compelling them to disgorge their hard-earned gains.

681. Stercorarius Spinicauda, Hardy ; Bp.

Cons. Av., p. 210 ; *Lestris Spinicauda ; Lestris Parasiticus,* apud G. R. Gray.

(A) Throughout, a dull sooty-black, slightly variegated on the back of the neck with brown ; tail slightly pointed.

(B) Throughout, mottled brown and fulvous, the latter predominating on the neck and head ; obsolete on the tail, which is much pointed.

(C) Back of head, wings, and tail, deep sooty brown ;

chin and under parts white; breast mottled with brown, forming a collar round the neck; vent also mottled with brown; centre tail-feathers not fully developed. Length, 16″; wing, 12″; tail, 6″.

Common in Table Bay in the summer months, in the three phases of plumage marked A, B, C. They chase the lesser gulls and terns, and make them disgorge their prey; but I have also seen them swimming on the water searching for their own food. Their flight is very swift, and sustained by powerful strokes of the wing, urging their bodies through the air with great rapidity : their rapid evolutions when in chase of the nimble terns are most graceful.

Messrs. G. R. Gray and Tristram identify specimens of the small *Lestris* of Table Bay sent home by me as examples of *L. Parasiticus.* I cannot myself detect any difference, but M. Bonaparte seems to do so, and has called our Cape bird *L. Spinicauda.*

Genus LARUS, Linn.

Bill more or less strong, as long as or shorter than the head, straight, and laterally compressed, with the culmen straight at the base, and arched to the tip; the gonys slightly angulated and advancing upwards; the nostrils lateral, with the opening near the middle of the bill, and longitudinal; wings lengthened and pointed, with the first quill the longest; tail moderate and even; tarsi nearly as long as the middle toe, strong, and covered in front with transverse scales; toes moderate, the anterior ones united by a full web; the hind toe short and elevated.

682. Larus Dominicanus, Vieil.; *L. Vetula,*
Baillon., Bp. Cons. Av., p. 214.

BACK, sooty-black; quill-feathers jet-black, tipped with white; near the end of the outermost a white spot; ends of the smaller wing-feathers also white; rest of the plumage pure white; bill yellow, with knob near the tip of the lower mandible scarlet. Length, 24″; wing, 17″; tail, 7¼″. Young: throughout mottled brown and white; bill almost black; irides rich-warm brown; legs livid flesh-colour.

The above descriptions are taken from specimens killed on the 29th of April, 1865, in Table Bay, by my son. It is an abundant species, frequenting all our coast line, and breeding in large numbers on the islands to the North-Westward. Their eggs are very variable in colour and markings, but are ordinarily of a darkish olive-green, profusely spotted with brown, chiefly in a band round the obtuse end: axis, 3″; diam., 2″.

I have had one in confinement for nearly two years, and he has only just assumed the full adult plumage. I do not think that it is put on till the third year. My friend "Jack" is quite tame, and stalks about the garden and into the house at pleasure: he feeds on any garbage, and generally takes his morsel to his water-tub, and washes or wets it before swallowing. His favourite sleeping-place, when not perched on the coal-heap, is an old tree-stump standing in the garden. He is a great adept at catching mice, which he swallows whole, after giving them a few preliminary raps on the ground. Indeed, they seem a great *bon bouche* with him; and he will hurry from the furthest end of the garden, if I do but hold one in my fingers, or show him the trap. He is very fond of worms, but will not eat slugs. He usually washes himself morning and evening, and sleeps much during the day.

683. Larus Poiocephalus, Swain., Nat. Lib.
Vol. 12, p. 245, Pl. 29 ; *Xema Phaeocephalum,* Strick. Jard., Cont. to Ornith., 1852, p. 160 ; *Cirrocephalus Minor,* Bp. Consp.

GENERAL colour of back, cinereous; a hood of the same colour, but lighter, covers the head, chin, and upper portion of the neck ; wing-feathers black, faintly tipped with white; a white spot on the two outermost near the tip, and a broad band across some of the others towards the centre ; the rest of the plumage fine white ; bill and feet, deep crimson ; legs flesh-colour ; irides white ; cere round the eye as in legs. Length, 16″; wing, 12½″; tail, 5″.

In the winter, according to Hartlaub, the hood disappears. I have, however, never seen it with the grey head in Table Bay, though it is abundant all the year round. I saw four specimens with the hood at Zoetendals Vlei in November (1865), and shot two.* The stomach of one killed in Table Bay on the 29th April (without the hood) contained nothing but quantities of a small crustacean common on sand.

I took three broken eggs, floating on the vlei, which I think can only belong to this species ; they are greenish-brown, profusely spotted with brown and purple : axis, 2″ 1‴; diam., 1″ 6‴.

Eggs said to belong to the white-headed birds have been brought to me from the Islands; they are light-green, spotted with dark-brown and purple, but vary in colour. In size they resemble those before described.

The Sub-Family, STERNINÆ, or Terns,

have the bill more or less lengthened, generally slender, straight, with the culmen sometimes curved at the tip, which is acute ; the nostrils basal, lateral, and linear; the wings very long and pointed ; the tail long, and more or less forked ; the tarsi usually short and slender ; the toes of various lengths, and more or less webbed ; the hind one long and slender.

---

* In the fresh killed specimens I was much struck with the exceeding beauty of the lovely delicate roseate tint which pervaded the whole of the under side; this faded very rapidly after death; but I think it was the most lovely colour I ever beheld. I have never noticed the faintest approach to this tint in any specimen killed in Table Bay, nor in the living example which I kept for a short time in my garden.

Genus STERNA, Linnæus.

Bill more or less long, strong, with the culmen slightly curved to the tip, which is acute; the gonys straight, and half the length of the bill; the nostrils lateral, placed towards the middle of the bill, and longitudinal, with the frontal plumes advancing close to or near the opening; wings very long and pointed, with the first quill the longest; tail more or less long, and generally forked; tarsi more or less long and slender; toes moderate, the two outer ones nearly equal, and the three anterior ones united by an indented web; the hind toe very short; the claws moderate, slightly curved, and acute.

684. Sterna Caspia, Pallas; Temm., Man. II, p 733; *Sterna Melanotis,* Swain.; *Sylochelidon Caspia,* Brehm; *The Caspian Tern.*

ABOVE, throughout, pale pearl-grey; quill-feathers of wings darkest; beneath pure white; head of male in full plumage, at the top, from the base of the bill, including the eye, and down the back of the neck, pure black; in the female, this portion is only profusely speckled with black; bill fine scarlet; points horn-coloured; legs black. Length, 21″; wing, 17″; tail to the centre, 5″, forked.

This noble Tern is not uncommon on our shores; but is, I think, a periodical visitant, appearing chiefly in the summer; though my son tells me he observed a pair during the day of the great gale, May 17th, 1865. It feeds on fish, and plunges from a considerable height, completely submerging itself. It is a solitary species, never appearing in flocks.

685. Sterna Dougalli, Temm.; McGillivray, Brit. Birds, Vol. 5, p. 648; *S. Douglasii,* Mont.; *Sterna Paradisea,* Keyser.

HEAD above, and upper part of hind neck, deep-black, tinged with greenish-blue; hind neck and sides of head, white; fore neck, and all the lower parts, roseate, fading into white; upper parts, pale bluish-grey; hind part of back and tail, lighter, almost pure white; the edge of the wings, the tips and inner edges of the quills white, as are their shafts; the outer three primaries greyish-black, with the inner margin white; tail long, deeply forked. Length, 15″; wing, 9¼″; tail, 7½″.—(McGillivray.)

The Common Tern of Table Bay and the adjoining coast has been identified by Dr. Hartlaulb as "McDougal's" or "The Roseate Tern." I confess I am sceptical on this point, and am inclined to think it accords better with Swainson's *S. Senegalensis*. (Nat. Lib , Vol. 12, p. 258.) I have killed many specimens at all times of the year, but have never detected the slightest rosy tint on any one of them. It is very common, seeking its prey along the edge of the surf or on the marsh at the mouth of Salt River.

686. Sterna Cantiaca, Gmel. Sys. Nat. 1, 606 ; *St Boysii*, Lath. Ind. Orn. II, 806 ; *Thalasseus Cantiacus*, Bonap. Consp. List, 61 ; *S. Africana*, Gmel. ; Gould's Birds of Eur., Pl 415 ; *Sandwich Tern*, Bewick's Birds of Eur., p. 177.

♂ : Above, throughout, pale pearl-grey, rather darker on the wing-feathers : tail white, and deeply forked ; on top of the head, from the bill, including the eye, and down the back of the head, a black hood, ending in a point on the back of the neck ; bill long and narrow, black, with the tip yellow horn-colour ; legs and feet black. ♀ as the male, but the black on the top of the head replaced by black speckles ; forehead white. Length, 15″ ; wing, 12″ 2‴ ; tail (to the end of centre feather), 4″.

Pretty common throughout the seaboard. In habits they resemble the next species, with which they associate.

687. Sterna Bergii, Licht. Verz. 1823, p. 80, No. 833 ; *Sterna Velox*, Rüpp. Atlas, t. 13, p. 31 ; nec. *St. Bergii*, Reichenbach ; nec. Hartlaub. ; Schlegel. Mus. des Pays-bas, Sternæ, p. 11.

♂ : General colour above, rather dark pearl-grey ; darkest on the wing-feathers ; under parts, and lower part of the neck, pure white ; *forehead white* ; top of head, including the eye, and crest in ♂, jet-black ; in the ♀ this is replaced by black speckles ; tail very deeply forked ; bill, rich lemon-coloured ; iris brown. Length (to centre of tail), 16½″ ; wing, 15″ ; tail 4½″ ; outside feathers, 8″.

The " Swift Tern " is, perhaps, our commonest species, appearing in large flocks in Table Bay throughout the year. It breeds on the islands to the North-Westward. Lays two eggs of a rich cream-colour, dotted, blotched, and streaked, with very dark-brown and purple : axis, 2″ 5‴ ; diam., 1″ 7‴. The running of the smaller spots into streaks at once seems to identify the eggs of this species from those of others that may breed here.

It feeds on small fish and crustaceans, the former of which it captures by dropping upon them from a considerable height ; the latter it either gathers in a graceful swoop, as it flies along the edge of the retiring wave, or composedly picks up while walking along the sand, on which it readily perches.

It is a singular sight to see the small craft in the bay covered by these birds and cormorants: each mast-head will be garnished with one, the bowsprit has a line of them, and along the sides and taffrail they sit as close as they can find room. When thus in repose, they will permit a close approach, well knowing that the police regulations of the bay prevent their being fired at. Elsewhere they are wary enough.

688. Sterna Galericulata, Licht. Verz. 1823' p. 81, No. 834; *St. Longirostris*, Less. Tr. d'Ornith, p. 621 (?); *St. Cristata*, Swain. Nat. Lib., Vol. 12, Pl. 30, p. 247.

UPPER plumage, but for a very slight tinge of grey hardly perceptible, might be termed as white as the under parts, the quills alone being grey, posted as it were with white; the deep black on the head and front includes half the lores and the eyes, but is interrupted by a white spot on the lower eyelid; the feathers of the hind head and nape are lengthened and pointed, so as to form a nuchal crest; the wings are long, and reach to the end of the tail, with a frosted white appearance on their outer webs; the inner webs are dark-grey on their inner half, and white on their outer, but the stripe of this colour almost disappears on the fourth and fifth quills, whose inner shafts are almost entirely dark-grey, excepting a narrow white edging; the tail, which is deeply forked, is white, both above and below; bill rich-orange; feet black and naked for nearly an inch above the tarsus. Length, $20''$; wings, $14''$; tail, base, $7\frac{1}{2}''$; depth of fork, $3\frac{2}{10}''$.

Hartlaub quotes this species, the description of which I have borrowed from Swainson (loc. cit.), as a native of South Africa. I have not yet seen it. It closely resembles *S. Velox*, but may easily be distinguished from it by the black of the head being continued down *to the bill*; whereas *S. Velox* has a *broad white forehead*.

689. Sterna Brachypus, Swain. Nat. Lib., Vol. 12, p. 252.

UPPER part of head and back of neck, black; general colour cinereous, both above and below, but paler and nearly white on the chin; upper and under tail-coverts, pure white; tail the same, but the outer web of the two external feathers is deep cinereous, almost sooty-black; quills blackish-cinereous; the shafts pure white, with the inner web half cinereous, half white; bill red; feet orange; tail one inch longer than the quills. Length, $15''$; wing, $11''$; tail from the base, 8; depth of fork, $4\frac{1}{2}''$; tarsus, $\frac{1}{2}''$.

During a drive late one evening across a marsh formed by the celebrated hot spring called "Brandt Vley," near the town of Worcester (South Africa), I observed a tern, new to me, flying in considerable numbers over a portion of open water. Owing to the lateness of the hour, I was unable to stop and procure specimens ; but a short time afterwards I received a tern from Tulbagh, a village at the end of the same valley (which is enclosed between high mountains), though distant about sixty miles, which I immediately recognised as belonging to the same species as those at Brandt Vley. It answers in all respects to Swainson's description (loc. cit.) of the short-footed tern, and as such I consider it.

Genus DROMAS, Paykull.

Bill longer than the head, and straight, with the culmen gradually sloping to the tip, which is acute ; the lateral margins straight, the sides compressed, and the gonys very long, ascending, and much angulated at the base ; the nostrils placed in a broad, short groove, and covered posteriorly by a membrane, leaving an oval opening exposed ; wings long, with the second quill the longest ; tail moderate and broad ; tarsi very long, compressed, and covered in front with transverse scales ; toes long, with the anterior ones palmated to the end of the toes, but much indented in the middle ; the hind toe long, free, and partly resting on the ground.

690. Dromas Ardeola, Paykull, Pl. Col., 362.;
Erodia Amphilensis, Salt's Trav. Abyss., IV., p. 60 ;
Lath. His. Birds, IX., Pl. 149.

GENERAL colour, white, with a black patch between the wings ; wing-feathers black ; head more or less mottled with black ; sometimes a grey tint on the back and shoulders ; bill black, very strong ; legs long, black ; feet semipalmated. Length, 1″ 3‴; wing, 8″; tail, 2″ 9‴.

This curious bird is said to be allied to the terns. It frequents the whole of the East Coast of Africa as low down as Natal ; but its favourite localities are sand-banks far out to sea. It feeds on small crustaceans, which it collects while walking. I never saw it take its prey on the wing. It runs with considerable swiftness.

The Second Family,* COLYMBIDÆ, or Divers,

have the bill more or less long, much compressed, straight and acute ; the nostrils placed in a longitudinal groove, with the opening basal, linear, or rounded ; the wings moderate, with the first quill longest ; the tail very short ; the tarsi short and much compressed ; the toes long, and the three anterior ones more or less united together by a membrane ; the hind toe short, and margined by a small membrane.

---

* Owing to an unfortunate displacement of the MSS., an error in the sequence of the Families occurred, and was not discovered until the preceding sheets were struck off.—E. L. L.

The Sub-Family, PODICIPINÆ, or Grebes,

have the bill long, straight, compressed at the sides, with the culmen slightly curved to the tip, which is acute and entire ; the gonys ascending ; the nostrils pierced in a groove, and oblong ; the wings short, with the first quill the longest ; the tail not apparent ; the tarsi short, and much compressed ; the toes long, the outer longer than the others ; the anterior ones broadly lobed on the sides, especially on the inner side ; the claws short, very broad, and obtuse.

Genus PODICEPS, Latham.

Bill more or less long, strong, straight, the culmen slightly curved at the tip, which is acute and entire; the sides much compressed, and the gonys short and advancing upwards to an acute point; the nostrils placed in a short groove, with the opening longitudinal and exposed; wings short and pointed, with the first or sometimes the second quill the longest, and slightly emarginated at the tips; tail short, not apparent; tarsi shorter than the middle toe, much compressed, the anterior and posterior edges covered with small scales, which are serrated posteriorly, and the sides with transverse scales; toes long, the outer the longest, depressed, margined on the sides, especially on the inner side, and united at the base to the middle toe; the hind toe short and strongly lobed; the claws short, very broad, flat, and obtuse.

691. Podiceps Cristatus. (Linn.) Lath., Pl.

Enl., 400 ; *Colymbus Cristatus*, Linn. ; *C. Cornutus*, Briss. ; *C. Urinator*, Linn. Ed. Birds, Pl. 360, f. 2 ; *The Crested Grebe*, Bewick's Brit. Birds, p. 146.

GENERAL colour above, greyish-black, tinged with rufous ; on the wing a large white patch ; under parts pure white ; greyish on the breast and flanks ; head ornamented with a double crest, and throat with a large ruff; top of the head, crest, back of neck, anterior and lower portion of neck-ruff, shining-black, tinged here and there with rufous ; chin white, changing into golden-yellow and rufous on the ruff ; below the ruff white, tinged with rufous ; tail spurious ; iris red. Length, 20"; wing, 7½".

The Crested Grebe is common on all our vleys. It breeds in companies, six or eight nests generally appearing within a few yards of each other ; these are built on the water—a mere flat form of sedge, *generally damp throughout* ; indeed, I remarked that every egg taken by us at Zoetendals Vley was *wet*. This might have been caused, how-

*U

ever, by the wet water-weed with which each bird carefully covered her eggs as we approached the nest. I watched three birds perform this manœuvre through my binoculars. They slid off their nests and rapidly picked up the floating weed, which they carefully disposed over the eggs, so as completely to hide them from view. We examined some fifteen or twenty nests, each one of which was thus covered. We never found more than three eggs in each nest, of a dirty chalky texture: axis, 2″; diam., 17‴. The bird feeds on small fish and water-insects.

692. Podiceps Auritus. (Linn.) Lath. Edw.
Birds, t. 96, f. 2 ; *Colymbus Auritus*, Linn. ; *The Eared Grebe.*

GENERAL colour, greyish-black, tinged on the sides with rufous ; beneath, from the centre of the breast to the vent, satiny-white ; behind the eye, over the ear, a tuft of longish reddish-yellow feathers ; patch on the wing white. Length, 13″; wing, 5″ ; tail spurious.

The Eared Grebe was unknown to me as an inhabitant of South Africa until the year 1859, when, having an opportunity of visiting Vogel Vley, in the Wellington district, I found it breeding in considerable numbers amid the rushes that border portions of that lake. Each pair seemed to keep guard over its special province, and never to stray to any distance from the haunt. The nest was constructed of sedge, and was a large compact structure ; the eggs, four or five in number, are chalky-white.

693. Podiceps Minor. (Gmel.) Lath. Pl. Enl.,
905 ; *Colymbus Minor*, Linn.; *Colymbus Fluviatilis*, Bris. ; *The Little Grebe*, Bewick's Brit. Birds, p. 173.

UPPER parts, top of head, chin, and back of neck, greenish-black ; rest of neck deep-rufous ; breast brownish-grey ; flanks the same, tinged with rufous ; under parts satiny-white ; all the plumage lustrous, and very dense, more like hair than feathers ; eyebrow and tip of bill, clear horn-coloured ; base, bright-green. Length, 10″; wing, 4″; tail spurious.

Not uncommon throughout the colony, frequenting still "reaches" (here called "*zeekoe gaten*," seacow-holes in the rivers) and all the vleys. Young birds have been brought to me. I never knew the nest to have been taken here ; but in Ceylon I have found those of an allied species (*P. Phillippensis*), constructed of sedge, and closely resembling those of *P. Cristatus* in all but size—the eggs pointed at both ends, and chalky white.

The Sub-Family, HELIORNINÆ, or Sungrebes,
have the bill long, straight, and compressed, with the tip slightly curved and emarginated ; the gonys of the lower

mandible short, and advancing upwards ; the wings moderate and rounded ; the tail long, and much rounded ; the tarsi short ; and the toes margined with a membrane, which is more or less united to the middle one.

Genus PODICA, Lesson.

Bill longer than the head, straight, sides compressed, with the culmen rather elevated, and gradually curved to the tip, which is acute and emarginated ; the gonys of the lower mandible short, and advancing upwards ; the nostrils placed in a broad groove, with the opening large, longitudinal, and near the middle ; wings moderate, with the third, fourth, and fifth quills the longest ; tail lengthened, rounded, and composed of long narrow feathers, with the shaft of each strong at the base ; tarsi shorter than the middle toe, and scutellated in front ; toes long, and margined with a broad, lobed membrane ; the lateral ones unequal ; the hind too long, and margined with a broad lobe ; the claws short and curved.

694. Podica Mosambicana, Peters; *Podica Petersii*, Hartb., Beitr. Fur. Orn., West. Af., p. 6 ; Abhandlung aus d. Geb. d. Naturwissensch., 1861, 2.

ABOVE brown ; beneath white ; chest, flanks, and vent, variegated fuscous ; head and neck above, black, with violet reflections ; beneath, and on the sides, shining-black ; shoulders violet-black, slightly spotted with white ; bill yellowish, with the culmen black ; legs yellow.

I have never seen this bird in any of the collections submitted to my inspection. Mr. Ayres has, however, procured it in Natal.

1855.—While walking along the banks of the River Zonder End, at Mr. Vigne's farm, I startled a bird from the cover of a thick bush which overhung the water. At first I took it for a coot, and abstained from firing ; but as the bird dropt its legs, I saw the feet were bright-orange, and apparently webbed : other peculiarities also convinced me that I had a stranger before me. I fired, and my bird dropt into the water, but instantly dived. It appeared again on the surface, and I gave it my second barrel—half a charge of dust-shot. I saw this strike all over it, but it again dived ; and though 1 hunted for more than an hour, I saw no more of my quarry. Since I have seen the Indian *Podica* in the collection of Mr. Webb, of Newstead Abbey, I am convinced this was our African bird.

February, 1867.—I saw a second specimen of this bird at the Berg River, at a farm belonging to Sir William Hodges. Though badly wounded, it dived and concealed itself amid the bushes drooping into the river, and I lost it. As I saw it swim across the river, and had a good view of it, I am sure of its identity.

The Third Family, ALCIDÆ, or Auks,

have the bill more or less strong, generally compressed on the sides, and the culmen usually curved to the tip, which is sometimes hooked; the wings generally short, and more or less imperfectly formed; the tail short and graduated; the tarsi usually short and compressed; the toes entirely webbed, with the hind toe small or wanting.

The Sub-Family, SPHENISCINÆ, or Penguins,

have the bill more or less long and straight, with the sides compressed and grooved; the culmen rounded and curved at the tip, which is acute; the nostrils placed in the lateral groove, and linear; the wings short and imperfect, being only covered with scale-like plumes; the tail more or less short, and composed of narrow rigid feathers; the tarsi very short and depressed; the toes moderate and depressed, with the anterior toes united by a web; the hind toe very small, and united to the side of the tarsus.

Genus SPHENISCUS, Brisson.

Bill moderate, much compressed, and strong, with the culmen rounded and curved at the tip, which is acute; the tip of the lower mandible suddenly truncated, and the gonys moderate and curved upwards; the nostrils rather rounded, and placed in the lateral groove near the middle of the bill; wings imperfect, and covered with scale-like plumes; tail very short; tarsi very short, thick, flattened, and covered with small scales; toes long, the lateral ones unequal, and united to the middle toe by a web; the hind toe very small, and united to the tarsus at the base of the inner toe; the claws long, compressed, and slightly curved.

695. Spheniscus Demersa. (Linn.) Temm., Pl.

Enl. 382; *Aptenodytes Torquata*, Sonn. (?) Edw. Birds, Pl. 94; *Common Penguin.*

GENERAL colour, blueish-grey; below white; a band, the colour of the back, extends from the front of each thigh, up the flanks and side, passes in front of the root of the wings, and forms an arch at the base of the neck; a patch of the same colour includes the eye, ear, chin, and a portion of the side of the neck. Length, 26"; wing, 7" 3"'; tail, 1".

The whole plumage is of a stiff character, resembling strips of whalebone. The wings are mere paddles for progression under water, and are incapable of flight—the feet placed so far back as to cause the bird to appear always falling backward if it attempts to stand on dry land.

This singular bird is very common on all our coasts, swimming to great distances out to sea. It breeds on rocky islets, in the months of August, September, and October.

Through the kindness of a dear and valued friend, the late Admiral Trotter, I once had the pleasure of visiting a breeding place of this bird, Seal Island, a lonely rock in the centre of False Bay. It can only be approached in certain winds, as the surf breaks on it with great violence. The penguins occupy the lower ledges by hundreds, having dug their holes among the rocks. As we walked over the stones, the birds rushed out at us, snapping at our legs, and drawing blood plentifully from the naked feet of the sailors, in defence of their one large solitary white egg, which is laid on the bare ground, without any protection from the damp—axis, 2″ 9‴; diam., 2‴—abruptly pointed at the small end. When not sitting on their eggs, they stand upright on the rocks in long rows, balanced on their stiff tails. They feed exclusively on fish, which they overtake beneath the surface by their swiftness in swimming.

The Sixth Family, PELICANIDÆ, or Pelicans,

have the bill more or less long, broad at the base, straight and compressed to the tip, which is sometimes hooked ; the nostrils linear, and sometimes scarcely visible ; the wings long ; the first quill the longest ; the tarsi short and robust ; the toes long, and all four connected together by a broad membrane ; the face and throat more or less naked, the latter sometimes furnished with a naked, dilating skin, or pouch, from the base of the lower mandible.

The Sub-Family, PLOTINÆ, or Darters,

have the bill lengthened, very slender and acute, with the lateral margins finely serrated, and the gonys long, and scarcely ascending ; the nostrils basal, and covered by a shield ; the wings long ; the tail lengthened, and widening towards the end ; the tarsi short, very strong ; the toes long ; the anterior ones united by a broad web ; the hind toe long, and united to the inner one by a broad web ; the claws short and curved.

Genus PLOTUS, Linn.

Bill longer than the head, straight, and very slender, with the sides much compressed to the tip, which is very acute ; the lateral margins finely serrated, and the gonys long, and slightly ascending ; the nostrils basal, linear, and scarcely visible ; wings long, with the second and third quills equal and longest; tail long, and broad towards the end, which is rounded; tarsi half the length of the middle toe, strong, and covered with small scales; toes rather long, all united by a broad web; the outer toe as long as the middle one; the claws short, curved, and acute.

696. Plotus Congensis, Cranch; *Plotus Levaillantii*, Temm., Pl. Col.; Lesson, Vol. 2, p. 380; *P. Rufus*, Licht., Pl. Enl., 107; *Sweet-water Duiker.*

GENERAL colour, black, shining green on the back and shoulders, longitudinally striped with fulvous, that colour occupying the centre of each feather; neck and head, rufous-brown; black lines extend from the back of the eye, down the sides of the neck, becoming fainter as they descend, and coalescing at the base of the neck; below this line, but only extending to one-third the distance, is a pure white line; tail and wing feathers, black: the former very stiff; the two centre ones being corrugated. Length, 36"; wing, 13½"; tail, 10" 9‴.

A female in my possession is of a rufous tint throughout, and wants the fine glossy black green body of the male, and the black line down the neck; the colours of the back are also less vivid.

The "*Anhinga,*" "*Snake-bird,*" or "*Darter,*" is not unfrequent in certain localities, among which may be mentioned chiefly: Verloren Vley and the Berg River generally, the River Zonder End, and Zoetendals Vley. In this latter place, I saw several individuals, evidently building, as they carried long trailing rushes in their bills; but I could not discover their nests. They may often be seen sitting on the "snags" projecting out of the water in the still reaches of the rivers, off which they dive with such ease and dexterity as hardly to leave a ripple to betray their departure. An egg said to belong to this bird was given me by Dr. Versfeld, of Stellenbosch, taken on the Berg River, is of a green ground colour, covered with white chalk: axis, 2" 2‴; diam., 1" 5‴.

The Sub-Family, PELICANINÆ, or Pelicans,

have the bill lengthened, slender, depressed on the culmen; the sides much compressed, the tip hooked, and more or less compressed; the nostrils placed in lateral grooves, and scarcely visible; the wings lengthened and pointed; the tail rather short, and generally wedge-shaped; the tarsi short and robust; the toes lengthened, with the outer one equalling the middle one, and all four united together by a membrane; the lower mandible and throat furnished beneath with a membranous pouch, more or less capable of extension.

Genus SULA, Brisson.

Bill longer than the head, robust, straight, broad at the base, with the sides compressed, and grooved towards the tip, which is slightly curved, and the lateral margins obliquely and unequally serrated ; the nostrils basal, lateral, linear, placed in a lateral groove, and almost invisible ; wings long, pointed, and tuberculated, with the first two quills the longest ; tail moderate and graduated ; tarsi short, one-third shorter than the outer toe, rounded anteriorly, and keeled posteriorly ; toes lengthened, the outer and middle ones nearly equal, and all four connected by a full membrane ; the claws moderate, and rather flat, with that of the middle toe serrated ; the hind claw rudimental ; beneath the base of the lower mandible is a naked space, reaching towards the breast, which is capable of expansion.

697. Sula Capensis, Licht. ; *S. Melanura,* Temm. ; *Malagash* of Colonists.

GENERAL colour throughout, white ; the larger feathers of the wings and tail, black-brown ; the shafts of the former grey ; those of the latter white ; head and neck, and particularly the back of the latter, ochreous-yellow ; space round and before the eye, bare, and of a dark-blue colour ; a bare stripe of the same extends from the angle of the mouth, on each side of the head, and from the chin, two-thirds of the way down the neck ; irides, pale-fulvous ; legs, dark livid-colour. Length, 36″ ; wing, 19″ ; tail, 10″.

The Common Gannet of South Africa frequents, in countless thousands, the whole of our coast line, breeding on the various islands scattered over the whole extent from St. Ann's River to the Eastward of Natal to the guano islands off Angra Pequina. It visits Table Bay in vast numbers in the months of April and May, in pursuit of the shoals of fish that then appear on the surface, upon which they pounce with almost unerring aim, from a great altitude, becoming entirely submerged by the violence of their descent. I have never seen this species far from land : they invariably have disappeared on the morning after the ship's departure, if standing off the land. Eggs of a blue-ground, covered with white chalk : axis, 3″ 5‴ ; diam., 2″.

Genus GRACULUS, Linnæus.

Bill moderate, straight, somewhat slender, with the culmen concave, and suddenly hooked at the tip ; the sides compressed and grooved ; the nostrils basal, lateral, linear, placed in the lateral groove, and scarcely visible ; wings moderate and

pointed, with the second and third quills the longest; tail
moderate, and rounded at its end; tarsi short, one-third less
than the middle toe, much compressed, and covered with
reticulated scales; toes long, with the outer rather longer
than the middle one, and all four united by a full web; the
base of the lower mandible is furnished with a coriaceous
pouch, which is capable of expansion.

698. Graculus Carbo; *Phalacrocorax Carbo,*

Linn., Pl. Enl., 927; *Carbo Cormorans,* Mey., Vieill.
Gal. des Ois., t. 407; Audub. B. of Am., Pl. 266;
Duiker of Colonists.

HEAD and back of neck, dark-brown, with a tinge of rufous;
back, sides, and flanks, black-green; wing-coverts and scapulars grey, edged with black-green, causing a scaled appearance, which is very beautiful; tail black, slightly frosted with
grey; chin, fore part of throat, and under parts, pure white.
Length, 27"; wing, 13"; tail, 7", fourteen feathers.

Not so common as the lesser *G. Capensis,* but still in considerable
abundance. Its chief haunt is the rocky, lonely shore at the base of
Cape Point. I have not succeeded in discovering where it breeds, and
fancy it must be somewhere to the North-West, among the islands off
Walwich Bay.

699. Graculus Capensis, G. R. Gray; *Phala-*

crocorax Capensis, Bp. Consp., 2, p. 170; *Pelicanus
Capensis,* Sparr., Mus. Carls., t. 61; *P. Gracula,* Lath.

GENERAL colour of adult, black-green; the wing-covers and
scapulars very dark-grey, narrowly edged with black-green;
tail black, slightly frosted with grey; chin bare and bright-
orange; irides green. Length, 22"; wing, 9⅓"; tail, 4".

Young: throughout of a dull rufous-brown, palest on the
neck, and darkest and least rufous on the wings and tail.

Extremely common along the whole coast, I have seen it in flocks
extending upwards of three miles in length and a quarter of a mile
wide, across the bay, the hindermost constantly flying forward and
plunging into the sea as soon as the outside of the army was reached.
In this way the flock gradually moved out of the bay, feeding as it
went. It breeds on Pomona Island, and on all the small rocks along
the coast, laying two eggs of a blue ground covered with white chalk:
axis, 2" 4'''; diam., 1" 6'''. It is eaten by the Dutch colonists; and
after soaking for a night in vinegar, or being buried for a few hours in
the ground, it loses its disagreeable odour, and is not bad food.

700. Graculus Africanus, G. R. Gray; *Peli-canus Africanus*, Gmel.; *Carbo Longicauda*, Swain., Nat. Lib., Vol. 12, Pl. 31, ♀, p. 255; Bp. Consp., 2, p. 178.

♂: GENERAL plumage black; back and wing feathers, light-grey, with a terminal spot, and sometimes a white tip; bill, and cere round the eye, bright-yellow; irides blue. The female figured by Swainson (loc. cit.) is white underneath. Length, 17''; wing, 8½''; tail, 6½''.

Of this elegant little Cormorant, but two examples have come under my notice: one was sent from Colesberg by Mr. Arnot; the other appeared in Mr. Chapman's collection, and was killed on the Lake N'Gami.

Genus PELICANUS, Linn.

Bill very long, straight; the culmen rounded at the base, and flat towards the tip, which is strongly armed with an acute, compressed, strong hook, the sides slightly enlarging towards the tip; the lower mandible broader at the base than the upper, and becoming slender towards the tip; the nostrils basal, lateral, linear, placed in the lateral groove longitudinally, and hardly visible; wings moderate, with the second quill the longest, and the secondaries nearly equalling the quills; tail short and rounded; tarsi the length of the outer toe, strong, compressed, and covered with reticulated scales; toes long, the middle one longer than the outer, and all four united by a full web; the lower mandible is furnished beneath, from the base to near the tip, with a naked membrane, capable of great extension, and advancing some way down the throat.

701. Pelicanus Onocrotalus, Linn., Pl. Enl. 87; *P. Roseus*, Eversman; *P. Minor*, Rüpp.; *Onocrotalus Phœnix*, Less.

PLUMAGE throughout, white, tinted with rose-colour; the larger wing-feathers frosted with grey, and some of the smaller with grey edges; on the breast a tuft of stiff bright straw-coloured feather; feathers of the head coming down in a point on the forehead; head crested; cheeks bare, and with the pouch, which extends half-way down the throat, light-yellow; centre of upper mandible, and base of lower, blue-black; the other parts yellow, with crimson markings; legs

*v

flesh-coloured.   Length, 6′ 2″ ; wing, 2′ 5″ ; tail, 1′ 3″ ; tip of bill to forehead, 18″.

Widely distributed, frequenting vleys and the mouths of rivers. Generally associates in small flocks ; soars to an immense attitude. I have never seen it plunge after its prey. It swims with great rapidity, and captures its food while sitting on the water.

702. Pelicanus Rufescens, Lath ; *P. Mitratus*, Licht.; *Onocrotalus Philippensis*, Briss., Ornith. VI, p. 527 ; *Pelicanus Cristatus*, Less. ; Rüpp. Atlas, t. 21.

GENERAL colour, white ; frosted with grey on the wings, tail, head, neck, and belly ; tinged with yellow on the stiff feathers of the breast, and with pink behind the thighs and all down the back ; wing-feathers dark-brown ; head crested ; pouch, yellow, with numerous perpendicular crimson lines.   Length, 4′ 7″ ; wing, 1′ 10″ ; tail, 10″.

This delicately-tinted Pelican has several times been shot in the neighbourhood of Cape Town, in company with the preceding, which it resembles in habits. I saw them both in considerable numbers at Zoetendals Vley in October, 1865. They have also been procured in Natal.

MEANING OF TERMS

ABDOMEN.—The belly: that part between the breast *(pectus)* and the vent *(crissum)*.

ALA.—The wing.

ALULA.—Spurious wing-feathers attached to the rudimentary thumb.

AUCHENIUM.—The posterior portion of the neck.

AURICULARS.—Feathers covering the ears.

CERE.—Bare skin covering the base of the bill in certain birds.

CAUDA.—The tail.

CERVIX.—The whole of the back of the neck, as opposed to *guttur*, the fore part.

COLLUM.—The neck.

COVERTS.—Either of wing or tail: small feathers covering the quills both on the upper and lower, or outer or inner sides.

CRISSUM.—The vent : sometimes applied to the under tail-coverts.

CULMEN.—The ridge of the upper mandible.

DIGITI.—The *fore-toes*, as distinguished from the hind-toe *(hallux)*.

DORSUM.—The entire back, from the *cervix* to the base of the tail *(uropigium)*.

EPIGASTREUM.—The anterior portion of the abdomen.

FEMUR.—The thigh : usually concealed.

FLEXURA.—The bend of the wing.

FRONS.—The forehead.

GONYS.—The ridge of the lower mandible.

GULA.—The anterior portion of the *guttur*, or fore-part of the throat.

GUTTUR.—The throat, or fore-part of neck.

HALLUX.—The hind toe.

HUMERUS.—The shoulder.

HYPOCHONDRIA.—The side, or space under the wing.

INTERSCAPULIUM.—The part between the shoulders, on which are the *scapulæ*, or scapular feathers.

JUGULUM.—The lower anterior part of the throat.

LORES.—Space between base of bill and eye.

MANDIBULÆ.—The upper and lower parts of the bill. They are said to be *depressed* when flattened horizontally ; *compressed* when flattened at the sides ; *cuneated* when wedge-shaped, &c. &c.

MAXILLA.—Upper and lower mandibles.

MENTUM.—The fore-part of the *gula*, chin.

NARES.—The nostrils.

NUCHA.—The anterior portion of neck, cervex.

ORBITS.—Bare spaces round the eyes.

OPTHALMICO REGIO.—The part about the eye.

PAROTICO REGIO.—The part about the ears : the feathers over the ears are also called *ear-coverts* and *auriculars*.

PECTUS.—The breast.

PILEUS.—The entire top of head, including *frons, vertex*, and *sinciput*.

RECTRICES.—Tail-feathers: the side-feathers are called *laterales;* the middle, *intermediæ.*

REMIGIES.—Wing-feathers : divided into *primariæ*, or large quill-feathers; *secundariæ*, or secondary quill-feathers, attached to the middle division of wing, corresponding to our fore-arm ; *tertiariæ*, or tertiary quills.

SCAPULARIÆ.—Feathers rising on shoulder-blades, and covering side of back.

SINCIPUT.—The back of the head.

TARSUS.—The lower joint of the leg next to the foot.

TECTRICES.—Wing and tail coverts : divided on the wings into *minores*, the lesser, *mediæ*, the middle, and *majores*, the greater wing-coverts ; and on the tail into *superiores*, upper, and *inferiores*, lower coverts.

TERGUM.—The lower portion of the back.

TIBIA.—The shank : that part of the leg immediately above the *tarsus*.

UROPIGIUM.—The rump : also applied to the upper tail-coverts.

VENTER.—The belly.

VERTEX.—The crown of the head.

VIBRISSÆ.—Hairs about the bills of certain birds, such as flycatchers, goatsuckers, &c.

INDEX.

[NOTE.—The names printed in Roman letters rank only as synonyms.]

*w

*x